Gerd Wagner · Frank Werner · Tuncer Oren ·
Floriano De Rango
Editors

Simulation and Modeling Methodologies, Technologies and Applications

International Online Conference (SIMULTECH 2021)

Springer

Editors
Gerd Wagner
Brandenburg University of Technology
Cottbus, Germany

Tuncer Oren
School of Electrical Engineering
and Computer Science
University of Ottawa
Ottawa, ON, Canada

Frank Werner
Otto von Guericke University of Magdeburg
Magdeburg, Sachsen-Anhalt, Germany

Floriano De Rango
University of Calabria
Rende, Cosenza, Italy

ISSN 2367-3370 ISSN 2367-3389 (electronic)
Lecture Notes in Networks and Systems
ISBN 978-3-031-23148-3 ISBN 978-3-031-23149-0 (eBook)
https://doi.org/10.1007/978-3-031-23149-0

This Springer imprint is published by the registered company Springer Nature Switzerland AG
The registered company address is: Gewerbestrasse 11, 6330 Cham, Switzerland

Lecture Notes in Networks and Systems 601

The series "Lecture Notes in Networks and Systems" publishes the latest developments in Networks and Systems—quickly, informally and with high quality. Original research reported in proceedings and post-proceedings represents the core of LNNS.

Volumes published in LNNS embrace all aspects and subfields of, as well as new challenges in, Networks and Systems.

The series contains proceedings and edited volumes in systems and networks, spanning the areas of Cyber-Physical Systems, Autonomous Systems, Sensor Networks, Control Systems, Energy Systems, Automotive Systems, Biological Systems, Vehicular Networking and Connected Vehicles, Aerospace Systems, Automation, Manufacturing, Smart Grids, Nonlinear Systems, Power Systems, Robotics, Social Systems, Economic Systems and other. Of particular value to both the contributors and the readership are the short publication timeframe and the world-wide distribution and exposure which enable both a wide and rapid dissemination of research output.

The series covers the theory, applications, and perspectives on the state of the art and future developments relevant to systems and networks, decision making, control, complex processes and related areas, as embedded in the fields of interdisciplinary and applied sciences, engineering, computer science, physics, economics, social, and life sciences, as well as the paradigms and methodologies behind them.

Indexed by SCOPUS, INSPEC, WTI Frankfurt eG, zbMATH, SCImago.

All books published in the series are submitted for consideration in Web of Science.

For proposals from Asia please contact Aninda Bose (aninda.bose@springer.com).

More information about this series at https://link.springer.com/bookseries/15179

Preface

The present book includes extended and revised versions of a set of selected papers from the 11th International Conference on Simulation and Modeling Methodologies, Technologies and Applications (SIMULTECH 2021), held as an online web-based event, due to the COVID-19 pandemic, from July 7 to 9, 2021.

SIMULTECH 2021 received 64 paper submissions from 30 countries, of which 17% were included in this book. The papers were selected by the event chairs, and their selection is based on a number of criteria that includes the reviews and suggested comments provided by the program committee members, the session chairs' assessments, and also the program chairs' global view of all papers included in the technical program. The authors of selected papers were then invited to submit a revised and extended version of their papers having at least 30% new material.

The purpose of the International Conference on Simulation and Modeling Methodologies, Technologies and Applications (SIMULTECH) is to bring together researchers, engineers, applied mathematicians, and practitioners interested in the advances and applications in all fields of modeling and simulation. The specific topics listed under each of these tracks highlight the interest of this conference in aspects related to computing, including Conceptual Modeling, Agent-Based Modeling and Simulation, Interoperability, Ontologies, Knowledge-Based Decision Support, Petri Nets, Business Process Modeling and Simulation, among others.

The papers selected to be included in this book contribute to the understanding of relevant trends of current research on Continuous and Discrete-time Modeling and Simulation applied to the following research topics: Modeling Computer Simulation Techniques, Fluid Dynamics, Dynamical Systems Models and Methods, Nonlinear Systems; Domain-Specific Tools, Verification and Validation; Nonlinear Systems, Optimization Issues, Rare Event Simulation and Co-Simulation; Dynamical Systems Models and Methods, Fluid Dynamics, Hydraulic and Pneumatic Systems, Industrial Processes, Mathematical Simulation, Multiscale Simulation, Performance Analysis, Discrete-Event Simulation; Conceptual Modeling, Laboratory Simulation Software, Performance Analysis, Petri Nets, Stochastic Modeling and Simulation, Verification and Validation, and Wireless Systems. This year, some accepted contributions covered on this Springer book series, also, some hop topics such as Clean Energy and Power Systems, Virtual Reality and Graphical Simulations, Biological and Social Systems Simulation, and Biologically and Nature-Inspired Systems Simulation.

We would like to thank all the authors for their contributions as well as to the reviewers who have helped ensuring the quality of this publication. We also thank the staff of INTICCC and Springer for their good efforts and cooperation.

July 2021

Floriano De Rango
Gerd Wagner
Frank Werner
Tuncer Ören

Organization

Conference Chair

Floriano De Rango	University of Calabria, Italy

Program Co-chairs

Gerd Wagner	Brandenburg University of Technology, Germany
Frank Werner	Otto-von-Guericke-Universität Magdeburg, Germany
Tuncer Ören (Honorary)	University of Ottawa, Canada

Program Committee

Saleh Abdel-Afou Alaliyat	Norwegian University of Science and Technology, Norway
Mikulas Alexik	University of Zilina, Slovak Republic
Vera Angelova	Institute of Information and Communication Technologies, Bulgarian Academy of Sciences, Bulgaria
Hamid Assadi	Brunel University London, UK
Simonetta Balsamo	University of Venezia Ca' Foscari, Italy
Isaac Barjis	City University of New York, USA
Mohamed Bettaz	Prague College, Czech Republic
Wolfgang Borutzky	Bonn-Rhein-Sieg University of Applied Sciences, Germany
Christos Bouras	University of Patras, Greece
Francesco Casella	Politecnico di Milano, Italy
Gopinath Chennupati	Los Alamos National Laboratory, USA
Franco Cicirelli	Università della Calabria, Italy
Flavio Correa da Silva	University of Sao Paulo, Brazil
Andrea D'Ambrogio	Università di Roma "Tor Vergata", Italy
Guyh Dituba Ngoma	Université du Québec en Abitibi-Témiscamingue, Canada
Karim Djemame	University of Leeds, UK
Sabeur Elkosantini	University of Carthage, Tunisia

Zuhal Erden	ATILIM University, Turkey
Paul Fishwick	University of Texas at Dallas, USA
Jason Friedman	Tel-Aviv University, Israel
José Manuel Galán	Universidad de Burgos, Spain
Charlotte Gerritsen	Vrije Universiteit Amsterdam, The Netherlands
Apostolos Gkamas	University Ecclesiastical Academy of Vella, Ioannina, Greece
Alexandra Grancharova	University of Chemical Technology and Metallurgy, Bulgaria
Giorgio Guariso	Politecnico di Milano, Italy
Mykola Gusti	International Institute for Applied Systems Analysis, Austria
Rainer Hegselmann	University of Bayreuth, Germany
Tsan-Sheng Hsu	Institute of Information Science, Academia Sinica, Taiwan, China
Xiaolin Hu	Georgia State University, USA
Nobuaki Ishii	Kanagawa University, Japan
Mhamed Itmi	INSA, Rouen, France
Anniken Karlsen	Norwegian University of Science and Technology, Norway
Okba Kazar	University of Biskra, Algeria
Peter Kemper	College of William and Mary, USA
Etienne Kerre	Ghent University, Belgium
Radek Koci	Brno University of Technology, Czech Republic
Juš Kocijan	Jozef Stefan Institute, Slovenia
Petia Koprinkova-Hristova	IICT - Bulgarian Academy of Sciences, Bulgaria
Vladik Kreinovich	University of Texas at El Paso, USA
Claudia Krull	Otto-von-Guericke University, Germany
Jean Le Fur	IRD (Inst. Res. Development), France
Willem le Roux	CSIR, South Africa
Antonio Lopes	University of Porto, Portugal
Johannes Lüthi	FH Kufstein Tirol, Austria
José Machado	Institute of Engineering, Polytechnic of Porto, Portugal
Nick Malleson	University of Leeds, UK
Romolo Marotta	University of L'Aquila, Italy
Carla Martin-Villalba	UNED, Spain
Moreno Marzolla	University of Bologna, Italy
Radek Matušu	Tomas Bata University in Zlin, Czech Republic
Nuno Melão	Instituto Politécnico de Viseu, Escola Superior de Tecnologia e Gestão de Viseu, Portugal
Adel Mhamdi	RWTH Aachen University, Germany

Bozena Mielczarek	Wroclaw University of Science Technology, Poland
Vikram Mittal	United States Military Academy, USA
Jairo Montoya-Torres	Universidad de La Sabana, Colombia
Nobuhiko Mukai	Tokyo City University, Japan
Bertie Müller	Swansea University, UK
Navonil Mustafee	University of Exeter, UK
Luis Gustavo Nardin	Ecole des Mines de Saint-Etienne, France
Giovanni Nardini	University of Pisa, Italy
Àngela Nebot	Universitat Politècnica de Catalunya, Spain
Lialia Nikitina	Fraunhofer Institute for Algorithms and Scientific Computing (SCAI), Germany
George Pavlidis	"Athena" Research Centre, Greece
Alessandro Pellegrini	University of Rome "Tor Vergata", Italy
Alexandr Petukhov	Lomonosov Moscow State University, Russian Federation
Tomas Potuzak	University of West Bohemia, Czech Republic
Marco Remondino	University of Genova, Italy
M. R. Riazi	Montreal Oil & Gas, Canada
José Risco-Martín	Universidad Complutense de Madrid, Spain
Ella Roubtsova	Open University of the Netherlands, The Netherlands
Jaroslav Rozman	Brno University of Technology, Czech Republic
William Spataro	University of Calabria, Italy
Giovanni Stea	University of Pisa, Italy
Mu-Chun Su	National Central University, Taiwan, China
Yevgeniya Sulema	National Technical University of Ukraine "Igor Sikorsky Kyiv Polytechnic Institute", Ukraine
Wenjie Tang	National University of Defense Technology, China
Klaus Troitzsch	University of Koblenz-Landau, Koblenz Campus, Germany
Alfonso Urquia	Universidad Nacional de Educación a Distancia, Spain
Durk-Jouke van der Zee	University of Groningen, Netherlands
Svetlana Vasileva-Boyadzhieva	Bulgarian Modeling and Simulation Association (BULSIM), Bulgaria
Maria Viamonte	Instituto Superior de Engenharia do Porto, Portugal
Antonio Virdis	University of Pisa, Italy
Friederike Wall	Alpen-Adria-Universität Klagenfurt, Austria
Frank Werner	Otto-von-Guericke-Universität Magdeburg, Germany

Billy Williams	North Carolina State University, USA
Kuan Yew Wong	Universiti Teknologi Malaysia, Malaysia
Yiping Yao	National University of Defense Technology, China
František Zboril	Brno University of Technology, Czech Republic
Feng Zhu	National University of Defense Technology, China

Additional Reviewer

Virginia Ahedo	Universidad de Burgos, Spain

Invited Speakers

Marlon Dumas	University of Tartu, Estonia
Peter Fritzson	Linköping University, Sweden
Gregory Monakhov	AnyLogic, USA

Contents

Development and Extended Validation of a Lumped Parameter Prediction Model for Analysing Injury Parameters in a Vehicle Crash

Gulshan Noorsumar(✉), Svitlana Rogovchenko, Kjell G. Robbersmyr, Dmitry Vysochinskiy, and Andreas Klausen

University of Agder, Jon Lilletuns vei 9, Grimstad, Norway
gulshan.noorsumar@uia.no

Abstract. We present a lumped parameter model (LPM) for improving vehicle crashworthiness analysis. The novel methodology divides the event into two phases: until maximum crush and when the vehicle starts pitching forward. We built a three degrees of freedom (DOF) model for the analysis of a crash event supporting the vehicle development process. The model has been validated against the National Highway Traffic Safety Administration (NHTSA) finite element (FE) simulation of a truck and a sedan. The LPM shows good correlation with the FE test data. A parameter variation study, changing the thickness of the metal parts by 10% and 20%, is presented to improve the vehicle crash performance resulting in the reduction in pitching of the vehicle. The Simulink based simulation captures the change in the performance confirming the reliability of the model to predict event kinematics.

Keywords: Mathematical model · Collision mitigation · Stiffness · Lumped parameter model · Vehicle pitching · Robustness and prediction

1 Introduction

Each year 1.23 million people are reported to die in road accidents and vehicle crashes have been among the major causes of mortality [8]. Even a larger number of people suffers from non-fatal injuries with many incurring a disability due to the injury. Production of vehicles that ensure safety for all road users including occupants is crucial to reduce the road related injuries.

Most vehicle safety regulations require crash testing at a specialized facility to determine the crashworthiness parameters. Car manufacturers conduct full vehicle or Vulnerable Road User (VRU) tests to ensure that the car design meets the regulations. Usually, crash-testing is time consuming and costly. Mathematical models are employed to represent crash dynamics, for example, in the case of a car impacting a barrier or another car. These models involve differential equations of motion describing the deformation

Supported by University of Agder.

G. Wagner et al. (Eds.): SIMULTECH 2021, LNNS 601, pp. 1–23, 2023.
https://doi.org/10.1007/978-3-031-23149-0_1

of the parts in the vehicle. The occupants inside the car can also be included in a mathematical model to predict injury values during a crash, models of a human present a valuable complement to other models, such as animal models and crash dummies. The vehicle front-end and side structures have been modified to improve energy absorption capability [9]. Finite Element Methods (FEM) have received impetus in vehicle crash modeling in the past decades. With improved computational speeds the models became more accurate and reliable for vehicle development. Benson et al. [4] presented the calculations of crashworthiness design laying the foundations for application of FEM in the automotive industry. However, development of FE models is time-consuming and needs CAD data which is not available during the early stages of vehicle design. Lumped parameter models (LPM) were first applied for modeling vehicle crash events in [13] where the vehicle was represented by three lumped mass components and eight resistances representing the deformable parts in the vehicle. The cited paper paved the way for many more studies using LPMs to represent the behaviour of a vehicle and occupants under impact. Recently, LPMs were used by Elkady et al. [9, 10] to develop a multi-DOF mathematical model to simulate a crash event with active vehicle dynamics control systems (VDCS). The model replicated a full frontal and offset impact between two vehicles comparing the performance of a baseline vehicle with a vehicle equipped with VDCS features. It also includes a 3-DOF occupant impact model derived using Lagrangian formulation. Munyazikwiye et al. [20] use a mass-spring-damper model with two lumped mass components to represent a full frontal impact with a rigid barrier. The study shows good correlation with test data suggesting that a simple LPM can replicate the impact kinematics successfully.

Occupant injury prediction is an important area of research where the vehicle-occupant interaction in a vehicle impact scenario is studied and the injury patterns of occupants in the car are determined with a help of mathematical models. Large vehicle deceleration has been identified as one of the main causes of head and chest injuries, and vehicle rotational motions in different axes also lead to occupant injuries [5].

In a full frontal impact, vehicle pitch and drop are significantly larger compared to rolling and yawing motions. Neck injury is one of the most common types of injury in vehicle accidents [15]. In a vehicle crash, unbelted occupants could interact with the vehicle interiors leading to severe injuries. In the recent past, the research focusing on unbelted occupants to meet Federal Motor Vehicle Safety Standards (FMVSS 208) requirements demonstrated that vehicle pitch and drop contributed to higher head and neck injury values.

The objective of a vehicle structure is not just to absorb energy and optimize crash pulses, but also to minimize vehicle pitch and drop [5, 28]. Chang et al. [7] have developed an FE model to study vehicle pitch and drop in body-on-frame vehicles. The model is correlated to barrier tests and predicts factors affecting vehicle pitch and drop in a crash event. This research points to the fact that design of vehicle rails plays an important role in the load distribution during an impact scenario for body-on-frame vehicles. The out-of-plane bending of the vehicle rails increases the role of a vertical component of the barrier force, causing an imbalance in the vehicle, leading to forward pitching on the vehicle. Wei et al. [27] have estimated the relationship between energy absorbing components and the crash pulse, establishing that the bumper and the front rails both significantly contribute to the energy absorption in a full frontal crash event.

Researchers use different methodologies to improve vehicle crashworthiness modifying the vehicle structure or materials used to manufacture different vehicle parts. Genetic algorithm to estimate and optimize the vehicle parameters for a vehicle-vehicle impact was used in [21]. Li et al. [16] used lightweight optimization and material modification to meet crashworthiness requirements balancing contradictory vehicle dynamics and fuel economy requirements. The design optimization using a DOE to develop surrogate models reducing the pitch and drop in an FE model that improves interactions between the occupant's head and vehicle interior parts was presented in [6]. Our paper is an extension of the work presented by the authors at the conference SIMULTECH 2021 [23]. The model developed has been extended to validate a sedan FE model. We establish the robustness of the model to predict the impact of changes in stiffness of the vehicle on the reduction of vehicle pitching. To this end, we simplify the system splitting the vehicle motion into two phases as in [23]:

– the horizontal linear motion, and
– the rotation of the vehicle body.

We replicate a full frontal vehicle crash event at 56 km per hour (kmph) employing an LPM with multiple DOFs to predict

– the maximum deformation in the vehicle to absorb energy, and
– the pitch angle of the vehicle due to the crash response.

The model has been validated with a 2014 pickup truck (Chevrolet Silverado) and a 2010 sedan (Toyota Yaris). FE model simulations of the two cars were used to compare the LPM results. The Toyota Yaris FE model has also been modified to study stiffness variations in the crashworthiness of the vehicle. The LPM is robust to predict the changes in stiffness of the vehicle making this model suitable for prediction of injury parameters in a vehicle crash.

2 Methodology

Literature documents that a crash event leads to pitching, rolling and yawing of the vehicle along with the deceleration of the vehicle and movement in horizontal and vertical directions. It is difficult to model the impact scenario in different axes and to generate the governing equations. It was observed that the time for the vehicle to attain minimum velocity after impact coincides with the maximum deformation on the vehicle.

In this study, we separate the horizontal translational motion from the vertical motion during the impact event. In a full frontal crash event the vehicle experiences forward pitching; whereas the effect of rolling and yawing can be neglected. Taking into account these assumptions we split the crash event into two phases:

– time interval until maximum deformation and minimum vehicle velocity after start of crash event t_1, and
– time interval after maximum deformation to the end of the crash event t_2.

2.1 FEM Simulations for Validation

We conducted a FE simulation for a 2014 Chevrolet Silverado and a 2010 Toyota Yaris running at 56 kmph and hitting a frontal barrier at 0% offset. These FE models were developed by National Crash Analysis Center (NCAC) in collaboration with NHTSA (National Highway Traffic Safety Administration) through the reverse engineering process [2].

Chevrolet Silverado Model. The FE model in Fig. 1 consisting of 1,476 parts, 2,741,848 nodes and 2,870,507 elements has been correlated to NHTSA Oblique Test and Insurance Institute for Highway Safety (IIHS) Small Overlap Front Test. The FE model weighs 2,582 kg which is close to the physical test vehicle weighing 2,624 kg. It replicates the material and geometrical properties of the physical vehicle [26].

Fig. 1. An FE Model of a 2014 Chevrolet Silverado developed by NCAC [23].

Toyota Yaris Model. The FE model replicates a 2010 four-door passenger sedan consisting of 917 parts, 1,480,422 nodes and 1,514,068 elements. The FE model weighs 1,100 kg which is close to the physical test vehicle weighing 1,078 kg. The validation is conducted against an NCAP frontal wall impact with actual data from NHTSA Tests 5677 and 6221. It replicates the material and geometrical properties of the physical vehicle [18]. The model was also validated against test data from other scenarios. The curves correlate well with the test data and the FE model has been used by several authors [1].

The FE models were run on LS-DYNA with 32 CPUs in an HPC environment and the corresponding curves generated were used for the parameter estimation and validation of the LPM in MATLAB Simulink. In the FE simulation, the acceleration of some nodes on the vehicle body are recorded by the solver LS-DYNA. These nodes are selected by the user at the pre-processing stage. This process was employed to determine the acceleration of the vehicle center of gravity (CG) as well as the barrier forces, employed for the validation. Figures 1 and 2 show the FE model used in the simulations. These FE model generated the piecewise linear curve data for the spring and damper

Fig. 2. An FE model of a 2010 Toyota Yaris developed by NCAC.

coefficients. Newton-Euler numerical integration is used to calculate the values and predict the time for maximum dynamic crush of the vehicle. The algorithm is discussed in the following section.

The Toyota Yaris FE model was further updated to increase the stiffness by changing the thickness of all the metal parts by 10% and 20%. This was achieved by changing the shell element thickness on the FE regardless of the parts being load bearing or deformable members. The parts undergoing thickness change are represented in Fig. 3. The increase in vehicle mass, mass moment of inertia and, change in vehicle CG were noted and updated in the LPM. The FE simulations were run and validated against the LPM to compare the performance.

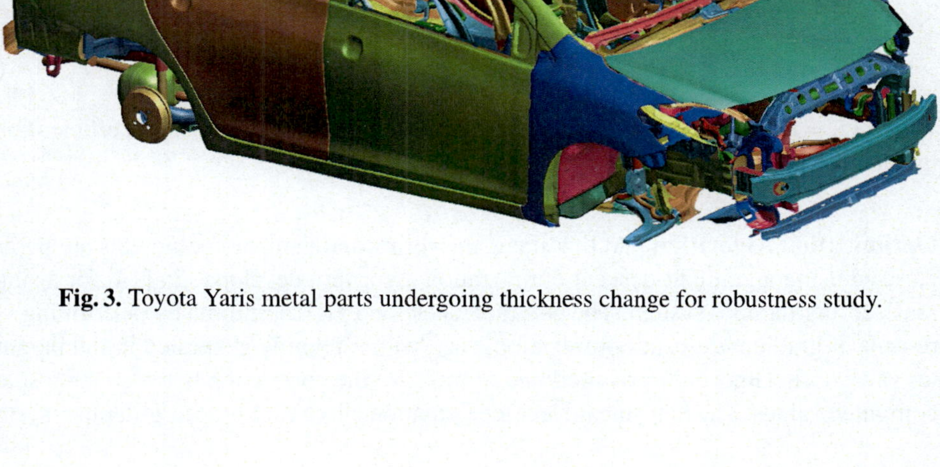

Fig. 3. Toyota Yaris metal parts undergoing thickness change for robustness study.

2.2 Description of the Lumped Parameter Model

As a single-mass system, the LPM developed utilizes a spring and damper system in the front side as the bumper and deformable system; this is commonly known as the Kelvin model. The front springs allow translational motion only in the direction of x-axis [12]. There are two pairs of springs and dampers in the suspension of the vehicle, allowing translation in the z axis and rotation around the y-axis. There are three degrees of freedom in this system, making it quite challenging to solve. The lumped mass body can move in the direction of horizontal (x) and vertical (z) axes along with the rotation around one (y) axis. The CG of the vehicle is located at a distance l_f from the front end and l_r from the rear end suspension points; l_0 represents the distance between the CG and the front occupant compartment zone. The two phases are described below.

2.3 Vehicle Crash Model - Phase I

First we model only the translational movement along the horizontal axis of the vehicle hitting the barrier at 0% offset. The LPM developed in Simulink replicates the maximum vehicle deformation until the time of maximum crush t_m. Additionally, this value corresponds to the instant the vehicle reaches its zero velocity or minimum speed. If the vehicle front end does not absorb energy by deforming plastically, then it is possible that it will not reach zero velocity by the time of maximum deformation. A single DOF equation with a spring-damper unit is used in the mathematical model. The stiffness of the spring is tuned to represent the maximum deformation of the vehicle at a particular speed. For this problem we have assumed the speed of 56 kmph (NHTSA regulations for frontal crash). The motion of suspension system in the model has been neglected during this phase of the event. Figure 4 represents the vehicle in a deformed state. The Simulink model predicts the time until the maximum deformation of the vehicle and the maximum displacement of the vehicle CG.

The prediction of the values of spring deformation coefficient k and damper coefficient c used in the general equation of motion has been a challenge for researchers in the past [17,24]. The stiffness of the vehicle front in a crash was estimated using various parameter estimation studies. Despite the highly nonlinear behavior of the front end, it was approximated by a piecewise linear relationship [3,19].

In our case the equation of motion assumes the form:

$$m\ddot{x} + c\dot{x} + kx = Q_i \tag{1}$$

where $Q_i = 0$ (i.e. no force component is added here); k is the spring coefficient; c is the damper coefficient for the bumper model.

Optimization Algorithm. At this stage, the spring and damper coefficients are parameterized using a gradient-descent optimization algorithm developed in [14] for a single mass-spring-damper system. The code searches for a global minima by performing 100 re-runs of gradient descent optimization, each with randomly generated initial parameter values. The algorithm was modified to improve the correlation between the test and computed values. The non-linear force-deformation curve for the spring-damper system

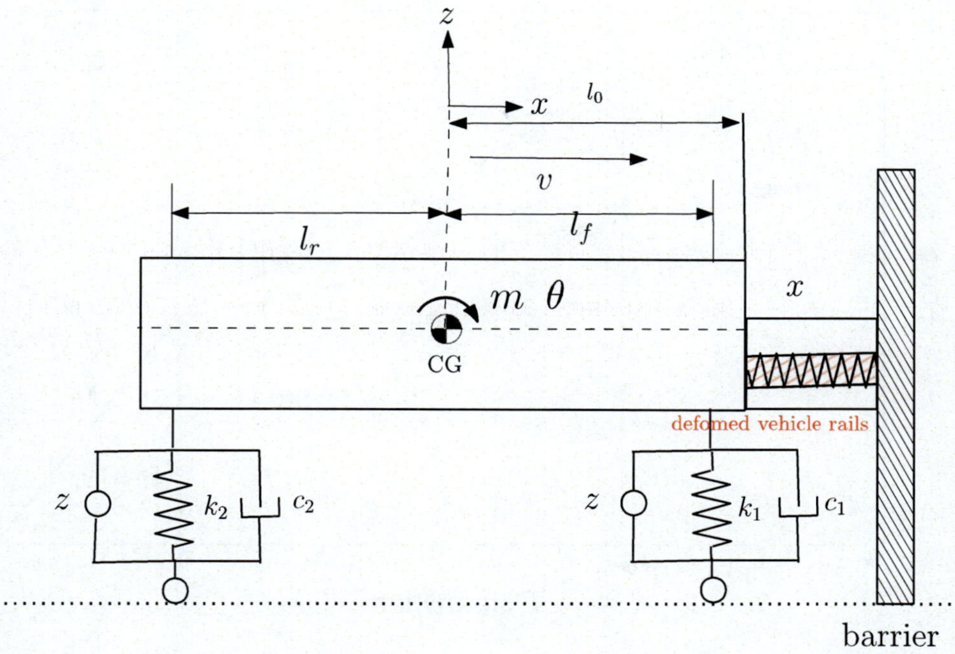

Fig. 4. Vehicle representation in Phase 1 of the event: Deformed front end [23].

is assumed to be piecewise-linear with six breakpoints in the curve. The forces on the spring are calculated using the general relationship between the force and deformation for a spring-damper system [9], see Fig. 5. The stiffness of the spring k and the spring force component F_k vary according to the deflection values in the spring.

The spring stiffness and damping coefficients in the model are defined as the piecewise-linear functions of x and $\dot{x}$, respectively:

$$k(x) = \begin{cases} \frac{(k_2-k_1)\cdot|\hat{x}|}{x_1} + k_1, & \text{for } |\hat{x}| \leq x_1, \\ \frac{(k_3-k_2)\cdot(|\hat{x}|-x_1)}{(x_2-x_1)} + k_2, & \text{for } x_1 \leq |\hat{x}| \leq x_2, \\ \frac{(k_4-k_3)\cdot(|\hat{x}|-x_2)}{(x_3-x_2)} + k_3, & \text{for } x_2 \leq |\hat{x}| \leq x_3, \\ \frac{(k_5-k_4)\cdot(|\hat{x}|-x_3)}{(x_4-x_3)} + k_4, & \text{for } x_3 \leq |\hat{x}| \leq x_4, \\ \frac{(k_6-k_5)\cdot(|\hat{x}|-x_4)}{(x_5-x_4)} + k_5, & \text{for } x_4 \leq |\hat{x}| \leq x_5, \\ \frac{(k_7-k_6)\cdot(|\hat{x}|-x_5)}{(C-x_5)} + k_6, & \text{for } x_5 \leq |\hat{x}| \leq C. \end{cases}$$

The damper characteristics are defined similarly to the spring characteristics:

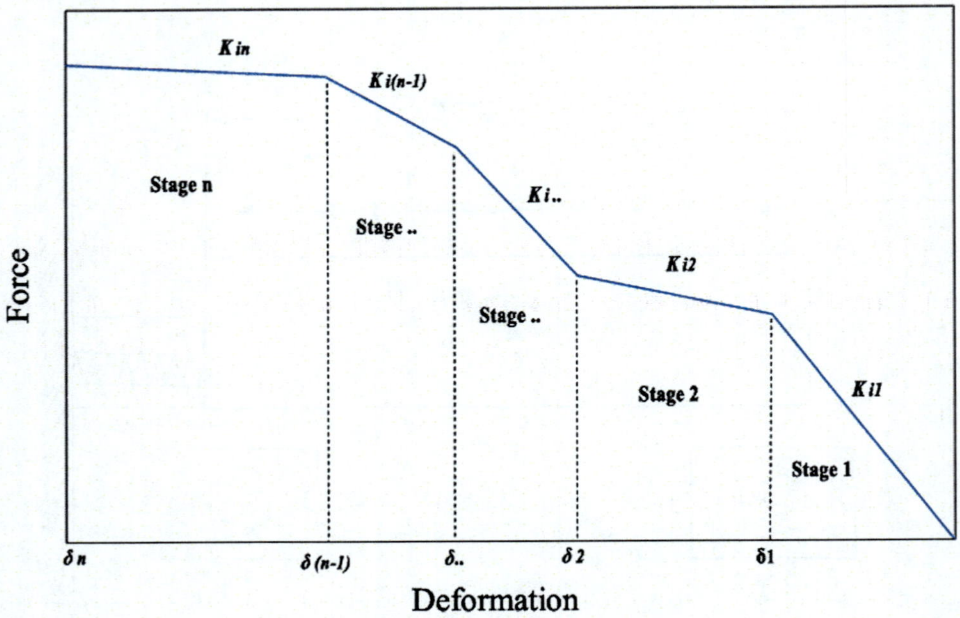

Fig. 5. General piecewise force-deformation characteristics [9,23].

$$c(\dot{x}) = \begin{cases} \frac{(c_2-c_1)\cdot|\hat{\dot{x}}|}{\dot{x}_1} + c_1, & \text{for } |\hat{\dot{x}}| \leq \dot{x}_1, \\ \frac{(c_3-c_2)\cdot(|\hat{\dot{x}}|-\dot{x}_1)}{(\dot{x}_2-\dot{x}_1)} + c_2, & \text{for } \dot{x}_1 \leq |\hat{\dot{x}}| \leq \dot{x}_2, \\ \frac{(c_4-c_3)\cdot(|\hat{\dot{x}}|-\dot{x}_2)}{(\dot{x}_3-\dot{x}_2)} + c_3, & \text{for } \dot{x}_2 \leq |\hat{\dot{x}}| \leq \dot{x}_3, \\ \frac{(c_5-c_4)\cdot(|\hat{\dot{x}}|-\dot{x}_3)}{(\dot{x}_4-\dot{x}_3)} + c_4, & \text{for } \dot{x}_3 \leq |\hat{\dot{x}}| \leq \dot{x}_4, \\ \frac{(c_6-c_5)\cdot(|\hat{\dot{x}}|-\dot{x}_4)}{(\dot{x}_5-\dot{x}_4)} + c_5, & \text{for } \dot{x}_4 \leq |\hat{\dot{x}}| \leq \dot{x}_5, \\ \frac{(c_7-c_6)\cdot(|\hat{\dot{x}}|-\dot{x}_5)}{(v_0-\dot{x}_5)} + c_6, & \text{for } \dot{x}_5 \leq |\hat{\dot{x}}| \leq v_0, \end{cases}$$

where k is the spring coefficient, c is the damper coefficient, $\hat{x}$ is the computed vehicle deformation, $\dot{x}$ is the vehicle velocity, $\hat{\dot{x}}$ is the computed vehicle velocity, v_0 is the velocity at the time of maximum dynamic crush, C is the maximum dynamic crush, F_k and F_c are the built-up spring and damping forces defined by the following equations

$$F_k = k(x) \cdot x, \tag{2}$$

$$F_c = c(\dot{x}) \cdot \dot{x}. \tag{3}$$

The proposed algorithm uses an optimization approach to minimize an objective function. The objective function to be minimized is the error function $E(\Theta, t)$ where Θ denotes the unknown variables in the mode. The error function is defined as follows: $E(\Theta, t) = E_1(\Theta, t) + E_2(\Theta, t) + E_3(\Theta, t)$ where

$$E_1(\Theta, t) = |(a_{FE} - a_{LPM})|, \tag{4a}$$

$$E_2(\Theta, t) = |(v_{FE} - v_{LPM})|, \tag{4b}$$

$$E_3(\Theta, t) = |(x_{FE} - x_{LPM})|, \tag{4c}$$

where a is the acceleration, v is the vehicle velocity, and x is the displacement. The error function $E(\Theta, t)$ determines the difference between the FE values and LPM values at every point, and the optimization algorithm minimizes the error values by altering $\Theta = [k_i, c_i], \quad \forall i \in [1, 7]$. The corresponding spring and damper coefficient values obtained from this minimization algorithm are discussed in the results section.

2.4 Vehicle Crash Model - Phase II

The second phase for the model describes what happens after the instant the vehicle achieves maximum dynamic crush and minimum velocity. At this instant the vehicle starts to pitch forward. Several studies were conducted to understand the vehicle pitching forward [5,22] suggesting that for the body-on-frame vehicles one of the reasons is the out-of-plane bending in vehicle rails which leads to the appearance of a vertical force component in the moment balance equation. This vertical force component is added to gravity force acting downwards and creates an imbalance of loading resulting in the vehicle pitching. The prediction of the pitching angle is important for determining the injury to occupants. Low pitching angles influences occupant protection design in a vehicle. This phase of the event is shown in Fig. 6 and Fig. 7. We consider here only vertical motion of the suspension springs and the rotation about the y-axis with angle θ.

The dynamics in the Lagrangian formulation is described by the equation [11]:

$$\frac{d}{dt}\frac{\partial L}{\partial \dot{q}_i} - \frac{\partial L}{\partial q_i} + \frac{\partial D}{\partial q_i} = Q_i \tag{5}$$

where, in the general case, $L = T - V$, T is the total kinetic energy of the system equal to the sum of the kinetic energies of the particles, $q_i\, i = 1, \ldots, n$ are generalized coordinates, Q_i is the external force acting on the system, which in this case is the vertical force component experienced by the vehicle at the time of maximum dynamic crush, and V is the potential energy of the system. For dissipation forces, a special function D must be introduced alongside L.

The equations for the kinetic, potential and dissipation energy are:

$$T = \frac{1}{2}J\dot{\theta}^2 + \frac{1}{2}m\dot{x}^2, \tag{6}$$

$$V = \frac{1}{2}k_1(x - l_f\theta)^2 + \frac{1}{2}k_2(x + l_r\theta)^2, \tag{7}$$

and

$$D = \frac{1}{2}c_1(\dot{x} - l_f\dot{\theta})^2 + \frac{1}{2}c_2(\dot{x} + l_r\dot{\theta})^2. \tag{8}$$

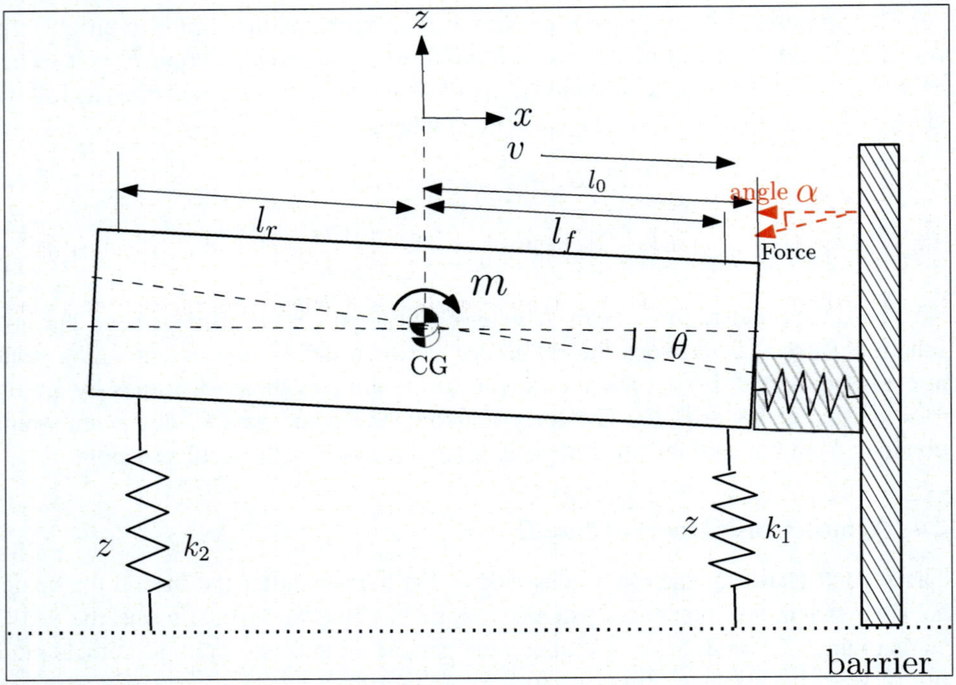

Fig. 6. Vehicle representation in Phase II of the event: Vehicle Pitching forward [23].

The values of standard automotive parameters k_1, k_2, c_1, c_2, l_f and l_r are taken from the literature, [25], see Table 1.

The values of the vehicle mass m and the moment of inertia J for the lumped mass system are calculated from the FE model of the vehicle. The governing equations of motion are [23]:

$$\begin{aligned} Q_i &= J\ddot{\theta} + (k_1 l_f^2 + k_2 l_r^2)\theta + (c_1 l_f^2 + c_2 l_r^2)\dot{\theta} \\ &\quad + (k_2 l_r - k_1 l_f)x + (c_2 l_r - c_1 l_f)\dot{x}, \end{aligned} \tag{9}$$

$$\begin{aligned} -Q_i &= m\ddot{x} + (k_1 + k_2)x + (c_1 l_f + c_2 l_r)\dot{\theta} \\ &\quad + (k_2 l_r - k_1 l_f)\theta + (c_1 + c_2)\dot{x}. \end{aligned} \tag{10}$$

2.5 Robustness Check

The LPM predicts important vehicle parameters, thus, contributing to analysis of vehicle crashworthiness. The model was validated to estimate the injury parameters for a truck and a sedan. The sensitivity of the model to stiffness is assessed by changing the thickness of the material; analyzing the spring and damper coefficient curves generated with the help of the optimization algorithm described in Sect. 2.3. The methodology for determining the vehicle parameters (maximum displacement, time for zero velocity and maximum pitch angle) is similar to the baseline model. It is interesting to observe

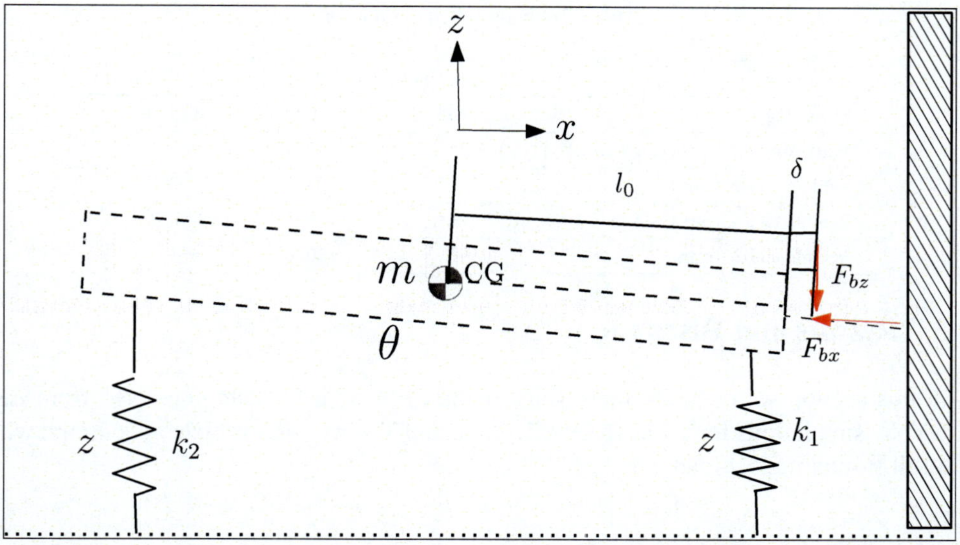

Fig. 7. Vehicle representation in Phase II with forces acting on the vehicle and suspension springs in play [23].

Table 1. Automotive parameters set [25].

Symbol	Value	Unit	Meaning
M	400	kg	Sprung mass
m_{ij}	50	kg	Unsprung masses (i = front, rear and j = left, right)
I_x	250	kg.m^2	Roll inertia
I_y	1400	kg.m^2	Pitch inertia
t	1.4	m	Front and rear axle
l_f	1.4	m	COG-front distance
l_r	1	m	COG-rear distance
r	0.3	m	Nominal wheel radius
h	0.7	m	Chassis COG height
k_f	30,000	N/m	Front suspension linearized stiffness (left, right)
k_r	20,000	N/m	Rear suspension linearized stiffness (left, right)
c_f	1500	N/m/s	Front suspension linearized damping (left, right)
c_r	3000	N/m/s	Rear suspension linearized damping (left, right)
k_t	200,000	N/m	Tire stiffness (front, rear and left, right)
β	50	rad/s	Suspension actuator bandwidth

the changes in vehicle pitching angle and acceleration by adding mass to the system in terms of elemental thickness to the metal parts. The changes in mass and moment of inertia for the model are presented in Table 2 below. These changes have been incorporated in the LPM to determine injury values.

Table 2. FE model specifications.

Parameter	Yaris baseline	10% stiffness	20% stiffness
Mass, (kg)	1253.5	1303.9	1353.6
Mass Moment of Inertia, (kgm^2) (I_{xx})	425128	445472	465782
Vehicle CG x, (mm)	1025	1033.6	1039
Vehicle CG y, (mm)	-3.0	-2.1	-1.5
Vehicle CG z, (mm)	557	560	563

3 Results and Discussion

In this section we compare the results of the LPM with FE data generated from LS-DYNA simulations for a Chevrolet Silverado and Toyota Yaris vehicle at 56 kmph with a full frontal impact loadcase.

3.1 Phase I

Baseline Chevrolet Silverado Model. First we simulate the time until maximum deformation of the vehicle; the spring and damper coefficients are determined using the Gradient Descent Optimization with an error function defined in Sect. 2.3. The computed and test (FE) values are plotted in Fig. 8; They show good correlation of results. The predicted values of the stiffness and damping coefficients are shown in Figs. 9 and 10.

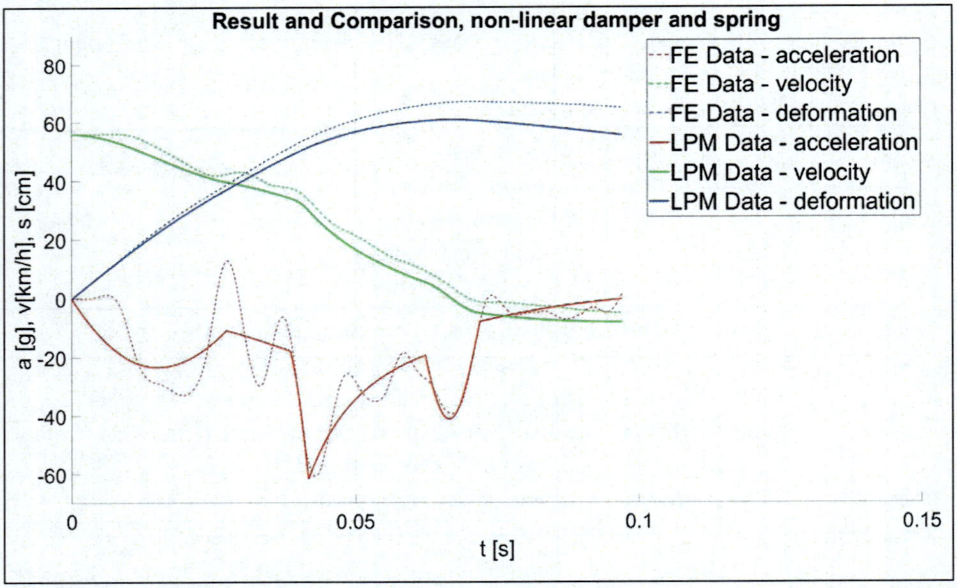

Fig. 8. Plot of computed and test values for parameter model for a Chevrolet Silverado Model.

The output from the Gradient Descent Optimization algorithm is used to predict the deformation and vehicle velocity in a MATLAB Simulink model.

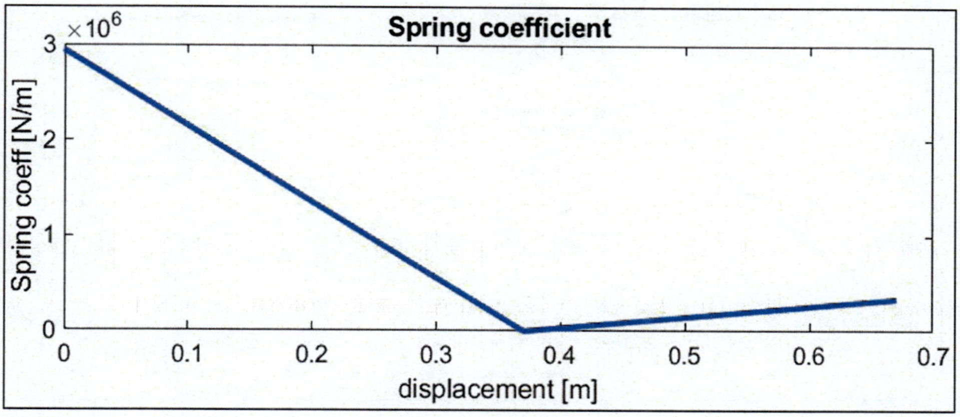

Fig. 9. Spring coefficient obtained from the algorithm for Chevrolet Silverado Baseline model.

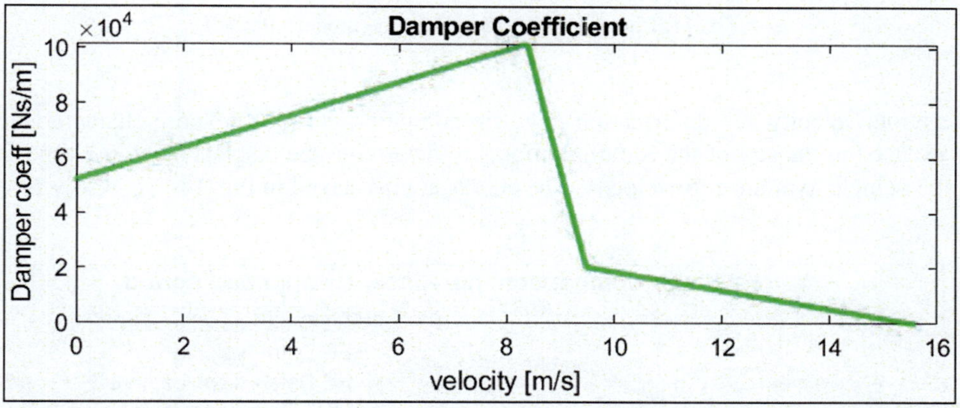

Fig. 10. Damper Coefficient obtained from the algorithm for Chevrolet Silverado Baseline model.

The plots of maximum FE vehicle deformation and LPM deformation in Fig. 11a show good correlation. A similar plot (Fig. 11b) was generated to compare the velocity of the vehicle at the CG; in the case of LPM at the lumped mass center. The LPM is represented by the mathematical model in the plots. The time the vehicle attains zero velocity is closely correlated in the plots but there is a small deviation after 40 ms. The reason for this deviation can be attributed to the spring and damper characteristics which are approximated in this study using a piece-wise linear function. The model can be improved using a non-linear function for the spring stiffness and damping characteristics. If the model is simulated beyond the time the vehicle attains zero velocity, a rebound is observed in the velocity. This velocity rebound could be due to the internal strain energy stored in the springs, and it would be interesting to investigate this further in the future.

Baseline Toyota Yaris Model. The baseline 2010 Yaris model FE simulations were used for estimating the front-end spring-damper characteristics shown in Fig. 13; accel-

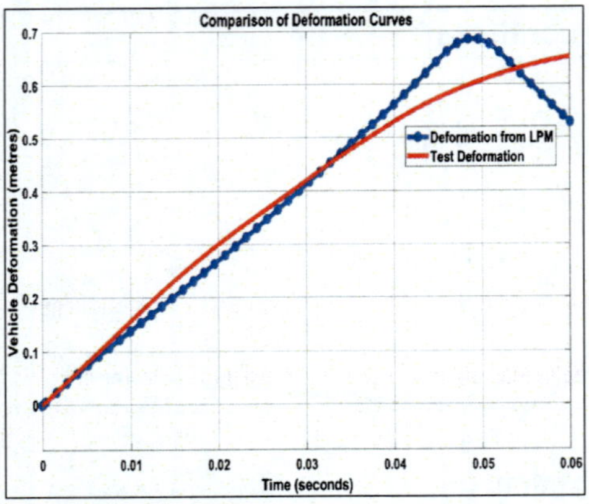

(a) Displacement of the vehicle CG curves comparison for LPM vs FE model

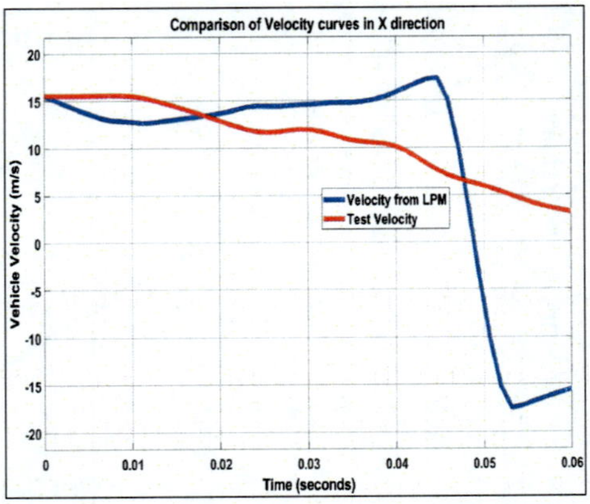

(b) velocity of the vehicle CG curves comparison for LPM vs FE model

Fig. 11. Chevrolet Silverado Baseline model - Phase I.

eration, velocity and deformation plots are compared in Fig. 12. These characteristics are used in Phase I of the Simulink model to determine the displacement and time for the vehicle to attain zero velocity. The curves are overlayed in Fig. 14.

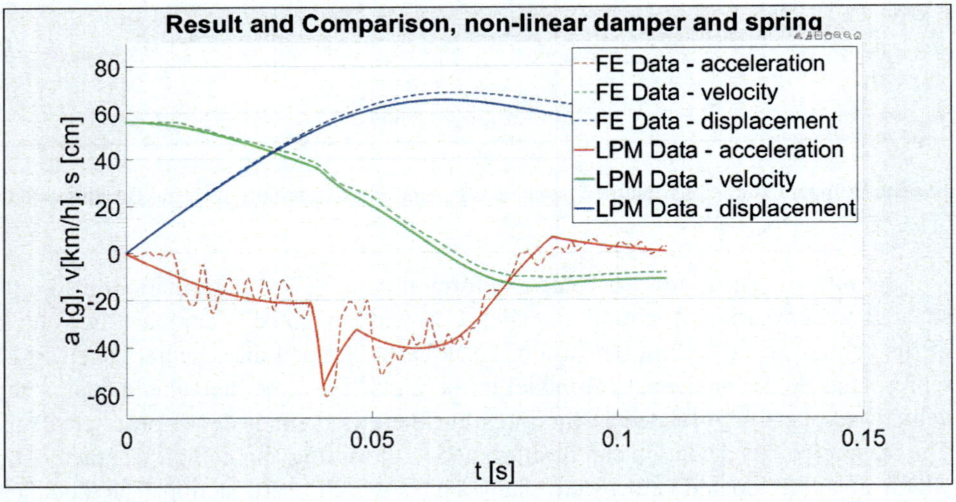

Fig. 12. Plot of computed and FE test values for lumped parameter model for a Toyota Yaris Model.

3.2 Phase II

The prediction for the second part of the lumped model using Simulink was conducted and plotted against the data from FE model. The quantity Q_i in the governing equations

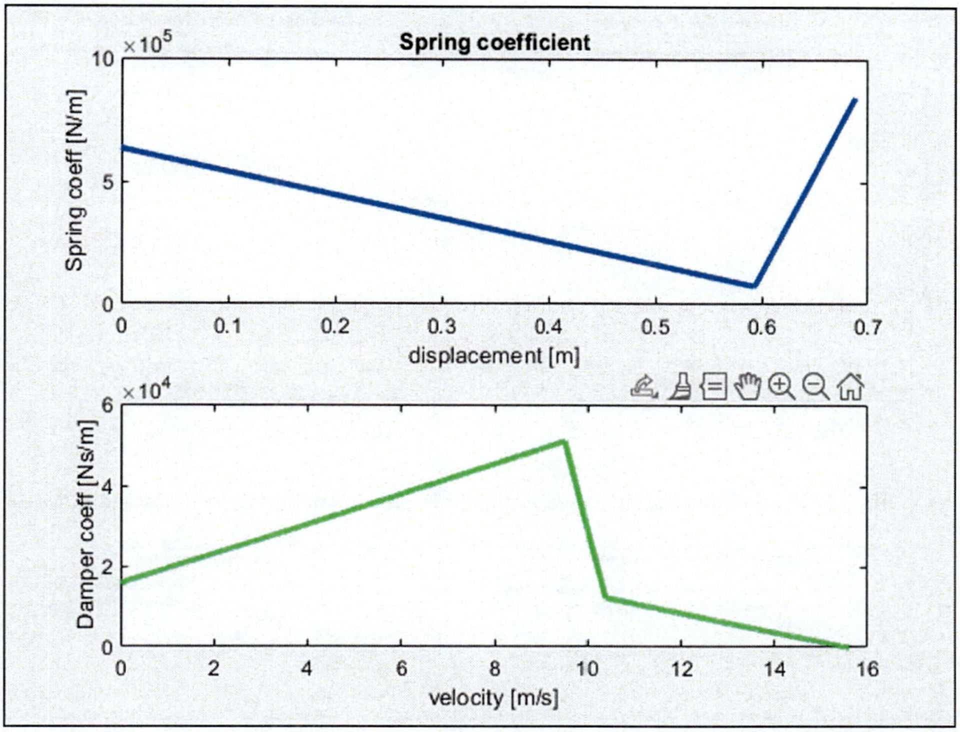

Fig. 13. Damper Coefficient obtained from the algorithm for Toyota Yaris Base-line model.

is the vertical component of the barrier force experienced by the vehicle in the crash. The force curve is derived from the FE model and inputted into the Simulink model to improve the prediction; it will be of interest to mathematically explain this force component in terms of residual impact energy after absorption. The Simulink model is run with numerical integration (variable timestep- ode 45) and the velocity of the lumped mass in z-direction along with the pitching angle is compared to the data from FE model.

Baseline Chevrolet Silverado Model. Figure 15a compares the z-velocity (vertical velocity) in the body with the curves generated from the FE data. The trend in the curve is similar but the peak values are not matching. One of the contributing factors to this deviation is the use of standard linear spring and damper coefficient values for the model. The use of the linear approximation for the spring and damper coefficients can lead to the difference in the values for this parameter as well. The values of l_f and l_r can also be further tuned to represent the Chevrolet Silverado (2014) model. However, we intentionally avoided fine tuning these values assuming that this data may not be available to vehicle development team at the start of the design process and it makes sense to use standard values for automotive parameters. Figure 15b compares the forward pitching angle for the FE model and the LPM developed in this study. The pitch angle comparison shows a similar trend observed in both curves. The vehicle starts to

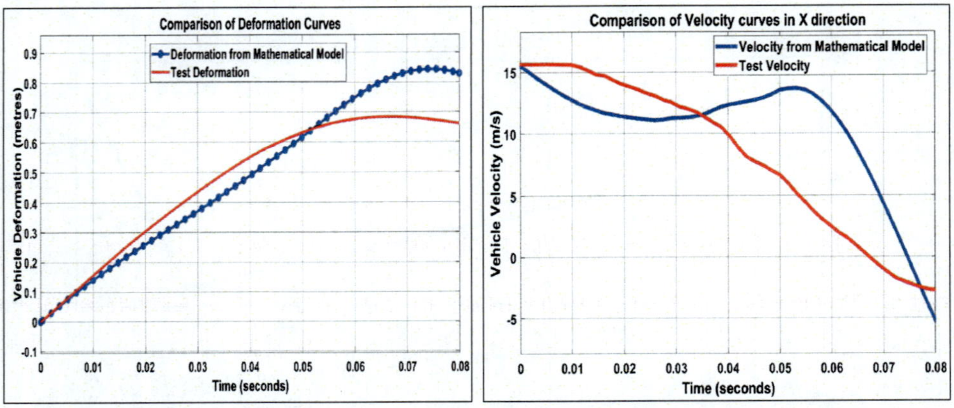

(a) Velocity curves overlaid **(b)** Displacement curves overlaid

Fig. 14. Toyota Yaris Baseline model - Phase I : Overlay of curves for LPM and FEM.

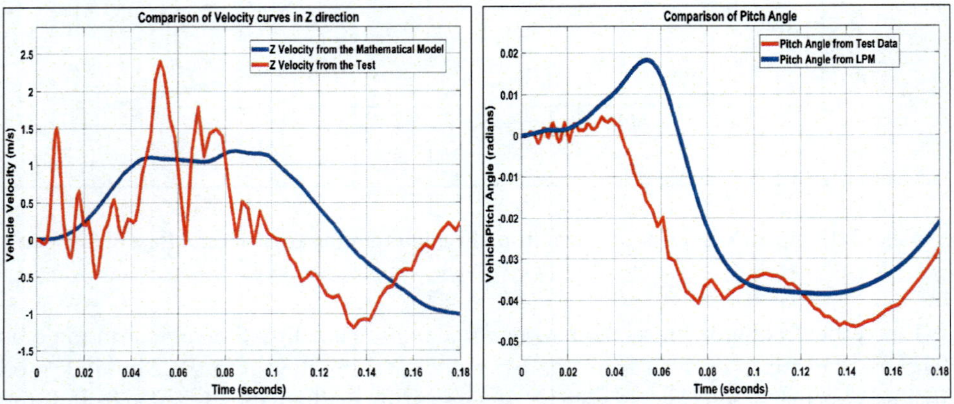

(a) Z-Velocity curve overlay for LPM vs FE model **(b)** Forward Pitch Angle curve overlay for LPM vs FE model

Fig. 15. Chevrolet Silverado Baseline Model - Phase II : Overlay of curves for LPM and FEM.

pitch around the same time during the crash event; this is crucial for designers planning airbag deployment in vehicles and other active protection features. The pitch angle curve for the simulation LPM peaks higher than the FE data at the start of the vehicle rotation but slowly follows the FE data curve showing comparable maximum pitch angle values. In addition, this is also a very important observation for vehicle safety designers. The difference between the curves can be explained by the linear approximation for the spring and damper coefficients and the barrier force definition. The study did not account for energy losses that may exist in the model.

Baseline Toyota Yaris Model. Similar to Phase I, the z velocity and pitch of the vehicle is overlaid for the Yaris model in Phase II of the impact presented in Fig. 16. The observations for the prediction of the injury parameters are consistent with the truck model prompting the reliability of the model for different vehicle platforms.

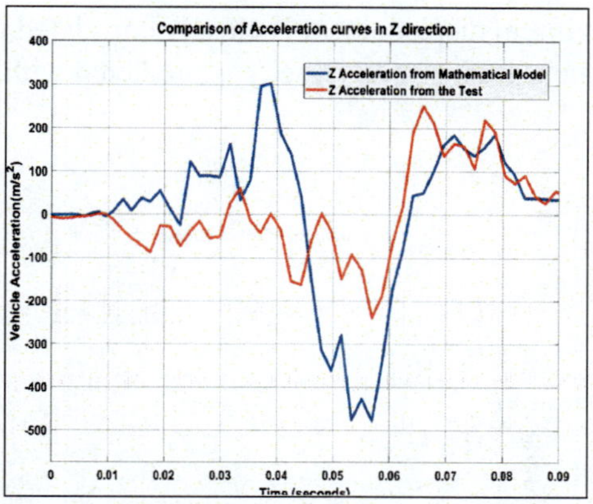

(a) Z- Acceleration curve

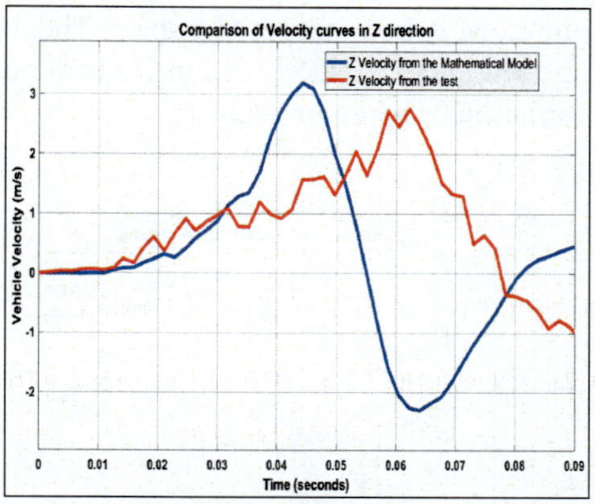

(b) Z- Velocity curve

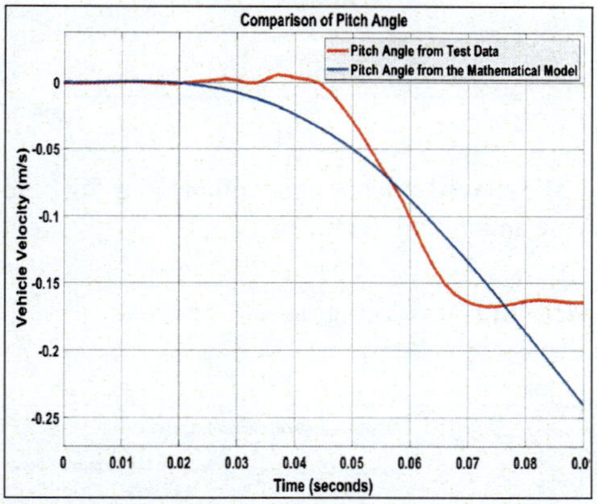

(c) Pitch curve

Fig. 16. Toyota Yaris Baseline - Phase II : Overlay of curves for LPM and FEM.

3.3 Robustness Check

According to Sect. 2.5 the LPM was used to predict stiffness changes in the model by changing the thickness of the model by

- increasing thickness of all metal parts by 10%
- increasing thickness of all metal parts by 20%

Figure 17 shows the acceleration and pitch curves for the baseline Toyota Yaris model and the modified models. It is observed that increasing the thickness of the parts reduces the peak acceleration values along with the vehicle pitching forward. However, the trend is non-linear indicating that only increasing the thickness is not a possible countermeasure to improve vehicle crashworthiness. There are other contributing variables which could help reduce the injury values in a crash.

Phase I - 10% Thickness. The spring and damper coefficient curves; mass/moment of inertia changes are updated in the Simulink model to determine the performance of the LPM in both phases of impact. The maximum displacement is closely correlated to the test data in Fig. 18a and the time the vehicle attains zero velocity is predicted with a variation of approximately 10 ms.

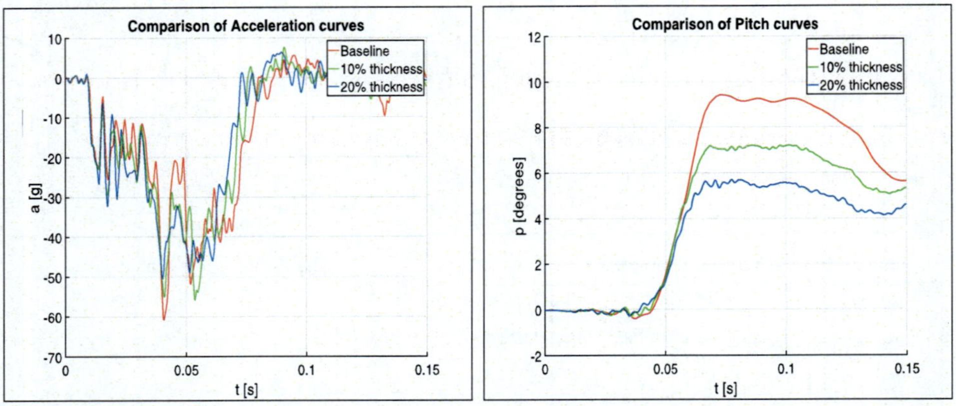

(a) X-acceleration comparison for baseline and modified models **(b)** Forward Pitch Angle comparison for baseline and modified models.

Fig. 17. Toyota Yaris Model - Stiffness Variation.

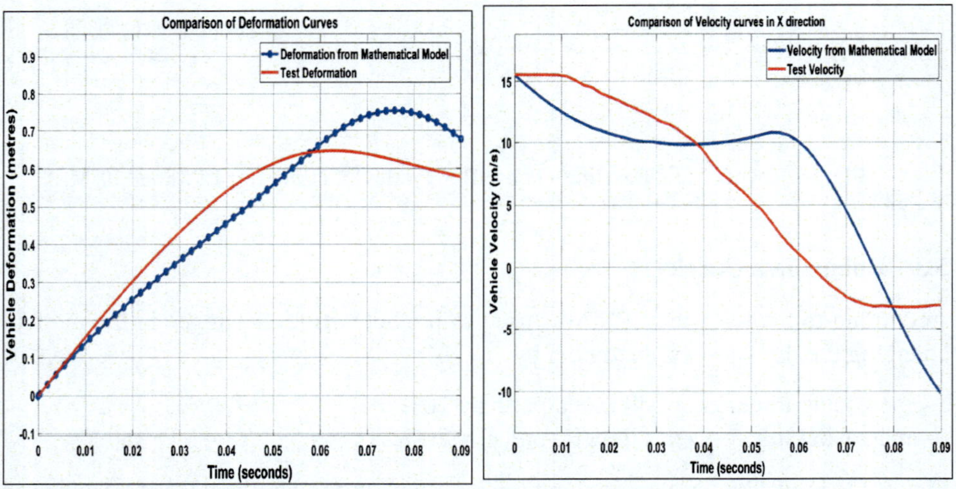

(a) Deformation curves overlaid **(b)** Velocity curves overlaid.

Fig. 18. Toyota Yaris Model - 10% Thickness Variation - Phase I : Overlay of curves for LPM and FEM.

Phase II - 10% Thickness. The simulation was repeated for the 10% stiffness model using Simulink to simulate the impact kinematics and predict the front-end deformation to absorb the energy of the impact; along with the forward pitching of the vehicle. The model predicts the maximum pitching angle and the z acceleration curve is closely replicated in the LPM simulations, see Fig. 19.

Phase I - 20% Thickness. The results of Phase I with thickness modification are presented in Fig. 20 and show good correlation with the results. The gap in correlation is consistent with the observations outlined with the Silverado model.

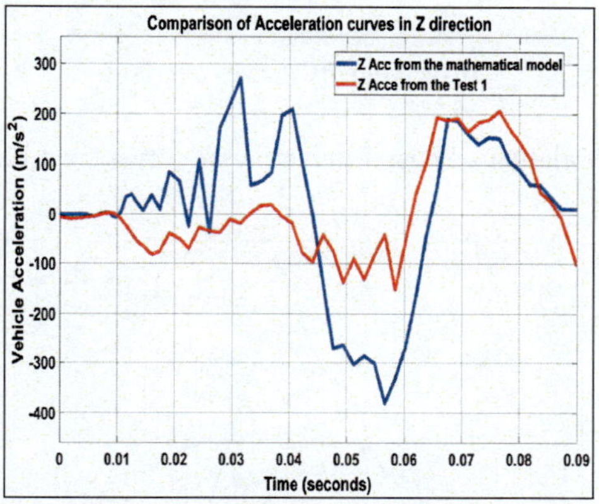

(a) Z Acceleration curves overlaid

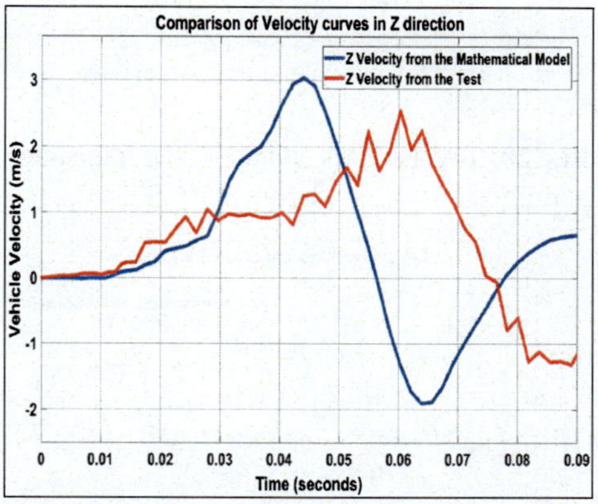

(b) Z Velocity curves overlaid

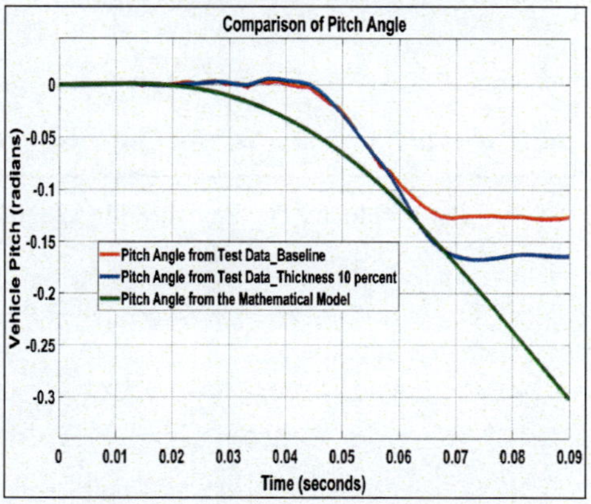

(c) Forward Pitching curves overlaid

Fig. 19. Toyota Yaris Model - 10% Thickness Variation - Phase II : Overlay of curves for LPM and FEM.

Phase II - 20% Thickness. The pitching curve in Fig. 21c follows the trend of the FE test data, however, the LPM simulation deviates from the test curve after the time of maximum pitching. This can be attributed to the constant stiffness of the suspension

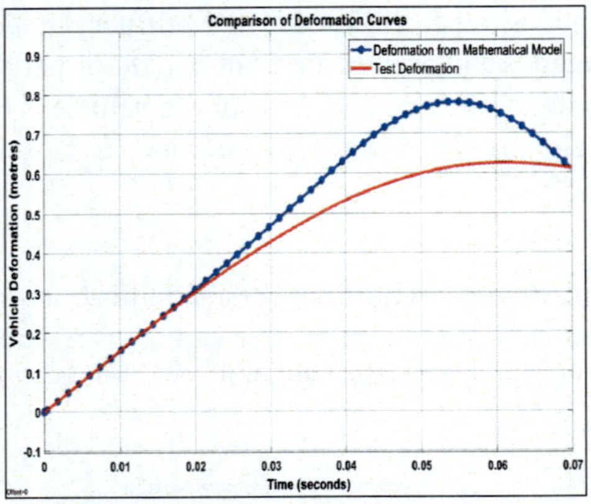

(a) Deformation curves overlaid

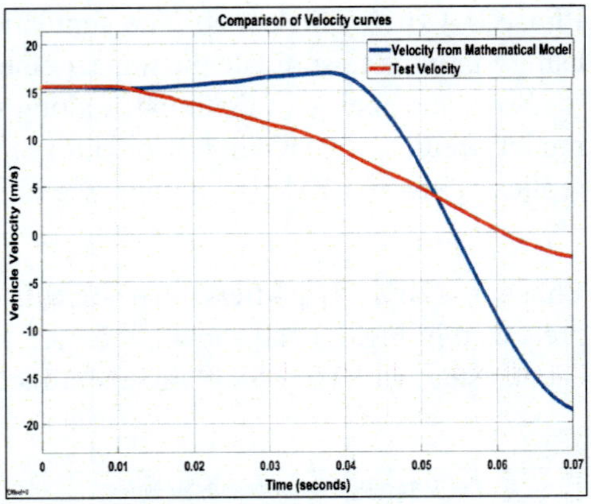

(b) x Velocity curves overlaid

Fig. 20. Toyota Yaris Model - 20% Thickness Variation - Phase I : Overlay of curves for LPM and FEM.

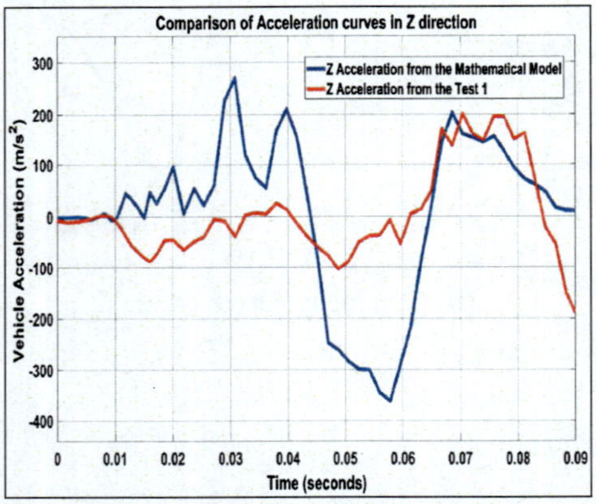

(a) z Acceleration curves overlaid

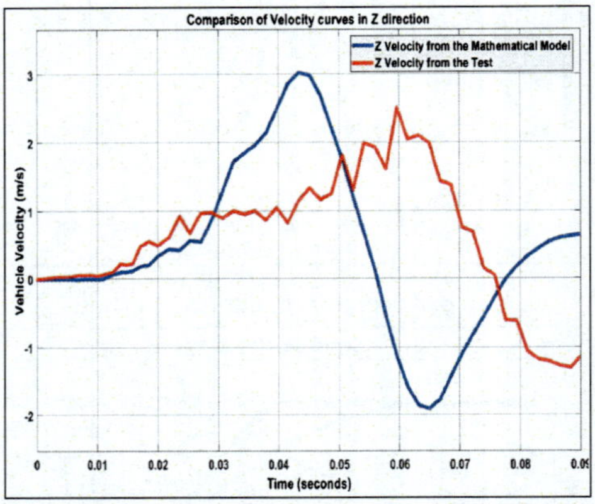

(b) z Velocity curves overlaid

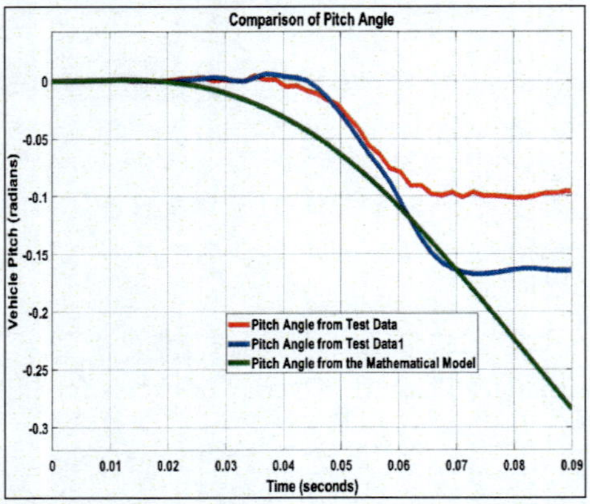

(c) Forward Pitch curves overlaid

Fig. 21. Toyota Yaris Model - 20% Thickness Variation - Phase II : Overlay of curves for LPM and FEM.

springs and damper coefficients. The LPM can be further improved by providing non-linear stiffness and damper characteristics representing the vehicle suspension system.

The change in stiffness is closely predicted in both models for the z acceleration and maximum pitching angle indicating a high reliability of the model. The maximum deformation and time to attain zero velocity is also correlated well in the LPM simulations; the differences in the result can be attributed to the assumption of linear characteristics for non-linear front-end spring data.

4 Conclusions

The novel technique developed in this paper for modeling a full frontal vehicle crash event successfully predicts the event kinematics. The study demonstrates that the two phase simulation model can describe a highly complex dynamical multiple DOF system with few equations and parameters, making the process of using LPMs very simple and reliable for safety design engineers. The robustness check and stiffness variation analysis indicates that the model is reliable and predicts variations in the parameters to determine injury values. The increase in thickness of the model by 10% and 20% improved the crashworthiness of the vehicle. Reducing the pitching angle would reduce the likelihood of injury to the occupants. The study highlights that parameter identification is an important part of the accident reconstruction process and influences the crashworthiness performance of the vehicle.

The assumptions used to arrive at a simpler LPM model providing reliable results include the following:

- The spring and damper characteristics are assumed to be piecewise-linear with six breakpoints although they are non-linear in physical systems.
- The suspension spring and damper coefficients were assumed same for the truck and sedan model used in the validation study.
- The vehicle acceleration is assumed to be zero at the time pitching starts in the crash event.
- Energy losses like friction and heat losses in the vehicle during the crash event are neglected to simplify the problem.
- Only vehicle rotations about the y-axis (pitching) are considered for modeling in the full frontal impact scenario; rotations about other axes are considered negligible and not impacting the occupant injuries.

The next steps include improving the prediction of deformation and pitching by using a non-linear force deformation curve for the spring stiffness curve with larger breakpoints along with including energy losses in the model. The LPM currently uses standard spring and damper coefficients for the suspension model which can be tuned to match a particular vehicle being studied. The model can be further validated by changing other vehicle parameters like design of the vehicle rails and defining the forces acting on the model with a mathematical expression. The simulation can be extended to include other vehicle crash scenarios to help design teams improve the vehicle crashworthiness without conducting complex time consuming simulations.

References

1. Center for Collision Safety and Analysis - 2010 Toyota Yaris. https://www.ccsa.gmu.edu/models/2010-toyota-yaris/
2. National Highway Traffic Safety Administration et al.: Crash simulation vehicle models (2016). https://www.nhtsa.gov/research-data/databasesand-software
3. Munyazikwiye, B.B., Vysochinskiy, D., Khadyko, M., Robbersmyr, K.G.: Prediction of vehicle crashworthiness parameters using piecewise lumped parameters and finite element models. Designs **2**(4), 43 (2018)
4. Benson, D., Hallquist, J., Igarashi, M., Shimomaki, K., Mizuno, M.: Application of dyna3d in large scale crashworthiness calculations. Technical report, Lawrence Livermore National Lab (1986)
5. Chang, J.M., Ali, M., Craig, R., Tyan, T., El-Bkaily, M., Cheng, J.: Important modeling practices in CAE simulation for vehicle pitch and drop. SAE Trans. 62–72 (2006)
6. Chang, J.M., Huang, M., Tyan, T., Li, G., Gu, L.: Structural optimization for vehicle pitch and drop. In: SAE Technical Papers. SAE International (2006). https://doi.org/10.4271/2006-01-0316. https://www.sae.org/publications/technical-papers/content/2006-01-0316/
7. Chang, J.M., Rahman, M., Ali, M., Tyan, T., El-Bkaily, M., Cheng, J.: Modeling and design for vehicle pitch and drop of body-on-frame vehicles. SAE Trans. 329–338 (2005)
8. Du Bois, P., et al.: Vehicle crashworthiness and occupant protection (2004)
9. Elkady, M., Elmarakbi, A.: Modelling and analysis of vehicle crash system integrated with different VDCS under high speed impacts. Open Eng. **2**(4), 585–602 (2012)
10. Elkady, M., Elmarakbi, A., MacIntyre, J.: Enhancement of vehicle safety and improving vehicle yaw behaviour due to offset collisions using vehicle dynamics. Int. J. Veh. Saf. **6**(2), 110–133 (2012)
11. Goldstein, H., Poole, C., Safko, J., Addison, S.R.: Classical mechanics. 3rd ed. Am. J. Phys. **70**(7), 782–783 (2002). https://doi.org/10.1119/1.1484149. http://aapt.scitation.org/doi/10.1119/1.1484149
12. Huang, M.: Vehicle Crash Mechanics. CRC Press (2002)
13. Kamal, M.M.: Analysis and simulation of vehicle to barrier impact. SAE Trans. 1498–1503 (1970)
14. Klausen, A., Tørdal, S.S., Karimi, H.R., Robbersmyr, K.G., Ječmenica, M., Melteig, O.: Mathematical modeling and optimization of a vehicle crash test based on a single-mass. In: Proceeding of the 11th World Congress on Intelligent Control and Automation, pp. 3588–3593. IEEE (2014)
15. Li, F., et al.: A review of neck injury and protection in vehicle accidents. Transp. Saf. Environ. **1**(2), 89–105 (2019). https://doi.org/10.1093/TSE/TDZ012. https://academic.oup.com/tse/article/1/2/89/5618803
16. Li, Z., Yu, Q., Zhao, X., Yu, M., Shi, P., Yan, C.: Crashworthiness and lightweight optimization to applied multiple materials and foam-filled front end structure of auto-body. **9**(8), 1–21 (2017). https://doi.org/10.1177/1687814017702806. https://journals.sagepub.com/doi/full/10.1177/1687814017702806
17. Marzbanrad, J., Pahlavani, M.: A system identification algorithm for vehicle lumped parameter model in crash analysis. Int. J. Model. Optim. **1**(2), 163 (2011)
18. Marzougui, D., Brown, D., Park, H.K., Kan, C.D., Opiela, K.S.: Development & validation of a finite element model for a mid-sized passenger sedan. In: 13th International LS-DYNA Users Conference Session: Automotive (2014)
19. Munyazikwiye, B.B., Karimi, H.R., Robbersmyr, K.G.: Application of genetic algorithm on parameter optimization of three vehicle crash scenarios. IFAC-PapersOnLine **50**(1), 3697–3701 (2017)

20. Munyazikwiye, B.B., Karimi, H.R., Robbersmyr, K.G.: Mathematical modeling of vehicle frontal crash by a double spring-mass-damper model. In: 2013 XXIV International Conference on Information, Communication and Automation Technologies (ICAT), pp. 1–6. IEEE (2013)
21. Munyazikwiye, B.B., Karimi, H.R., Robbersmyr, K.G.: Optimization of vehicle-to-vehicle frontal crash model based on measured data using genetic algorithm. IEEE Access **5**, 3131–3138 (2017). https://doi.org/10.1109/ACCESS.2017.2671357
22. Noorsumar, G., Robbersmyr, K., Rogovchenko, S., Vysochinskiy, D.: Crash response of a repaired vehicle - influence of welding UHSS members. In: WCX SAE World Congress Experience. SAE International (2020). https://doi.org/10.4271/2020-01-0197
23. Noorsumar, G., Rogovchenko, S., Robbersmyr, K.G., Vysochinskiy, D., Klausen, A.: A novel technique for modeling vehicle crash using lumped parameter models. In: Proceedings of the 11th International Conference on Simulation and Modeling Methodologies, Technologies and Applications, SIMULTECH 2021, pp. 62–70 (2021). https://doi.org/10.5220/0010529200620070
24. Pawlus, W., Karimi, H.R., Robbersmyr, K.G.: Development of lumped-parameter mathematical models for a vehicle localized impact. J. Mech. Sci. Technol. **25**(7), 1737–1747 (2011)
25. Savaresi, S.M., Poussot-Vassal, C., Spelta, C., Sename, O., Dugard, L.: Semi-active Suspension Control Design for Vehicles. Elsevier (2010)
26. Singh, H., et al.: Structural countermeasure/research program mass and cost increase due to oblique offset moving deformable barrier impact test. Technical report (2018)
27. Wei, Z., Karimi, H.R., Robbersmyr, K.G.: Analysis of the relationship between energy absorbing components and vehicle crash response. Technical report, SAE Technical Paper (2016)
28. Woitsch, G., Sinz, W.: Influence of pitching and yawing during frontal passenger vehicle crash tests on driver occupant's kinematics and injury. Int. J. Crashworthiness **18**(4), 356–370 (2013)

Integrating GPU-Accelerated Tetrahedral Mesh Editing and Simulation

Daniel Ströter[1(✉)], Andreas Halm[2,3], Ulrich Krispel[2,3], Johannes S. Mueller-Roemer[1,4], and Dieter W. Fellner[1,3,4]

[1] Interactive Graphics Systems Group, TU Darmstadt, Darmstadt, Germany
{daniel.stroeter,d.fellner}@gris.tu-darmstadt.de
[2] Fraunhofer Austria Research GmbH, Vienna, Austria
{andreas.halm,ulrich.krispel}@fraunhofer.at
[3] Institute of Computer Graphics and Knowledge Visualization, TU Graz, Graz, Austria
[4] Fraunhofer IGD, Darmstadt, Germany
johannes.mueller-roemer@igd.fraunhofer.de

Abstract. The use of computer-aided methods for the design of parts that must meet functional or stability requirements typically consists of an iterative cycle of design, physical simulation and testing or analysis, followed by redesign, etc. Each step is often performed with a domain-specific tool, e.g., a specific CAD modeling suite. This results in the need to convert the model representation between steps, such as meshing for finite element simulation for example.

In recent work, a distributed application framework has been proposed that allows for the interactive modification and simulation of tetrahedral meshes derived from existing CAD models, e.g., to create customized versions of parts that were designed for mass production. This shortens the design cycle by eliminating the need for conversion and switching between tools. In this paper, we present a more detailed description and improvements to this architecture by using GPU parallelization not only for simulation but also for mesh editing, which leads to even shorter iteration cycles.

Keywords: Computer aided design · Massively parallel geometry processing · Simulation environments · Client server architectures · Massively parallel and high-performance simulations

1 Introduction

This paper is an extended version of the work "TEdit: A Distributed Tetrahedral Mesh Editor with Immediate Simulation Feedback" [32], presented at the 11th International Conference on Simulation and Modeling Methodologies, Technologies and Applications (SIMULTECH 2021). TEdit [32] is a novel tetrahedral mesh editor with immediate simulation feedback, i.e., without tool switches or manual intervention and with minimal delay.

By making use of the fact that high-quality tetrahedral meshes and corresponding simulation load cases already exist for mass-produced parts, TEdit [32] significantly

G. Wagner et al. (Eds.): SIMULTECH 2021, LNNS 601, pp. 24–42, 2023.
https://doi.org/10.1007/978-3-031-23149-0_2

shortens the product development cycle for customized parts. Previously, these parts were limited to those that serve a purely cosmetic function, as simulations of the modified part are otherwise required to ensure it continues to fulfill its function correctly. Parts that do have to fulfill mechanical requirements lead to the usual iterative product development cycle of modifying a part in computer aided design (CAD), remeshing it, and simulating it in a computer aided engineering (CAE) tool, which is uneconomic in the context of individualized components. By directly editing the tetrahedral mesh—not the CAD model—and using its triangular surface for 3D printing, editing of the CAD model followed by full remeshing is avoided completely.

In this extended version, we shorten the design cycle even more by utilizing the graphics processing unit (GPU) not only for simulation, but also for geometry processing when performing mesh editing operations. Furthermore, we provide additional details on how TEdit maintains persistent surface IDs, as well as the architectural adjustments applied due to the use of GPU-accelerated mesh processing.

GPU-accelerated simulation provides direct feedback (see Sect. 2), while client hardware requirements are minimized using a multi-tier, distributed client-server architecture (see Sect. 3). We contribute massively parallel algorithms for mesh editing. As a result, we achieve a speedup of up to 12× for hole closing and up to 6× for erosion in comparison to the original TEdit implementation [32].

2 Related Work

In this section, we discuss the related work, with a focus on the extensions beyond the original TEdit framework [32].

2.1 GPU-Accelerated Mesh Processing

As mentioned in the introduction, we extend TEdit [32] with massively parallel mesh processing algorithms for the GPU, which result in improved runtimes and mesh quality (see Sect. 5). GPU-accelerated mesh editing is a research topic that has recently seen increasing interest. For example, Mahmoud et al. [14] recently presented RXMesh, a static triangular mesh data structure that enables GPU-parallel geometry processing tasks such as smoothing of surface meshes. However, most physical simulations using the finite element method (FEM) require volumetric meshes.

As tetrahedral meshes can be generated robustly (see, e.g., Hu et al. [11]), we focus on tetrahedral meshes. Mueller-Roemer et al. [17, 18] present a mesh data structure for volumetric meshes that allows for GPU-acceleration of geometry processing tasks as well as stiffness matrix assembly. We use this data structure in our work, because of the ability to couple geometry processing with stiffness matrix assembly is vital for integrating GPU-parallel mesh editing and simulation.

Ströter et al. [33] present massively parallel optimization techniques that allow for quick element shape optimization and enable application-specific boundary treatment. As FEM simulation requires sufficient element quality [27], we use these techniques for fast mesh optimization.

2.2 Tetrahedral Mesh Editing

While GPU-accelerated mesh editing is a new research topic, we highlight related work in the field of tetrahedral mesh editing on the CPU as well. Stoll et al. [31] allow users to interactively modify volumetric meshes using visual handles. They perform differential rotation updates for deformation. Mezger et al. [16] deform tetrahedral meshes with the use of physically based simulation. As they use the FEM model, they avoid significant changes in element volume. As our system is intended for virtual prototyping, deformation alone is not sufficient and topological changes to the mesh are required as well. The focus of our work is a set of editing operations that enable model customization including topological changes.

A related editing framework in the context of engineering was presented by Serna et al. [26], who propose an embodiment of the FEM mesh for modification of semantic features. Their framework offers extrusion, dragging and rounding operations on semantic features. While they focus on semantic features, we extract face groups from CAD for direct customization of models that originate from CAD, thereby shortening design cycles. Another tetrahedral mesh editing framework for FEA was presented by Xian et al. [39]. Their framework decomposes the model into local surface features representing the semantic meaning of particular components. Like our editing operations, their framework preserves element quality throughout model modification. In contrast to their framework, our system performs GPU-accelerated editing operations, which can be executed on the best-suited machines due to the distributed nature of our system. Like TEdit, Graphite [12] allows users to modify tetrahedral meshes, e.g., filling holes or deleting individual facets. Graphite provides an automatic hole detection that is controlled by a user-specified number of maximum boundary loop vertices. While users work on individual mesh facets in Graphite, we focus on interactive 3D modeling with face groups to accelerate virtual prototyping.

2.3 GPU-Accelerated Finite Element Analysis

Simulating a model with the FEM can be a compute-intensive task as well. For this reason, Weber et al. [37] perform GPU-accelerated finite element analysis (FEA) based on a merged kernel modified preconditioned conjugate gradient (MPCG) solver. Their MPCG solver is a GPU-optimized variant of Baraff and Witkin's MPCG [2] that reduces kernel launch overheads while enabling simulation of various boundary conditions without reassembly of the stiffness matrix. To achieve fast response times, we use the Weber et al.'s [37] MPCG solver in TEdit's FEM simulation.

2.4 Distributed CAD/CAE

Some CAD/CAE environments employ a distributed architecture to share model data and facilitate collaboration [5,40]. Wang and Nnaji [36] present a distributed CAD system that offers a thin client for visualization functions and outsources compute intensive tasks to suitable machines. In contrast to these distributed CAD systems, our system focuses on shortening design cycles by avoiding tool switches and remeshing after CAD-based editing instead of collaboration.

For the implementation of the front end, we use Unity [34]. Although the Unity project [34] started out as a game engine, its features and capabilities have developed beyond the field of computer games, and it is now used a general development platform for 3D visualization [22]. While its competitors are comparable in features [6], we choose Unity due to previous experience, as well as the support for web-based applications.

3 Distributed Editing and Simulation

The system architecture description in this section, including Sect. 3.1, 3.3 and 3.4, is based on Sect. 3 of the conference paper [32], with minor corrections and updates for the addition of GPU-accelerated editing. We have added Sect. 3.2 and 3.5 to describe how persistent surface IDs are maintained between editing operations and how speculative simulation is used to further reduce perceived delay, respectively.

We considered the following points in the design of the architecture of the proposed system: The resource demand of the complete application is high, as mesh processing of large meshes is compute- and memory-intensive, and the fast FEA requires a capable GPU. The requirements for the front end consist largely of what is needed for visualization and interaction. Thus, our approach is to create a distributed application instead of a monolithic one by grouping related components into services. These services communicate using a network protocol, which allows running components on different machines and distributing the workload to computers with adequate hardware. This approach also ensures a degree of encapsulation between existing components. In addition, we overlap simulation with user interaction and perform it speculatively, further reducing perceived latency. Communication takes place using WebSockets, in order to facilitate easy integration into various front end implementations such as web browsers and to facilitate the use and distribution of visualizations [13]. Figure 1 gives an overview of the services and communication between them.

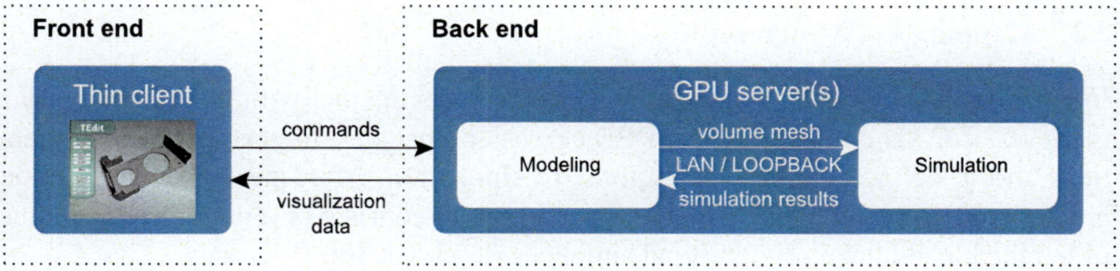

Fig. 1. As in TEdit [32], the system consists of three main services that may reside on different machines: the modeling, simulation, and user interface services. This separates the interactive visualization front end from compute-intensive geometry modification and simulation tasks. However, since both the modeling and the simulation service make use of GPU acceleration, it may be desirable to run both on the same GPU server. This allows the use of a more efficient loop-back connection, at the cost of potential resource contention between modeling and simulation (c.f. Sect. 3.4).

3.1 Workflow and Communication

The front end service presents the model to the user interactively. The user may select parts of the model and perform editing operations. The front end sends modification commands and their parameters to the modeling back end, which performs the topological modifications and sends the results back to the front end. At the same time, it sends an updated mesh to the simulation service, triggering simulation of the previously selected load case. When the user requests a simulation result, the front end queries the selected field from the back end.

Although simulation has to be performed on the tetrahedral representation, the outer surface triangles of the tetrahedral mesh are sufficient for visualization. Thus, from the front end point of view, the main data for upstream communication consists of modeling commands with model part selections, downstream data consists of surface triangles as well as simulation results. On the back end side, the modeling service sends tetrahedral meshes to the simulation service, and receives simulation results, such as scalar fields on the mesh. The modeling service organizes all the needed representations of the data—the tetrahedral mesh for simulation, the triangular surface for visualization and grouping information to translate between the two—and translates between them as needed. To minimize (de)serialization and bandwidth overhead, we use an efficient binary protocol based on Protocol Buffers [10] for communication between the different services.

One challenge that arises with a distributed modeling system is synchronizing model state between the services. In our implementation, the front end service communicates only with the modeling service, which acts as the central source of truth. On the back end side, the modeling service communicates with the simulation service using a tetrahedral mesh representation (see Fig. 1). Operations and visualization updates are carried out asynchronously, to keep the user interface responsive during modeling and simulation. A user may issue an editing command, while still being able to navigate and inspect the model, before a response is received. Additionally, simulation is started speculatively after editing to further reduce perceived latency (see Sect. 3.5).

3.2 Mapping CAD Surfaces

In existing CAD tools and CAE preprocessors, users interact with model geometry at the level of CAD entities, i.e., NURBS curves and patches, to perform editing operations and to define boundary conditions for simulation. After meshing, these entities are not part of the discretized model, which instead consists of a much larger number of tetrahedra and lower-dimensional simplices (triangles, lines, vertices).

Due to the large number and small size of the boundary triangles, user interaction at that level would both be tedious and result in low framerates, and would therefore be time-consuming and error-prone. Therefore, as mentioned in our conference publication [32] and described in more detail here, TEdit uses CAD surface IDs to associate every boundary triangle with its originating CAD surface. Some hierarchical meshers such as Gmsh [8], which we use to generate the initial input meshes, provide this information directly. Specifically, Gmsh MSH files store nodes/vertices in per-entity blocks and the bounding entities of higher-dimensional entities, e.g., 2D surfaces that bound a 3D volume (the entire model), 1D curves that bound a 2D surface, and so on [9].

Based on the data extracted from the input MSH file, we assign integer identifiers to each boundary triangle corresponding to the originating CAD surface (conceptually similar to Schmidt's triangle groups [24]). This allows the user to interact on the level of semantic faces, instead of individual boundary triangles that can be arbitrarily small, depending on meshing requirements for simulation. Besides user interaction, we use these surface IDs throughout TEdit, including the definition of boundary conditions for simulation.

However, the mapping between originating CAD surfaces and surface IDs is only a 1-to-1 mapping directly after loading a new mesh. As soon as the user begins editing the mesh, existing surfaces can be changed, split, or removed completely. All these changes only affect the tetrahedral mesh and not the original CAD model, but the relations between both need to be maintained even over modification steps, similar to the persistent naming problem in parametric CAD [7, 15]. To preserve the semantics of user selections and boundary condition definitions between edits, all editing operations must keep track of changes to the surface and communicate them to the front end. These changes are a 1-to-n mapping from each previous surface ID to zero (removal), one (re-indexing), or more (split/addition) new surface IDs.

We transmit this mapping in the form of a compressed sparse row (CSR)-like data structure (a description of the standard CSR matrix format can be found in the numerics literature [23]). In particular, we transmit an array of $k + 1$ offsets into an array of indices $i \in [0, k')$ corresponding to the CSR column array, where k is the number of surfaces before and k' after applying an editing operation. We omit the CSR value array. This means, that every surface ID $s \in [0, k)$ is mapped to the new surface IDs $\texttt{indices}[\texttt{offsets}[s]\texttt{..offsets}[s+1]]$ (if $\texttt{offsets}[s]$ equals $\texttt{offsets}[s+1]$, the surface was removed).

For example, assume an input mesh of a square plate with 7 surfaces $[0, 7)$, where 1 and 2 are the front and back faces, $[3, 6]$ are the sides, and 0 is a cylindrical hole connecting 1 and 2. A close hole operation on surface 0 would result in a new mesh with 6 surfaces and the following mapping:

$$\texttt{offsets} = [0, 0, 1, 2, 3, 4, 5, 6]$$
$$\texttt{indices} = [0, 1, 2, 3, 4, 5]$$

Essentially, surface 0 is removed, and the other surfaces IDs are shifted down by one. In the original implementation [32], the closing surfaces of the hole would not have been merged into the surrounding faces, which would instead have resulted in:

$$\texttt{offsets} = [0, 2, 3, 4, 5, 6, 7, 8]$$
$$\texttt{indices} = [0, 7, 1, 2, 3, 4, 5, 6],$$

where surface 0 is now replaced by the two new surfaces closing the hole: surface 0 and the newly inserted surface 7. The other surfaces maintain their indexing.

3.3 Front End

Game engines have gained in popularity for the development of 3D interfaces in many areas outside of games, as they provide solid frameworks for rapid development of interactive 3D applications. We chose the Unity platform [34] to implement our front end service. The viewer consists of an interactive 3D rendering and navigation interface in which the user inspects and modifies the model, or inspects simulation load cases and simulation results (see Fig. 2).

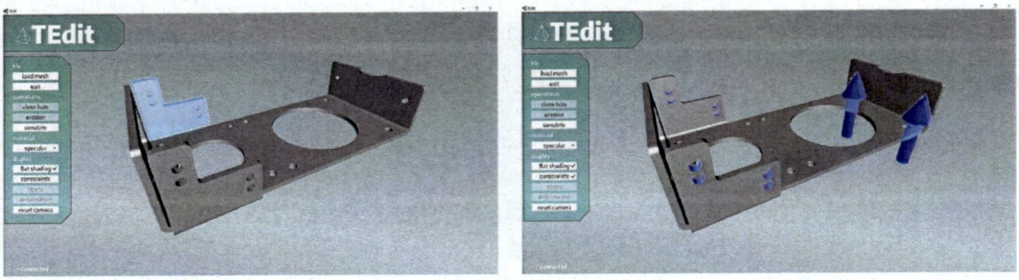

Fig. 2. A user may quickly select relevant model parts, as a selection interaction automatically selects triangle groups, such as ones that originate from a single CAD surface *(left)*. Furthermore, simulation load cases are visualized by highlighting surfaces, such as constrained surfaces, or surfaces with applied forces *(right)* [32].

Navigation within the 3D scene uses the trackball navigation concept [4], which revolves an object around a center point. The center point is initialized to be the center of the 3D object. To facilitate selection of semantic groups for quick interaction with the model, the front end also manages a mapping of visible triangles to groups, such as those originating from a single CAD surface or editing operation, as described in Sect. 3.2.

After sending a command message, the front-end service enters a waiting mode; The command buttons are disabled until a response is received, but the user can still rotate or move the model so that responsiveness of the application is maintained. Additionally, a busy indicator (a rotating stylized arc) is displayed in the title of the application while commands are carried out by the back end services to further support feedback to the user.

3.4 Modeling and Simulation Services

To enable optimal use of hardware for tetrahedral mesh editing and simulation, which requires an NVIDIA GPU in our case (see Sect. 2), the back end is further split into a modeling service and a simulation service. This results in a distributed, multi-tier client-server architecture, as the modeling service acts as a server for the front end, while acting as a client for the simulation service. The client therefore only communicates with the modeling service directly (see Fig. 1).

While the front end only needs the boundary mesh and the simulation results for selection and visualization, both the modeling and simulation services operate on tetrahedral meshes. Therefore, the front end can be remote and connected via a high-latency, low-bandwidth connection (latency-free interaction remains possible due to asynchronous operation). However, both back end services should run in the same data center for low-latency, high-bandwidth communication, as they exchange full volumetric meshes.

Due to the addition of GPU-accelerated modeling operations, the modeling service now requires an NVIDIA GPU as well (see Sect. 4). As the multi-tier architecture allows for flexible assignment of services to servers, the modeling and simulation services can be run on the same GPU server, resulting in the use of an even lower-latency loopback connection. However, this can lead to GPU resource contention, as both the simulation and the modeling service require a significant amount of GPU memory.

3.5 Speculative Simulation

Besides allowing limited user interaction, such as camera navigation or surface selection while an edit or simulation is running, TEdit uses speculative simulation to reduce perceived latency even further. Whenever an editing operation completes, the modeling back end sends a simulation command to the simulation back end in addition to sending the new boundary surface to the UI front end (see Fig. 3). This allows the simulation back end to receive the new simulation mesh and speculatively begin simulating before the user explicitly triggers a new simulation, under the assumption that they likely want to simulate their new model after editing it. Therefore, as soon as the user interacts with the "Simulate" button, the simulation has already started (or even completed) and the result can be provided with reduced latency. If the user performs another edit instead, the back end cancels the running simulation and starts a new one. Altenhofen et al. [1] previously used by a similar approach in a parametric customization UI with simulation.

4 GPU-Accelerated Geometry Editing

We present a set of editing operations for model customization: volumetric hole filling (Sect. 4.2) and erosion (Sect. 4.3). In contrast to the CPU-based algorithms presented in the conference paper [32], we present massively parallel algorithms for surface mesh extraction and erosion leading to faster processing times.

As the front end and the modeling service only exchange surface mesh data, the service must maintain a mapping from surface vertex indices to vertex indices in the tetrahedral mesh and vice versa. Whenever the service receives a request for an editing operation, it receives surface triangles that can be converted to tetrahedral mesh triangles using the mapping. The response then includes the new surface triangles, face groups, and the mapping described in Sect. 3.2. For this purpose, we extract the surface mesh of the altered model and check for each face group if its associated triangles are still included in the surface. For each triangle, we map surface vertex indices to vertex indices of the previous tetrahedral mesh, the resulting tetrahedral mesh, and finally the new surface mesh. New surface triangles of the tetrahedral mesh are identified by the editing operation as one or more face groups and can be converted to surface triangles using the mapping of tetrahedral vertex indices to surface vertex indices.

4.1 Surface Mesh Extraction

As the tetrahedral mesh is repeatedly altered by user-defined editing operations, fast extraction of the triangular surface mesh annotated with the data required for mapping

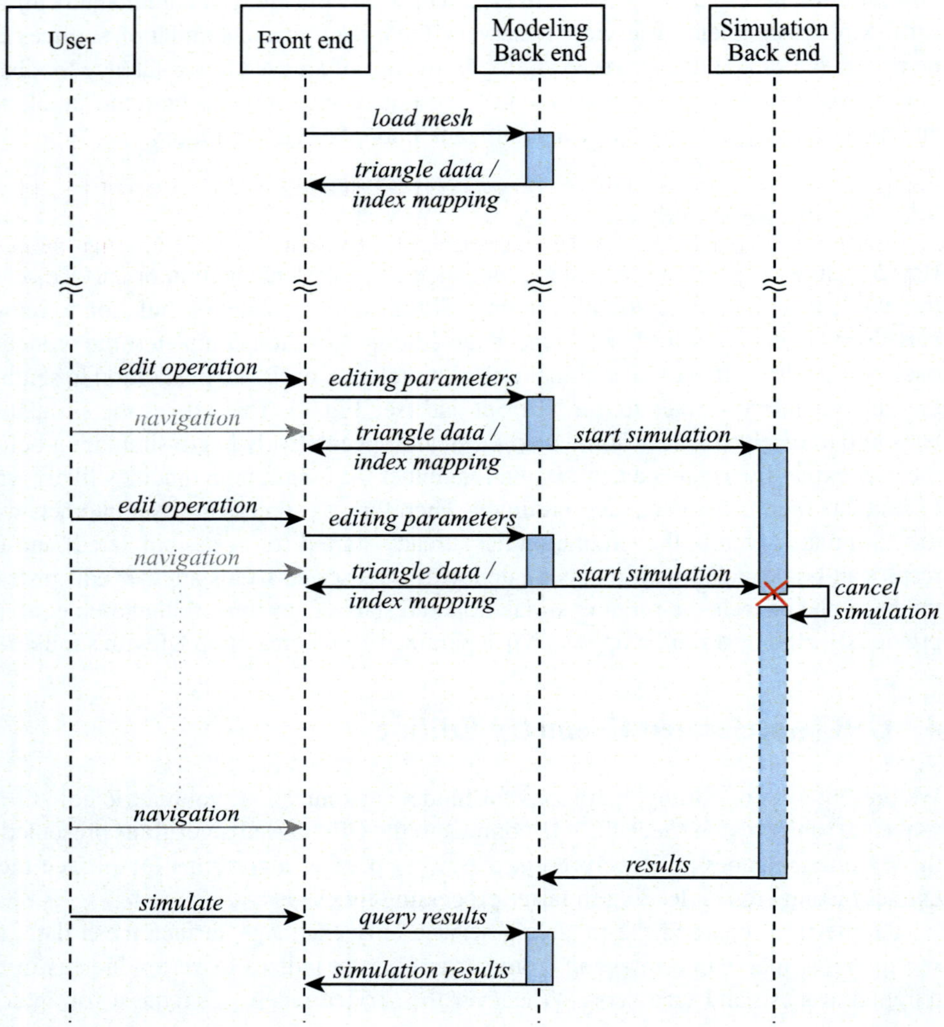

Fig. 3. User interaction and service communication sequence diagram. When the front end is started, it connects to the back end and sends a message to load mesh data as well as the corresponding simulation configuration. Triangle data as well as mapping information is sent back to the front end to allow the user to act on the level of semantic surfaces. After a mesh editing operation completes, the back end sends the modified triangle/mapping data and speculatively starts the simulation as early as possible to have the results ready as soon as possible—possibly even before the user requests them. The back end cancels running simulation tasks when it receives a new mesh/simulation command. While the modeling and/or the simulation service are busy handling requests, navigation and interaction remain possible in the front end. When the user requests a simulation, the modeling service queries the simulation service for results. As soon as the simulation service finishes the calculations, it returns the resulting data to the front end. This may happen immediately if the simulation has already finished.

the triangular primitives to the face groups is a necessity for short response times. Our experiments revealed that extraction of the surface mesh with computation of mappings from the surface mesh to the volumetric mesh and vice versa often becomes the bottleneck for editing operations such as hole closing. For instance, compare the relative runtimes shown in Fig. 4 for the hole closing operation shown in Fig. 5(a) and (b).

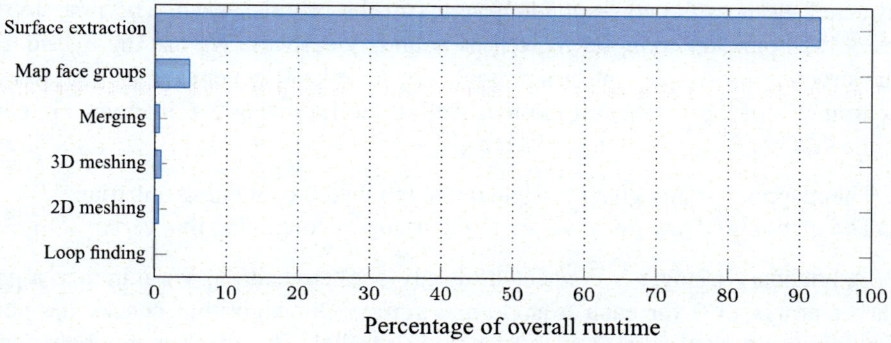

Fig. 4. Relative runtimes of the individual steps of the hole closing operation in Fig. 5(a) and (b). The figure shows that surface mesh extraction is the bottleneck.

Therefore, we present a massively parallel algorithm for quick extraction of the triangular surface coupled with computation of mapping data. Our algorithm computes the following data structure to facilitate maintenance of face groups:

```
struct {
    int num_boundary_triangles;
    int num_boundary_vertices;
    float3 * boundary_vertices;
    int3 * boundary_triangles;
    uchar * vertex_markers;
    int * tet_mesh_idx_to_boundary_idx;
    int * boundary_idx_to_tet_mesh_idx;
} TetMeshBoundary;
```

We save the numbers of boundary vertices and triangles and the surface mesh itself as arrays of three contiguous floats and integer indices per vertex and triangle, respectively. Each vertex of the tetrahedral mesh is annotated with a flag to indicate, if it is a boundary vertex or not. If the vertex is a boundary vertex, we additionally encode whether the vertex lies on a geometrical face, edge, or corner. This enables downstream element quality optimization algorithms to optimize boundary vertices without changing the model shape. We use the GPU-accelerated harmonic optimization algorithm by Ströter et al. [33] due to its fast runtime and tendency to lower the element count, reducing the computational cost of simulation. For mapping purposes, the `tet_mesh_idx_to_boundary_idx` array includes the boundary vertex index for each vertex index of the tetrahedral mesh. By construction, this array only includes valid values for vertices that are boundary vertices, i.e., whose markers indicate a boundary vertex. For each boundary vertex, the `boundary_idx_to_tet_mesh_idx` includes the corresponding index in the tetrahedral mesh.

Our algorithm extracts the triangular boundary of the tetrahedral mesh using GPU-accelerated sub-mesh extraction similar to the method presented by Wald [35]. A key difference to Wald's [35] method is that we are able to map from the vertex of the extracted mesh to the corresponding vertex in the original mesh. The first step is to detect the boundary triangles as well as boundary vertices. We use the hybrid TCSR data structure for simplex meshes presented by Mueller-Roemer et al. [17, 18] to extract the triangles of the tetrahedral mesh as well as the following two mappings in parallel on the GPU:

1. The mapping of any given triangle to the tetrahedra containing this triangle
2. The mapping of any given vertex to the triangles containing this vertex

A boundary triangle is connected to only one tetrahedron. We initialize a pair of marker arrays to 0 for each triangle and vertex. Our algorithm checks the number of neighboring tetrahedra of each triangle in parallel. The check marks boundary triangles and their vertices with the value 1. Subsequently, our algorithm performs a parallel prefix scan on the marker array in order to obtain offset positions as well as `num_boundary_triangles` and `num_boundary_vertices`. As the prefix sum over the boundary vertex markers can be used to map a boundary vertex index to the corresponding index in the tetrahedral mesh, we keep this prefix sum in the `tet_mesh_idx_to_boundary_idx` array. Using the prefix offset positions we write boundary vertices to the `boundary_vertices` array and boundary triangles to the `boundary_triangles`. Unlike Wald's method [35], we do not require a permutation array to update the indices, because the prefix positions are sufficient for a simple copy. After determining the boundary vertices, we classify them into one of the following three categories:

- **FACE:** The boundary triangles surrounding the vertex are coplanar, forming a geometric face.
- **EGDE:** The boundary triangles surrounding the vertex form a geometric edge.
- **CORNER:** The vertex is neither classified as FACE nor EDGE.

We classify boundary vertices in parallel on the GPU. We first retrieve the tetrahedral mesh index of the boundary vertex with the subsequently calculated mapping. Our algorithm then iterates over the triangles containing the vertex and calculates the normals of boundary triangles. If the dot products of the normalized surface normals are within a threshold $\varepsilon_n = 1\text{e} - 2$ of 1, i.e., nearly parallel, the boundary vertex is classified as FACE. If only two groups of surface normals exhibit a pairwise difference larger than ε_n, the boundary vertex is classified as EDGE. Otherwise, the vertex is classified as CORNER.

In a final parallel pass over the vertices of the tetrahedral mesh, our algorithm sets the `boundary_idx_to_tet_mesh_idx` mapping. Our algorithm only needs to check the boundary marking of the vertex and lookup the boundary vertex index in the `tet_mesh_idx_to_boundary_idx` array in order to write the tetrahedral mesh vertex index to the entry at the boundary vertex index position.

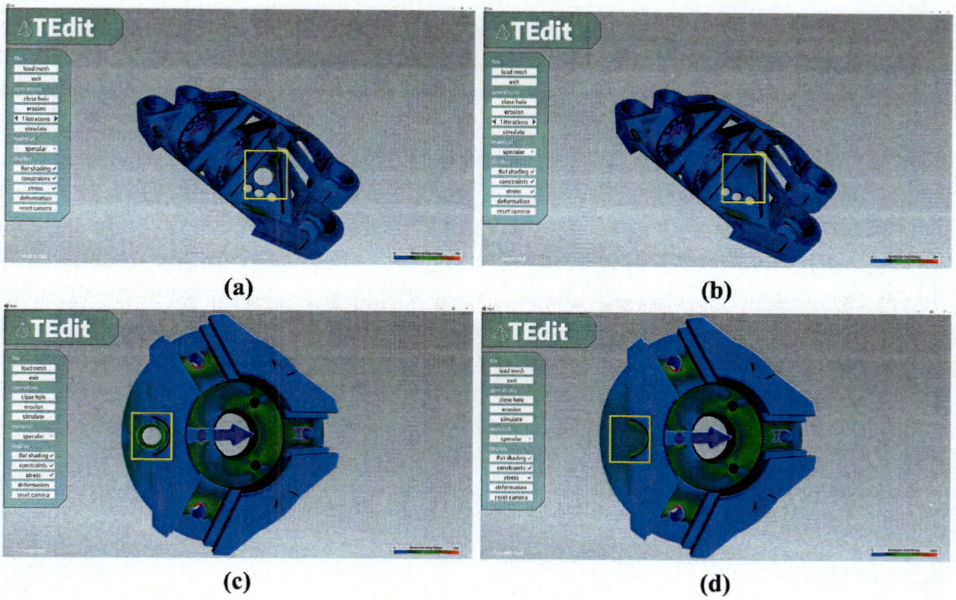

Fig. 5. Experiments for hole closing. The yellow rectangles indicate the area of the closed hole. Figure (a) shows the selected hole on the "SimJEB 633" model and Figure (b) shows the results. Figures (c) and (d) are from the conference paper [32]. Figure (c) shows the selected hole of a mechanical part and (d) the resulting model.

4.2 Volumetric Hole Filling

Mechanical parts often include holes, e.g., for mounting or weight relief, that users may want to remove in a modification step. However, beyond closing the hole at both ends of the cylinder formed by the hole's inner surface, the hole in the tetrahedral mesh must be filled volumetrically. For this purpose, our framework includes a volumetric hole closing algorithm. Our algorithm performs the following four steps previously described in our conference paper [32]:

1. Search the received triangular mesh of the hole shell for boundary loops
2. For each boundary loop: Close the loop by applying two-dimensional constrained meshing using Shewchuck's Triangle library [28]
3. Fill the resulting manifold by volumetrically meshing it using Tetgen [29]
4. Merge the resulting tetrahedral mesh with the tetrahedral mesh of the model

Our experiments have revealed that sequential execution on the CPU provides sufficiently fast runtimes for quick perceived responses. The number of surface triangles in the shell of a volumetric hole is typically small enough such that parallelization is not expected to achieve a significant impact on the perceived response time. In the case of a shell mesh with many surface triangles, the tetrahedral meshing step (step three) may be worth parallelizing, but no GPU-parallel constrained tetrahedral meshing algorithms are available yet. Consequently, we use the previously presented sequential volumetric hole filling algorithm [32], but obtain a substantial speedup by using the new GPU-accelerated boundary extraction approach (see Sect. 4.1).

As user-friendly maintenance of face groups requires reasonable assignment of newly added boundary faces to either persisting or new face groups, we extend TEdit's [32] volumetric hole filling algorithm to check if the newly added faces shall be assigned to a surrounding face group or not. Due to the hierarchical nature of our hole closing algorithm which meshes each boundary loop individually, we know which sets of newly added boundary faces are potentially added to an existing face group. We iterate through the boundary faces obtained by meshing each boundary loop and mark all the edges of the newly added surface triangles. In another loop, we iterate through all previously existing boundary triangles and check, if a triangle contains any of the marked edges. If a triangle contains a marked edge, we record the face group of the triangle. After this loop, we count the recorded face groups. If we only recorded one face group, we add the triangles closing the boundary loop to this face group. Otherwise, the newly added boundary triangles represent a new face group. Although parallelization of our face group assignment check algorithm is straightforward, we did not encounter any situation in which sequential execution of this algorithm was the bottleneck.

4.3 Erosion

The erosion operation can be used to remove parts of the model. See Fig. 6 for examples of model customizations with erosion. We build upon TEdit's [32] erosion scheme to develop a massively parallel erosion method for tetrahedral meshes. Our erosion method receives a set of boundary vertices belonging to the user selected surface triangles as an input. We use the mesh data structure presented by Mueller-Roemer et al. [17, 18], as it enables massively parallel computation of connectivity relationships with efficient memory use.

We determine the remaining part of the mesh after erosion by marking tetrahedra as 0 or 1, denoted as *to-be-removed* and *remaining*, respectively. Initially, every tetrahedron of the mesh is unmarked. The first step of our erosion scheme is to perform a parallel pass over the input boundary vertices and mark every tetrahedron including any of the input vertices as *to-be-removed*. While users intend to thin down or remove a part of the model and expect erosion to return a consistent mesh, direct removal of every marked tetrahedron can leave "islands" of tetrahedra that are not connected to the rest of the mesh. For this reason, we determine the remaining mesh by propagating a marking through the mesh starting from the unselected surface triangles in parallel on the GPU.

To initiate the propagation, we perform a parallel pass over every boundary triangle that checks the tetrahedron associated with each boundary triangle. If the tetrahedron is not marked as *to-be-removed* already, it is marked as *remaining*. We now propagate the *remaining* marking by advancing into the interior of the mesh. To this end, we check the tetrahedra sharing an interior face with a currently unmarked tetrahedron in parallel on the GPU. If an unmarked tetrahedron is connected to a tetrahedron marked as *remaining*, we mark it as *remaining* as well. We perform several parallel passes advancing the remaining part until no further tetrahedron can be marked as *remaining*. Every tetrahedron without a marking is not connected to the mesh and is marked as *to-be-removed*. With the use of the previously calculated marking, we write the remaining tetrahedra and vertices to a buffer using Wald's [35] GPU-parallel re-indexing method.

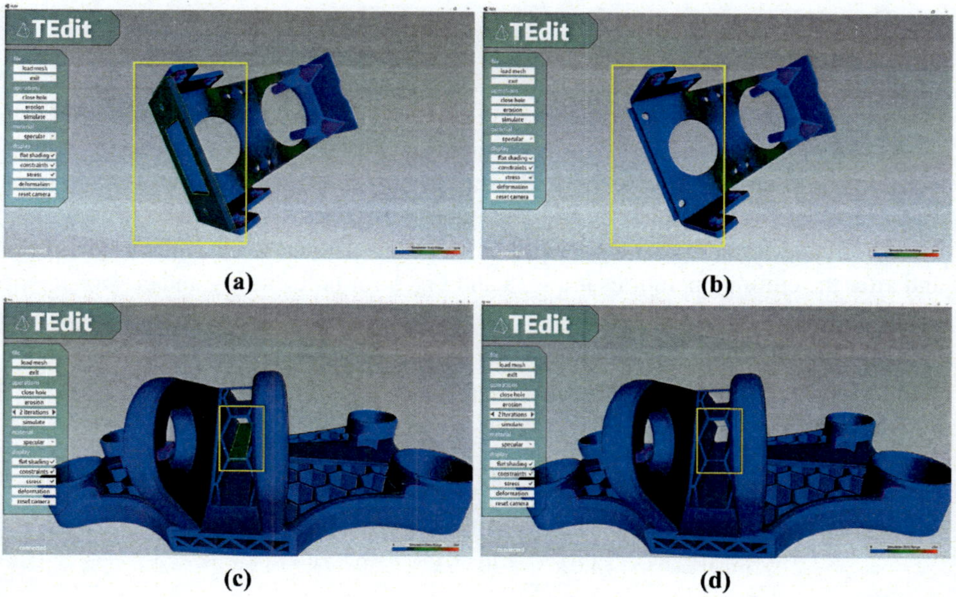

Fig. 6. Experiments for erosion. The yellow rectangles indicate the eroded parts of the model. Figures (a) and (b) are from the conference paper [32]. Figure (a) shows the selected surfaces of a bracket and (b) shows the resulting model. Figure (c) shows the selected strut of the "SimJEB 220" model and (d) shows the results.

After the user-specified part of the mesh is removed, a coarse "zigzag" surface appears at the surface newly introduced due to erosion. However, users expect a smooth surface without protruding triangles. Therefore, we perform GPU-accelerated smoothing. We smooth the new surface using the discrete Laplace-Beltrami operator [19], as it is efficient and provides a unique solution for a compact surface [30]. As simultaneous update of every vertex is more efficient on the GPU than updating every vertex individually, we smooth all surface vertices simultaneously.

We calculate the Laplace-Beltrami gradient for each surface vertex in parallel. Since our mesh editing algorithms aim to produce models suitable for the FEM, they must prevent element inversions. We perform a binary search to find an update step size λ such that relocating the vertices along their gradients does not produce any inverted tetrahedra. As soon as such a λ is determined, we perform parallel Laplace-Beltrami smoothing by relocating every vertex of the new surface. After a smoothing step, the element quality of adjacent tetrahedra typically deteriorates. In order to improve the element quality of the tetrahedral mesh and enable successive smoothing passes without introducing inverted elements, we perform mesh quality optimization after surface smoothing. For fast run times, we use the GPU-parallel mesh optimization algorithm by Ströter et al. [33]. In order to allow for appropriate boundary treatment, we extract the boundary with the GPU-parallel algorithm from Sect. 4.1 and feed the boundary vertex classification into the optimization algorithm. We perform a user-specified number of iterations of alternating passes of smoothing and mesh optimization until termination.

5 Results

In this section, we evaluate the impact of GPU-acceleration on processing times, using the models used for the experiments in the conference publication [32] and additional models from the SimJEB data set [38] as shown in Table 1. The evaluation machine is equipped with an Intel i7-3930K CPU and an NVIDIA RTX 3090 GPU. The impact of GPU-acceleration is quantified by comparing runtimes of the editing methods in TEdit [32] and the editing methods presented in this paper. We show the models before and after the editing operations in Fig. 5 and Fig. 6 for hole closing and erosion, respectively. Figure 7 shows the speedups of our novel GPU-accelerated editing operations compared to the CPU-sequential operations.

Table 1. Numbers of vertices, tetrahedra, and boundary triangles for the meshes used for the experiments.

Model	#Vertices	#Tetrahedra	#Boundary triangles
Bracket	30.9 k	108.0 k	50.5 k
SimJEB 633	74.0 k	303.9 k	97.7 k
Part	102.9 k	536.8 k	67.6 k
SimJEB 220	293.9 k	1.2 M	353.2 k

The experiments for the hole closing operations reveal the impact of GPU-acceleration on surface mesh extraction. When we close a hole in the "SimJEB 633" model to reduce surface stress, we observe a significant speedup of 6.3×. Although the "Part" model includes fewer surface triangles, we observe an even larger speedup of 12×, because more triangles have to be extracted from the tetrahedral mesh and be classified into either interior or boundary. Since surface mesh extraction is the main bottleneck for hole closing, our massively parallel algorithm leads to processing times shorter than 500 ms for complex models. As a result, our system allows for hole closing at interactive rates.

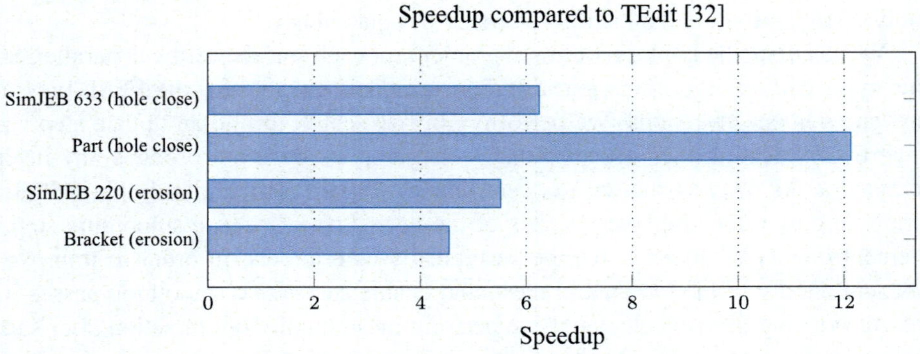

Fig. 7. This figure shows the speedups due to the application of our massively parallel algorithms compared to the sequential algorithms used in TEdit [32].

Our experiments for erosion show significant speedups due to the use of the GPU. Eroding a strut in the "Bracket" model (see Fig. 6(a) and (b)) with our massively parallel erosion and surface mesh extraction methods leads to a speedup of 4.6×. As in the conference publication [32], we performed six iterations of alternating surface smoothing and volumetric mesh optimization. When we erode a strut in the significantly more complex "SimJEB 633" model with two iterations of smoothing and mesh optimization, we observe a speedup of 5.5×. This leads to a significant reduction of processing time, and thus waiting time for end users.

The use of Ströter et al.'s [33] massively parallel optimization technique results in significantly better element qualities compared to the erosion algorithm in the conference paper [32]. We compare the 5%-quantiles of dihedral angles after erosion of both methods in Fig. 8. As our erosion technique applies full boundary optimization at the smoothed surface, we significantly improve element quality, while achieving significant speedups.

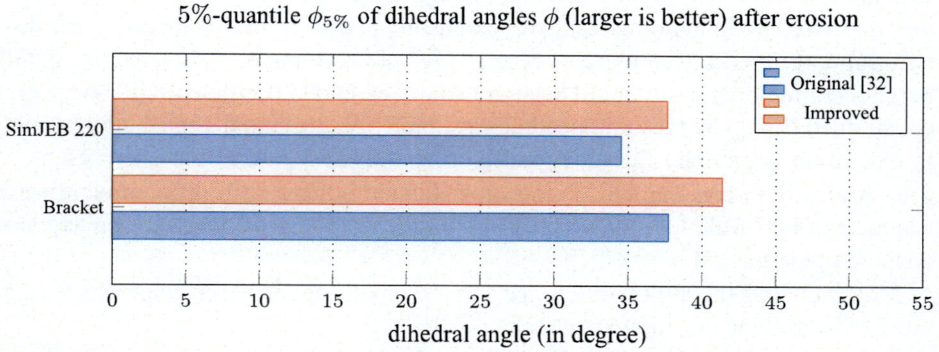

Fig. 8. We compare resulting 5%-quantiles of dihedral angles ϕ (larger is better) of meshes edited with the previous erosion method (blue) and our new erosion method (red) using the optimization technique by Ströter et al. [33]. The new erosion method achieves significantly better $\phi_{5\%}$.

6 Conclusions

In summary, we have presented significant improvements to the novel tetrahedral mesh editor with immediate simulation feedback TEdit [32] that shortens the iterative process of (re)design, analysis and inspection. By directly operating on CAE meshes, modifications can be performed and the resulting models simulated, while avoiding switching to a CAD tool and remeshing the entire domain. As 3D printing can support and often only supports discrete triangular surfaces, this approach completely avoids the necessity to feed back discrete modeling results into a CAD tool. While the current set of editing operations is somewhat limited, our improved prototype demonstrates how the iterative product design loop can be shortened for individualized versions of mass-produced parts with functional requirements.

The use of GPU-accelerated mesh editing operations presented in this paper and a GPU accelerated FEA solver ensure short iteration times, while the distributed architecture minimizes user hardware requirements. Due to the use of Unity [34] and WebSockets, the front-end can be deployed directly as both a desktop and web application.

Bandwidth requirements are low, as only surface triangles are transmitted between the front end and the modeling service (the modeling and simulation service should reside in the same data center or on the same server, as they exchange volumetric meshes).

We also presented a method for maintaining correspondence between individual bounding triangles and their original CAD surface IDs (or a newly created contiguous surface ID). This allows users to interact at a much higher level of abstraction than individual surface triangles. Furthermore, it also allows for automatic reassignment of surface-based constraints and loads for simulation. In addition, mesh optimization ensures that the mesh quality is maintained.

6.1 Future Work

In order to allow collaboration of different engineering teams, our system could be further improved by integrating it into an extensible platform for collaborative 3D design with support for GPU acceleration, such as NVIDIA's Omniverse [20]. Our set of GPU-parallel editing operations could be significantly extended if efficient GPU-accelerated constrained Delaunay meshing becomes available. A substantial contribution would be to extend unconstrained Delaunay meshing approaches such as the one presented by Cao et al. [3] to constrained 3D meshing. Recent advances in differentiable surface triangulation by Rakotosaona et al. [21] could lead to faster optimizer convergence if adapted to tetrahedral meshes.

Additionally, the following suggestions from the conference paper [32] remain interesting avenues for future research:

- Extending our editing operations further to support filling non-coplanar holes or to erode thicker structures that go beyond thin plate-like structures.
- Adapting geometric morphological operations such as the methods of Sellán et al. [25] to support tetrahedral meshes could significantly improve flexibility of our mesh editing toolset.

References

1. Altenhofen, C., Loosmann, F., Mueller-Roemer, J., Grasser, T., Luu, T., Stork, A.: Integrating interactive design and simulation for mass customized 3D-printed objects - a cup holder example. In: 2017 International Solid Freeform Fabrication Symposium (2017). https://doi.org/10.26153/16941
2. Baraff, D., Witkin, A.: Large steps in cloth simulation. In: Proceedings of the 25th Annual Conference on Computer Graphics and Interactive Techniques, SIGGRAPH 1998, pp. 43–54 (1998). https://doi.org/10.1145/280814.280821
3. Cao, T.T., Nanjappa, A., Gao, M., Tan, T.S.: A GPU accelerated algorithm for 3D Delaunay triangulation. In: Proceedings of the 18th Meeting of the ACM SIGGRAPH Symposium on Interactive 3D Graphics and Games, I3D 2014, pp. 47–54 (2014). https://doi.org/10.1145/2556700.2556710
4. Chen, M., Mountford, S.J., Sellen, A.: A study in interactive 3-D rotation using 2-D control devices. ACM SIGGRAPH Comput. Graph. **22**(4), 121–129 (1988). https://doi.org/10.1145/378456.378497

5. Denis, L., Gardan, Y., Perrin, E.: A framework for a distributed CAD system. Comput. Aided Des. **36**(9), 761–773 (2004). https://doi.org/10.1016/j.cad.2003.09.004
6. Dickson, P.E., Block, J.E., Echevarria, G.N., Keenan, K.C.: An experience-based comparison of Unity and Unreal for a stand-alone 3D game development course. In: Proceedings of the 2017 ACM Conference on Innovation and Technology in Computer Science Education, ITiCSE 2017, pp. 70–75 (2017). https://doi.org/10.1145/3059009.3059013
7. Farjana, S.H., Han, S.: Mechanisms of persistent identification of topological entities in CAD systems: a review. Alex. Eng. J. **57**(4), 2837–2849 (2018). https://doi.org/10.1016/j.aej.2018.01.007
8. Geuzaine, C., Remacle, J.F.: Gmsh: a 3-d finite element mesh generator with built-in pre- and post-processing facilities. Int. J. Numer. Meth. Eng. **79**(11), 1309–1331 (2009). https://doi.org/10.1002/nme.2579
9. Geuzaine, C., Remacle, J.: Gmsh 4.9.5 - MSH file format (2022). https://gmsh.info/doc/texinfo/gmsh.html#MSH-file-format. Accessed April 2022
10. Google: Protocol buffers (2022). https://developers.google.com/protocol-buffers/. Accessed April 2022
11. Hu, Y., Schneider, T., Wang, B., Zorin, D., Panozzo, D.: Fast tetrahedral meshing in the wild. ACM Trans. Graph. **39**(4), (2020). https://doi.org/10.1145/3386569.3392385
12. Inria: Graphite (2019). http://alice.loria.fr/index.php?option=com_content&view=article&id=22. Accessed April 2022
13. Krispel, U., Settgast, V., Fellner, D.W.: Dynamo - dynamic 3D models for the web: a declarative approach to dynamic and interactive 3D models on the web using x3dom. In: Proceedings of the 23rd International ACM Conference on 3D Web Technology, Web3D 2018 (2018). https://doi.org/10.1145/3208806.3208812
14. Mahmoud, A.H., Porumbescu, S.D., Owens, J.D.: RXMesh. ACM Trans. Graph. **40**(4), 1–16 (2021). https://doi.org/10.1145/3450626.3459748
15. Marcheix, D., Pierra, G.: A survey of the persistent naming problem. In: Proceedings of the Seventh ACM Symposium on Solid Modeling and Applications, SMA 2002, pp. 13–22 (2002). https://doi.org/10.1145/566282.566288
16. Mezger, J., Thomaszewski, B., Pabst, S., Straßer, W.: Interactive physically-based shape editing. Comput. Aided Geom. Des. **26**(6), 680–694 (2009). https://doi.org/10.1016/j.cagd.2008.09.009
17. Mueller-Roemer, J.S., Altenhofen, C., Stork, A.: Ternary sparse matrix representation for volumetric mesh subdivision and processing on GPUs. Comput. Graph. Forum **36**(5), 59–69 (2017). https://doi.org/10.1111/cgf.13245
18. Mueller-Roemer, J.S., Stork, A.: GPU-based polynomial finite element matrix assembly for simplex meshes. Comput. Graph. Forum **37**(7), 443–454 (2018). https://doi.org/10.1111/cgf.13581
19. Nealen, A., Igarashi, T., Sorkine, O., Alexa, M.: Laplacian mesh optimization. In: Proceedings of the 4th International Conference on Computer Graphics and Interactive Techniques in Australasia and Southeast Asia, pp. 381–389 (2006). https://doi.org/10.1145/1174429.1174494
20. NVIDIA Corporation: NVIDIA Omniverse (2022). https://www.nvidia.com/en-us/omniverse/. Accessed April 2022
21. Rakotosaona, M.J., Aigerman, N., Mitra, N.J., Ovsjanikov, M., Guerrero, P.: Differentiable surface triangulation. ACM Trans. Graph. **40**(6), 1–13 (2021). https://doi.org/10.1145/3478513.3480554
22. Rhyne, T.M.: Computer games and scientific visualization. Commun. ACM **45**(7), 40–44 (2002). https://doi.org/10.1145/514236.514261
23. Saad, Y.: Iterative Methods for Sparse Linear Systems. Society for Industrial and Applied Mathematics, Philadelphia (2003). https://doi.org/10.1137/1.9780898718003

24. Schmidt, R.: Designing for AM: integrating mesh-based modelling techniques with parametric CAD. In: SIAM Conference on Geometric and Physical Modeling, SPM 2016 (2016). https://d2f99xq7vri1nk.cloudfront.net/SIAMTalk_Oct2015.pptx
25. Sellán, S., Kesten, J., Sheng, A.Y., Jacobson, A.: Opening and closing surfaces. ACM Trans. Graph. **39**(6), 1–13 (2020). https://doi.org/10.1145/3414685.3417778
26. Serna, S.P., Stork, A., Fellner, D.W.: Tetrahedral mesh-based embodiment design. In: International Design Engineering Technical Conferences and Computers and Information in Engineering Conference (2010). https://doi.org/10.1115/detc2010-28971
27. Shewchuk, J.: What is a good linear finite element? Interpolation, conditioning, anisotropy, and quality measures (preprint). University of California at Berkeley 73, 137 (2002)
28. Shewchuk, J.R.: Triangle: engineering a 2D quality mesh generator and Delaunay triangulator. In: Lin, M.C., Manocha, D. (eds.) WACG 1996. LNCS, vol. 1148, pp. 203–222. Springer, Heidelberg (1996). https://doi.org/10.1007/BFb0014497
29. Si, H.: TetGen, a Delaunay-based quality tetrahedral mesh generator. ACM Trans. Math. Softw. **41**(2), 1–36 (2015). https://doi.org/10.1145/2629697
30. Solomon, J., Crane, K., Vouga, E.: Laplace-Beltrami: the Swiss army knife of geometry processing. In: Symposium on Geometry Processing Graduate School, Cardiff, UK, vol. 2 (2014)
31. Stoll, C., de Aguiar, E., Theobalt, C., Seidel, H.P.: A volumetric approach to interactive shape editing. Technical report, MPI-I-2007-4-004, Max-Planck-Institut für Informatik (2007)
32. Ströter, D., Krispel, U., Mueller-Roemer, J., Fellner, D.: TEdit: a distributed tetrahedral mesh editor with immediate simulation feedback. In: Proceedings of the 11th International Conference on Simulation and Modeling Methodologies, Technologies and Applications, SIMULTECH 2021, pp. 271–277 (2021). https://doi.org/10.5220/0010544402710277
33. Ströter, D., Mueller-Roemer, J., Weber, D., Fellner, D.W.: Fast harmonic tetrahedral mesh optimization. Vis. Comput. Springer Science and Business Media LLC (2022). https://doi.org/10.1007/s00371-022-02547-6
34. Unity: Unity real-time development platform. [Online; accessed Apr-2022] (2022), https://unity.com/
35. Wald, I.: GPGPU-parallel re-indexing of triangle meshes with duplicate-vertex and unused-vertex removal (2021). https://doi.org/10.48550/arXiv.2109.09812
36. Wang, Y., Nnaji, B.O.: Document-driven design for distributed CAD services in service-oriented architecture. J. Comput. Inf. Sci. Eng. **6**(2), 127–138 (2005). https://doi.org/10.1115/1.2194911
37. Weber, D., Bender, J., Schnoes, M., Stork, A., Fellner, D.W.: Efficient GPU data structures and methods to solve sparse linear systems in dynamics applications. Comput. Graph. Forum **32**(1), 16–26 (2013). https://doi.org/10.1111/j.1467-8659.2012.03227.x
38. Whalen, E., Beyene, A., Mueller, C.: SimJEB: simulated jet engine bracket dataset. Comput. Graph. Forum **40**(5), 9–17 (2021). https://doi.org/10.1111/cgf.14353
39. Xian, C., Gao, S., Zhang, T.: Tetrahedral mesh editing with local feature manipulations. In: 2011 12th International Conference on Computer-Aided Design and Computer Graphics, pp. 130–137 (2011). https://doi.org/10.1109/CAD/Graphics.2011.58
40. Zissis, D., Lekkas, D., Azariadis, P., Papanikos, P., Xidias, E.: Collaborative CAD/CAE as a cloud service. Int. J. Syst. Sci. Oper. Logist. **4**(4), 339–355 (2016). https://doi.org/10.1080/23302674.2016.1186237

Performance Study of Vertical Submersible Pump in Terms of Induced Loads and Vibrations

Patrick Zito Malonda[1], Guyh Dituba Ngoma[1(✉)], Walid Ghié[1], Fouad Erchiqui[1], Python Kabeya[2], and Francis Kifumbi[1]

[1] School of Engineering, University of Quebec in Abitibi-Témiscamingue, Rouyn-Noranda, Canada
{zito.malonda,guyh.dituba-ngoma,walid.ghie,fouad.erchiqui, francis.kifumbi}@uqat.ca

[2] Polytechnic Faculty, University of Kinshasa, Kinshasa, Democratic Republic of the Congo
python.kabeya@unikin.ac.cd

Abstract. In this research work a reliable numerical approach is developed to determine the axial and the radial forces, the strains, the stresses and the induced vibrations in vertical submersible pumps. The use of this approach in the manufacture of these pumps allows to further improve their performances in terms of head, shaft power and efficiency while increasing the reliability and the service life of the pump shaft bearings and/or bushings. Sizing starts from the design point characterized by a flow rate of 141 m^3/h, a head of 92 m and a rotating speed of 3600 rpm. After determining the various geometrical parameters of the impellers, the diffusers and the shaft, a 3D model of the pump is obtained in the solid and fluid domains. Numerical simulations are carried out and the results achieved are validated comparing them with the experimental results from the Technosub company. Thus, the axial and the radial forces and the torques resulting from the analysis of the liquid flow in the impeller and the diffuser induce the loads on the pump shaft and make it possible to determine the stresses and the strains including the vibration amplitudes on this shaft. Moreover, the impacts of the outer diameter of the impellers, the number of submersible pump stages and the rotating speed on the pump performances are carried out in order to improve the design of the vertical submersible pumps.

Keywords: Multistage centrifugal pump · Axial and radial forces · Stress · Strain · Vibrations · CFX · Static structural analysis · Harmonic response

1 Introduction

The high pressure vertical submersible pumps are used extensively in numerous industrial, mining applications and in the mining sites in construction for the dewatering and the control of the water level. The working of these high pressure pumps implies strong solicitations on the shaft and its antifriction bearings/plain bearings. Thus, the design process is a big challenge due to the performances to reach. To account for the essential parameters of the pump components such as the diameter, the width of the blades, the

G. Wagner et al. (Eds.): SIMULTECH 2021, LNNS 601, pp. 43–69, 2023.
https://doi.org/10.1007/978-3-031-23149-0_3

angles of the blades, the thickness of the blades of the impeller and the diffuser [1, 2], is primordial to assure an optimal manufacture. Indeed, the relative complexity to the analysis of the liquid flow through a vertical submersible pump leads to the use of the numerical tools in the goal to determine the performances of pump notably the head, the brake horsepower and the efficiency of the pump, but also the applied loads on the shaft for a good dimensionality of the antifriction bearings/plain bearings. Furthermore, the relevant literature review in connection with this research reveals that:

a) for the axial and the radial forces in the centrifugal pumps and the submersible pumps, the pump designers are confronted to problems of unbalance of the radial and axial loads on the impeller due of the distribution of the static pressure on the shrouds of the impeller. In the case of a submersible pump, the radial force doesn't modify the good operating of the pump appreciably whereas the axial force influences considerably on the running of the pump [3]. The use of an axial thrust bearing is ideal for the balancing of the axial force in the single-stage pumps and for weak rotating speed. The methods of the holes of balancing and the radial blades can be used to reduce force acting on the rear shroud of the impeller [4–6]. In the multi-stages pumps, considering the complexity of the calculation of the axial force from the distribution of the pressure on the shrouds of the impeller, the dimensioning of the balancing device of the axial force and the thrust bearing is often defined on the basis of the values by force measured at the time of the tests of the pump [7]. The design of the volute has an influence on the radial force. This last is minimal to the point of working of the pump for a volute of simple design. The inverse occurs for a circular volute with a maximal force to the point of good operating whereas a volute to double partition generates an appreciably uniform force [8];
b) concerning the stresses and the strains in the centrifugal pumps and the submersible pumps, in practice, the regions of the stress concentration are caused by grooves, the pins and the cracks that lead to an increase of the stresses in the pieces [9]. At the time of the pump operating, the impellers in rotation transmit the mechanical energy to the fluid. So the load of pressure of the fluid and the load of inertia due to the rotating speed cause some stresses on these impellers. The fluid pressure introduces a stress and a more important maximal strain than the one due to the force of inertia. But with the increase of the thickness of the blades, the stress and the maximal strain caused by the load of the inertia force increase progressively, while the one of the load of the pressure of the fluid decreases [10]. It agrees to underline that the increase of the diameter of the impeller also increases the stress and the strain in a centrifugal pump [11];
c) regarding the plain and the antifriction bearings in the centrifugal pumps and the submersible pumps, the adequate choice of the bearings depends on the dynamical behavior of the shaft, of the rotating speed and the factors as the bearing positions and the pump applications [7, 12, 13]. During the functioning of the pump; the plain and the antifriction bearings take in charge the axial displacement and the lateral deviation of the shaft. The capacity of the bearings to work correctly is damaged by wear, fatigue or the deterioration of the lubricant. The penetration of particles in the bearings also causes the elevated stresses and a premature rupture by fatigue. These particles also cause wear of the bearings reducing their life span [13]. In a

centrifugal pump, a thrust bearing must be used to balance the axial force completely in all operating conditions [10];

d) concerning the mechanical vibrations in the centrifugal pumps and the submersible pumps, it is noted that, during the pump operating, the components as the impellers, the bearings, the seals and the shaft undergo the influence of the hydraulic forces coming from the work process [14]. So, the circulation of the liquids in the centrifugal pumps generates some turbulence. Besides, the displacement between the impeller and the diffuser creates an unsteady interaction phenomenon that leads to some vibrations [15]. To get a reduction of the vibrations, the pump designers must take into account the analysis of this factor in order to avoid the vibrations in normal operating conditions. The modal analysis is essential in the design of machines subject to dynamic loads [16]. This analysis allows to determine the modes and the natural frequencies that are important characteristics of the vibrations of the components of the machine notably, whereas the harmonic analysis permits to insure inter alia that the design of a machine will be susceptible to overcome the resonance effects.

In the light of past research works described previously, in this study, it is about developing reliable and precise numerical approach to better determine the axial and the radial forces, the stresses, the strains and the vibrations induced in the vertical submersible pumps accounting for the effect of the pressure generated by the impeller in rotation on the pump shaft. This reseach work is indeed the extended results of [17].

2 Model Description

The six-stage vertical submersible pump considered as the reference pump for this work is illustrated in Fig. 1. It is composed, inter alia, of a shaft, six impellers and six diffusers. The solid and the fluid models of this pump are shown in Fig. 2.

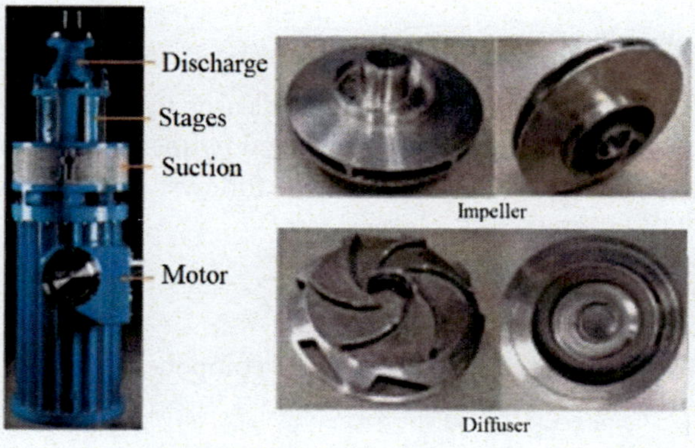

Fig. 1. Six-stage vertical submersible pump [17, 18].

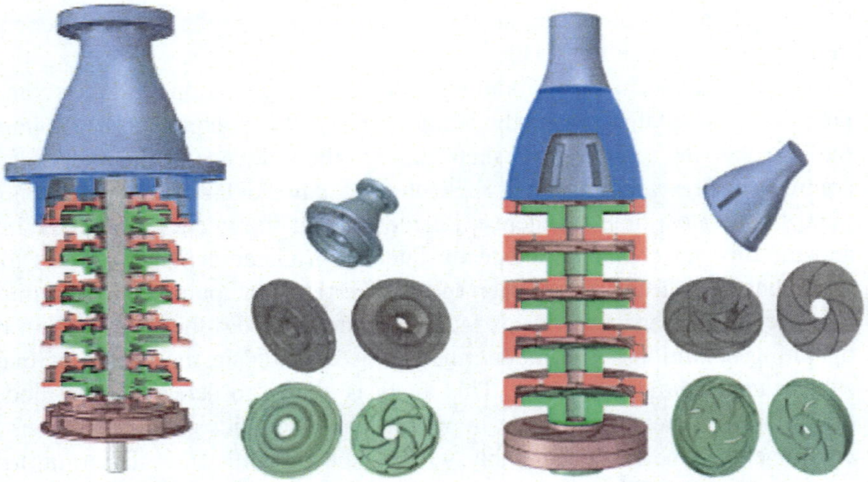

Fig. 2. Model of a six-stage vertical submersible pump [17].

3 Mathematical Formulation

To determinate the field of the liquid flow velocity, the field of the pressure, the stress and the strain in a multistage vertical submersible pump, the following hypotheses are considered for the liquid flow and the solid mechanics [2, 15]: (a) a steady state, three-dimensional flow; (b) the liquid being considered incompressible; (c) it is a Newtonian liquid; and (d) the liquid's thermophysical properties are constant with the temperature; (e) the material is considered continuous, doesn't have cracks, nor cavities; (f) the material is homogeneous and presents the same properties in all points; (g) the material is considered as isotropic; and (h) no internal force acts in the material before the application of the external loads.

3.1 Liquid Flow Velocity and Pressure [17]

The equations of continuity and Navier-Stokes are used to obtain the fields of liquid flow velocity and pressure. These equations are solved by means of the ANSYS CFX-code [19]. The equation of continuity is expressed as follows:

$$\frac{\partial u}{\partial x} + \frac{\partial v}{\partial y} + \frac{\partial w}{\partial z} = 0 \tag{1}$$

where u(x, y, z), v(x, y, z) and w(x, y, z) are the components of the liquid flow velocity U(u, v, w).

Accounting for the gravity, the equations of Navier-Stokes can be formulated by:

$$\begin{aligned}
\rho\left(u\frac{\partial u}{\partial x}+v\frac{\partial u}{\partial y}+w\frac{\partial u}{\partial z}\right) &= \mu_{eff}\left(\frac{\partial^2 u}{\partial x^2}+\frac{\partial^2 u}{\partial y^2}+\frac{\partial^2 u}{\partial z^2}\right)-\frac{\partial p}{\partial x}+\rho(\omega_z^2 r_x+2\omega_z v)+\rho g_x \\
\rho\left(u\frac{\partial v}{\partial x}+v\frac{\partial v}{\partial y}+w\frac{\partial v}{\partial z}\right) &= \mu_{eff}\left(\frac{\partial^2 v}{\partial x^2}+\frac{\partial^2 v}{\partial y^2}+\frac{\partial^2 v}{\partial z^2}\right)-\frac{\partial p}{\partial y}+\rho(\omega_z^2 r_y-2\omega_z u)+\rho g_y \\
\rho\left(u\frac{\partial w}{\partial x}+v\frac{\partial w}{\partial y}+w\frac{\partial w}{\partial z}\right) &= \mu_{eff}\left(\frac{\partial^2 w}{\partial x^2}+\frac{\partial^2 w}{\partial y^2}+\frac{\partial^2 w}{\partial z^2}\right)-\frac{\partial p}{\partial z}+\rho g_z
\end{aligned} \tag{2}$$

where g (g_x, g_y, g_z) is the gravity acceleration, p is the pressure; ρ is the density; μ_{eff} is the effective viscosity accounting for turbulence, it is defined as $\mu_{eff} = \mu + \mu_t$. μ is the dynamic viscosity and μ_t is the turbulence viscosity. It is linked to turbulence kinetic energy and dissipation [2].

3.2 Axial and Radial Forces

Figure 3 illustrates the axial and the radial forces and the pressure distribution on an impeller due to the liquid flow through a six-stage vertical submersible pump. These forces are determined using the ANSYS CFX-code. The axial forces are the result of unbalanced impeller forces acting in the shaft axial direction. Moreover, the radial force on the impeller results from a non-uniform distribution of pressure on the circumference of the impeller. The non-uniform pressure distribution can be caused by: the geometrical form of the diffuser for the multistage centrifugal pumps; the non-symmetrical impeller inflow; or the pump operating regime. It is to highlight that the radial force depends on the time. Its components are the static radial force and the dynamic radial force. Generally, the static radial force is greater than the dynamic radial force [1, 7, 15, 20–25].

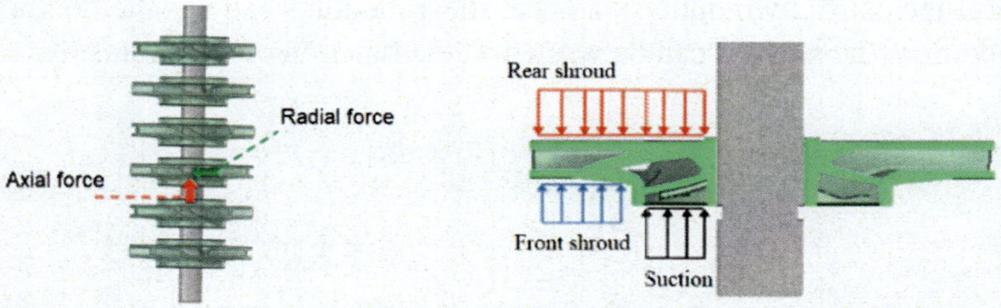

a) Axial and radial forces [17]. b) Pressure distribution on a pump impeller.

Fig. 3. Axial and radial force in a model of the six-stage vertical submersible pump.

3.3 Stresses and Strains [17]

The normal and the shear stresses on the pump shaft are determined by means of the equilibrium equations of elasticity in terms of stress neglecting the forces per unit of volume [26]. These equations are given by:

$$\begin{aligned}
&\frac{\partial \sigma_x}{\partial x} + \frac{\partial \tau_{yx}}{\partial y} + \frac{\partial \tau_{zx}}{\partial z} = 0\\
&\frac{\partial \tau_{xy}}{\partial x} + \frac{\partial \sigma_y}{\partial y} + \frac{\partial \tau_{zy}}{\partial z} = 0\\
&\frac{\partial \tau_{xz}}{\partial x} + \frac{\partial \tau_{yz}}{\partial y} + \frac{\partial \sigma_z}{\partial z} = 0
\end{aligned} \tag{3}$$

The normal and the shear strains are formulated as follows using the displacements (u, v, w) respectively in the directions of x, y and z:

$$\begin{aligned}
&\varepsilon_x = \frac{\partial u}{\partial x};\ \varepsilon_y = \frac{\partial v}{\partial z};\ \varepsilon_z = \frac{\partial w}{\partial z}\\
&\gamma_{xy} = \frac{\partial u}{\partial y} + \frac{\partial v}{\partial x};\ \gamma_{yz} = \frac{\partial w}{\partial y} + \frac{\partial v}{\partial z};\ \gamma_{zx} = \frac{\partial u}{\partial z} + \frac{\partial w}{\partial x}
\end{aligned} \tag{4}$$

Furthermore, the relationships between the stresses and the strains is given by:

$$\begin{aligned}
&\varepsilon_x = \frac{1}{E}\left[\sigma_x - \nu(\sigma_y + \sigma_z)\right]\\
&\varepsilon_y = \frac{1}{E}\left[\sigma_y - \nu(\sigma_z + \sigma_x)\right]\\
&\varepsilon_z = \frac{1}{E}\left[\sigma_z - \nu(\sigma_x + \sigma_y)\right]\\
&\gamma_{xy} = \frac{\tau_{xy}}{G};\ \gamma_{yz} = \frac{\tau_{yz}}{G};\ \gamma_{zx} = \frac{\tau_{zx}}{G}
\end{aligned} \tag{5}$$

where E is the elasticity modulus, G is the shear modulus and ν is the Poisson ratio.

In addition, the stresses can be written as the function of the strains by:

$$\begin{aligned}
&\sigma_x = \frac{E}{(1+\nu)(1-2\nu)}\left[(1-\nu)\varepsilon_x + \nu(\varepsilon_y + \varepsilon_z)\right]\\
&\sigma_y = \frac{E}{(1+\nu)(1-2\nu)}\left[(1-\nu)\varepsilon_y + \nu(\varepsilon_z + \varepsilon_x)\right]\\
&\sigma_z = \frac{E}{(1+\nu)(1-2\nu)}\left[(1-\nu)\varepsilon_z + \nu(\varepsilon_x + \varepsilon_y)\right]\\
&\tau_{xy} = G\gamma_{xy};\ \tau_{yz} = G\gamma_{yz};\ \tau_{zx} = G\gamma_{zx}
\end{aligned} \tag{6}$$

The stress of von Mises selected for the yield criteria can be expressed by:

$$\sigma' = \sqrt{\frac{1}{2}\left((\sigma_1 - \sigma_2)^2 + (\sigma_2 - \sigma_3)^2 + (\sigma_3 - \sigma_1)^2\right)} \tag{7}$$

where σ_1, σ_2 and σ_3, are the principal stresses in the directions of 1, 2 and 3 according to $\sigma_1 > \sigma_2 > \sigma_3$. These stresses can be determined as follows [26]:

$$\begin{aligned}
\sigma_1 &= \sigma_0 + 2\left(\left|\frac{J_2}{3}\right|\right)^{\frac{1}{2}} \cos\left(\frac{1}{3}\arccos\left(-0.5J_3\left(\left|\frac{J_2}{3}\right|\right)^{-1.5}\right)\right)\\
\sigma_2 &= \sigma_0 - 2\left(\left|\frac{J_2}{3}\right|\right)^{\frac{1}{2}} \cos\left(\frac{1}{3}\arccos\left(-0.5J_3\left(\left|\frac{J_2}{3}\right|\right)^{-1.5}\right)+\frac{\pi}{3}\right)\\
\sigma_3 &= \sigma_0 - 2\left(\left|\frac{J_2}{3}\right|\right)^{\frac{1}{2}} \cos\left(\frac{1}{3}\arccos\left(-0.5J_3\left(\left|\frac{J_2}{3}\right|\right)^{-1.5}\right)-\frac{\pi}{3}\right)
\end{aligned} \tag{8}$$

where

$$\begin{aligned}
\sigma_0 &= \frac{1}{3}(\sigma_x+\sigma_y+\sigma_z)\\
J_2 &= s_xs_y + s_ys_z + s_zs_x - \tau_{xy}^2 - \tau_{yz}^2 - \tau_{zx}^2\\
J_3 &= -\left(s_xs_ys_z - s_x\tau_{yz}^2 - s_y\tau_{zx}^2 - s_z\tau_{xy}^2\right)\\
s_x &= \sigma_x - \sigma_0;\ s_y = \sigma_y - \sigma_0;\ s_z = \sigma_z - \sigma_0
\end{aligned} \tag{9}$$

3.4 Mechanical Vibrations in the Vertical Submersible Pumps

To study the mechanical vibrations in the system defined by submersible pumps, the general shape of the equations of movement of the system are given by [15]:

$$[\mathrm{M}]\{\ddot{\mathrm{x}}\} + [\mathrm{C}]\{\dot{\mathrm{x}}\} + [\mathrm{K}]\{\mathrm{x}\} = \{\mathrm{F}\} \tag{10}$$

where [M] is the modal mass matrix, [C] is the damping matrix, [K] is the stiffness matrix, {x} is the displacement vector, $\{\dot{x}\}$ the vector velocity vector, $\{\ddot{x}\}$ is the vector acceleration, and $\{F\}$ is the force excitation vector.

Additionally, the displacement vector can be written as:

$$\{\mathrm{x}\} = \{\phi\}_\mathrm{m}\cos(\mathrm{w_m\, t}) \tag{11}$$

where $\{\phi\}_\mathrm{m}$ is the eigenmode vector and ω_m is the natural frequency.

The velocity vector is expressed by:

$$\{\dot{\mathrm{x}}\} = -\mathrm{w_m}\{\phi\}_\mathrm{m}\sin(\mathrm{w_m\, t}) \tag{12}$$

and the vector acceleration can be formulated as follows

$$\{\ddot{\mathrm{x}}\} = -\mathrm{w_m^2}\{\phi\}_\mathrm{m}\cos(\mathrm{w_m\, t}) \tag{13}$$

Finally, the force excitation vector is given by:

$$\{\mathrm{F}\} = \{\mathrm{F_o}\}_\mathrm{m}\cos(\omega_\mathrm{m}\mathrm{t}) \tag{14}$$

To obtain the amplitudes and the frequencies of the vibrations, the ANSYS-code for the harmonic response is used according to Eq. 10.

3.5 Diffuser Equations [17]

The diffuser equations [1] are used in this research to calculate the main parameter of the diffusers of the six-stage vertical submersible pump as illustrated in Fig. 4.

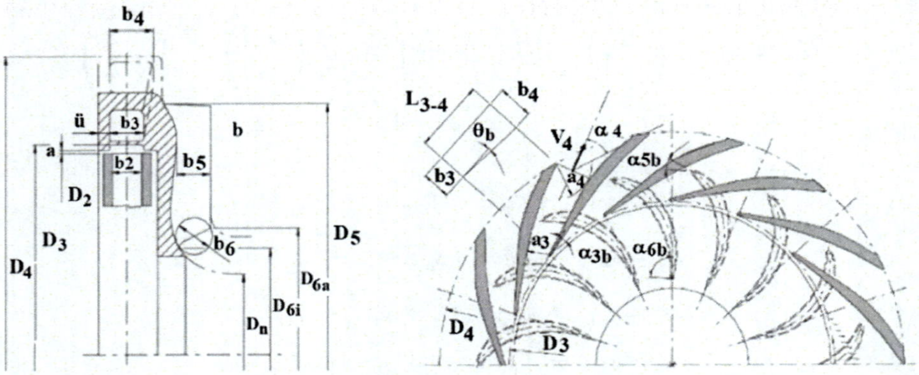

Fig. 4. Diffuser parameters [1].

These equations can be written as follows:

$$b_3 = (1.05 \text{ to } 1.3) \cdot b_2 \tag{15}$$

$$\alpha_{3b} = \alpha_3 \pm 3^\circ \text{ and } \alpha_3 = \tan^{-1}\left(\frac{V_{m3}}{V_{u3}}\right) \tag{16}$$

$$V_{m3} = \frac{Q \cdot \tau_3}{\pi \cdot D_3 \cdot b_3} \tag{17}$$

$$V_{u2} = \frac{gH}{\eta_h U_2} + \frac{U_{1m} V_{1u}}{U_2} \tag{18}$$

$$a_3 = f_{a3} \cdot \left(\frac{D_3}{2}\right) \cdot \left\{\exp\left(\frac{Q}{b_3 \cdot V_{2u} \cdot \left(\frac{D_2}{2}\right) \cdot Z_{Le}}\right) - 1\right\} \tag{19}$$

where $1.1 \le f_{a3} \le 1.3$, the vane number of the diffuser Z_{Le} is chosen according to Table 1.

Table 1. Number of blades required for the diffuser [1].

Z_b	5			6	7			
Z_{Le}	7	8	12	10	9	10	11	12	(15)

$$e_3 = (0.01 \text{ à } 0.015) \cdot D_2 \tag{20}$$

$$D_4 = ((1.05 \text{ à } 1.15) + 0.01 \cdot n_q) \cdot D_2 \tag{21}$$

where n_q is the specific speed.

$$\vartheta_b = \tan^{-1}\left(\frac{0.5 \cdot (a_4 - a_3)}{L_{3-4}}\right) \tag{22}$$

$$b_4 = b_3 + (\tan(\vartheta_b) \cdot L_{3-4}) \tag{23}$$

$$b_5 = b_6 = \frac{Q}{\pi \cdot D_6 \cdot V_{m6}} \tag{24}$$

$$V_{m6} = (0.85 \text{ à } 0.9) \cdot V_{m1} \tag{25}$$

$$\alpha_5 = \tan^{-1}\left(\frac{V_{m5}}{V_{u5}}\right) \tag{26}$$

$$V_{m5} = \frac{Q}{\pi \cdot D_5 \cdot b_5} \tag{27}$$

$$V_{u5} = V_{u4} \cdot \left(\frac{D_4}{D_5}\right) \tag{28}$$

$$V_{u4} = V_{u3} \cdot \left(\frac{D_3}{D_4}\right) \tag{29}$$

$$\alpha_{6b} = \alpha_6 \pm 5^\circ \tag{30}$$

4 Numerical Implementation, Submersible Pump Modeling and Simulation Steps

The differential equations result from mathematical modelings of the equations of continuity and Navier-Stokes including the model of turbulence are solved using the ANSYS-CFX module, and the equations of stresses, strains and vibrations are solved by the ANSYS-code while applying the module static structural, the modal and the harmonic response. In sum, Fig. 5 shows the steps of the numerical resolution with the ANSYS software. Besides, Fig. 6 illustrates the modeling and the simulation steps for a vertical submersible pump using the Inventor and the ANSYS softwares (modules: Spaceclaim, CFX-Pre, CFX-Solver and CFX-Post) and accounting for the boundary conditions.

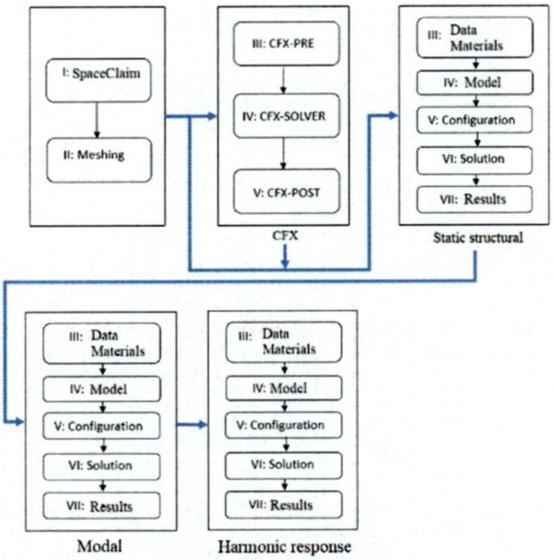

Fig. 5. Steps of the numerical resolution with the ANSYS software.

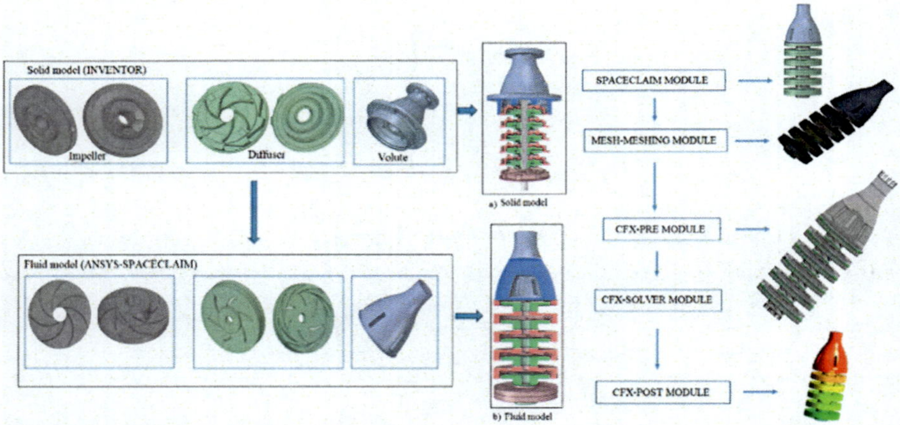

Fig. 6. Modeling of the reference submersible pump et simulation steps.

Additionally, Fig. 7 indicates the boundary conditions used in this research. Indeed, at the pump inlet, the static pressure is specified, while the flow rate is given at the pump outlet. For the interaction of the impeller-diffuser, the frozen rotor condition is applied.

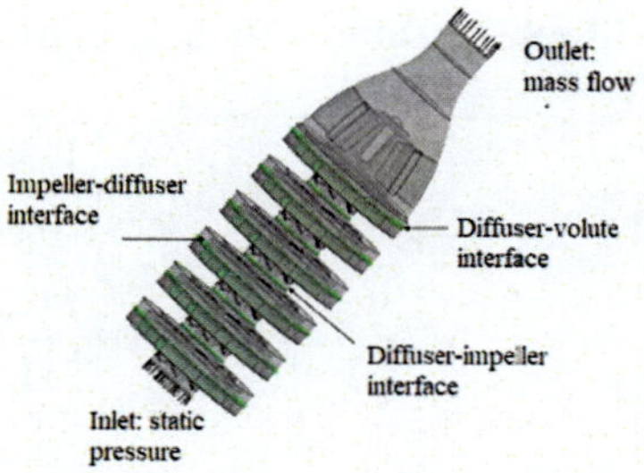

Fig. 7. Boundary conditions for the six-stage vertical submersible pump [17].

5 Results and Discussion

To study the shaft behavior in terms of axial and radial forces, stresses, strain and vibrations, five cases were selected: (a) the flow rate, (b) the intensity of the impeller trimming (100%, 91.7%, and 84.65%); (c) the number of the stages (3, 5 and 6 stages); (d) the shaft rotating speed (1800 rpm and 3600 rpm); and (e) the amplitudes and the frequencies of vibration. According to the numerical results obtained, different numbers of mesh elements are used in each case study to achieve the mesh-independent solution tests. The reference data applied for the shaft, the impeller, and the diffuser are given in Tables 2, 3, 4 and 5 [17].

Table 2. Pump shaft data.

Length L [mm]	655.15
Diameter d [mm]	45.06

Table 3. Impeller data.

Inlet blade height b_1 [mm]	30.17
Outlet blade height b_2 [mm]	14.48
Hub diameter D_{h1} [mm]	44,45
Inlet diameter D_{h2} [mm]	107.95
Outlet diameter D_2 [mm]	241
Inlet blade angle β_{b1} [°]	16
Outlet blade angle β_{b2} [°]	27.5
Blade thickness e [mm]	3.17
Blade number Z_b	7

Table 4. Diffuser (front side) data.

Inlet blade height b_3 [mm]	17.46
Outlet blade height b_4 [mm]	40.64
Inlet diameter D_3 [mm]	243,84
Outlet diameter D_4 [mm]	311.15
Inlet blade angle α_{3b} [°]	10
Blade thickness e_3 [mm]	3.175
Blade number Z_{Le}	8

Table 5. Diffuser (rear side) data.

Return vane number Z_R	6
Outlet return vane height b_5 [mm]	24.4
Diameter at the inlet of the return vane D_3 [mm]	311.15
Blade angle at the inlet of the return vane α_5 [°]	95
Blade angle at the outlet of the return vane α_6 [°]	18
Blade thickness of the return vane e_3 [mm]	6.04

Furthermore, the properties of the 17-4PH steel and the water considered are given in Tables 6 and 7 [17].

Table 6. Properties of the 17-4PH steel.

Module of the Young [Pa]	1.96×10^{11}
Poisson ratio	0.3
Compressibility module [Pa]	1.63×10^{11}
Shear modules [Pa]	7.53×10^{10}
Resistance coefficient [Pa]	9.2×10^{8}
Ductility coefficient [Pa]	10^{9}
Yield strength [Pa]	7.93×10^{8}
Ultimate tensile strength [Pa]	1.103×10^{9}
Density [kg/]	7750.4

Table 7. Properties of water in 25 °C.

Density [kg/]	Thermal expansion coefficient [K^{-1}]	Kinematic viscosity [m^2/s]
997	2,57 × 10^{-1}	0.884 × 10^{-6}

Effect of the Flow on the Axial and the Radial Forces, and the Strain for a Single-Stage Submersible Pump. The analysis of the stresses and the strains is achieved by means of the solid model while integrating the internal pressures of the impellers, the axial and the radial forces and the torques obtained from the simulations of the fluid model according to the flow rate and the rotating speed of rotation of the pump. Figures 8, 9 and 10 illustrate the pump head, the brake horsepower, the efficiency, the axial and the radial forces, the stresses and the strains according to the conditions of working in terms of flow rate and rotating speed. In Fig. 8, it is observed, when the head decreases, the axial force also diminishes whereas the radial force falls off then until a flow rate of 110.22 m^3/h believes in a continuous manner to more elevated flow rates. In accordance with Fig. 9, the rise of the brake horsepower and the efficiency of the pump also entails the growth of continuous manners of the stresses on the pump shaft. The corresponding curves for the axial and the radial forces, and the strain is depicted in Fig. 10. Therefore, the choice of the materials is of key importance for the stresses and the strains that will be generated by the pump in operating at a high flow rate.

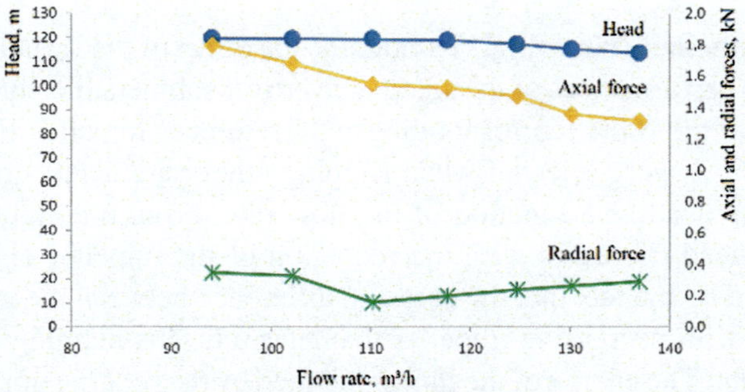

Fig. 8. Pump head, axial and radial forces versus flow rate.

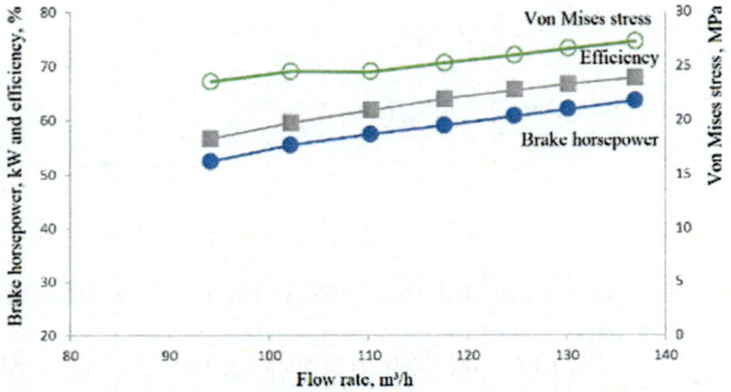

Fig. 9. Von Mises stress, brake horsepower and efficiency versus flow rate.

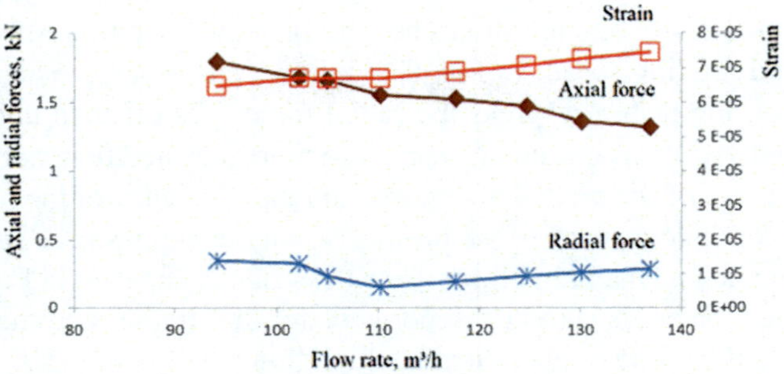

Fig. 10. Strain, axial and radial forces versus flow rate.

Effect of the Impeller Trimming. To examine the effect of the variation of the impeller diameter on the performances of the six-stage vertical submersible pump, three impeller diameter ratios of the 100% (corresponding to the reference impeller diameter: 241 mm), 91.7% and 84.65% were selected when keeping other parameters constant. Figure 11 shows the pump head as a function of the flow rate. From this figure, it can be seen that the pump head decreases with the reduction of the impeller diameter ratio. This can be explained by the fact that the pressure difference between the impeller outlet and inlet decreases with decreasing impeller diameter ratio maintaining the diffuser inner diameter constant. The energy of the fluid generated by the rotating impeller was affected by the impeller trimming which modifies the blade angle. In addition, Fig. 12 indicates that the brake horsepower diminishes with the reduction of the impeller diameter ratio due to the requested diminution impeller shaft torque relative to the size of the impeller diameter keeping the diffuser inner diameter constant. Furthermore, the corresponding efficiency curves as a function of the flow rate illustrated in Fig. 13. It is remarked that the efficiency is the best for the lowest impeller trimming. Moreover, Figs. 14, 15, 16 and 17 show that the impeller trimming decreases the axial force, the radial force, the stress, and the strain on the pump shaft. It may be due to the interaction between the impeller and the diffuser, which is reduced by the decrease of the impeller diameter.

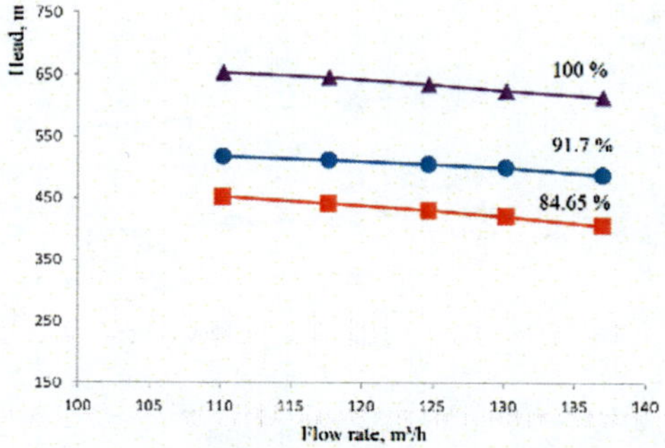

Fig. 11. Head versus flow rate [17].

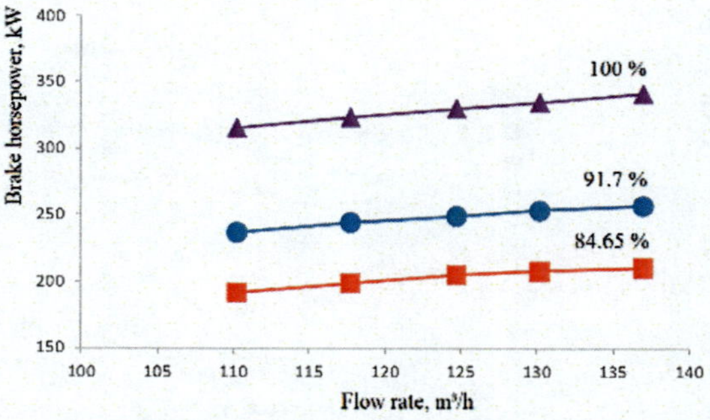

Fig. 12. Brake horsepower versus flow rate [17].

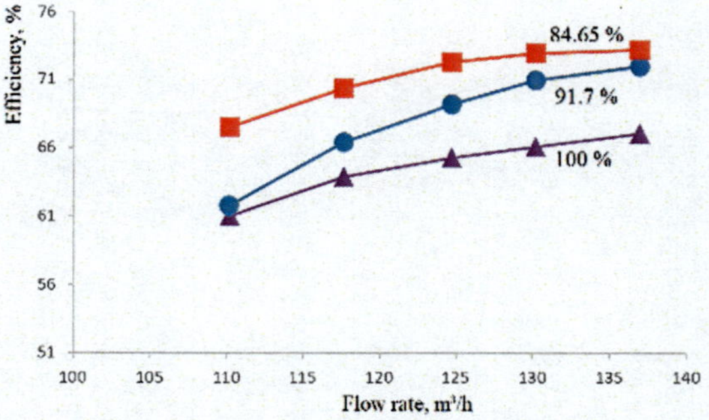

Fig. 13. Efficiency versus flow rate [17].

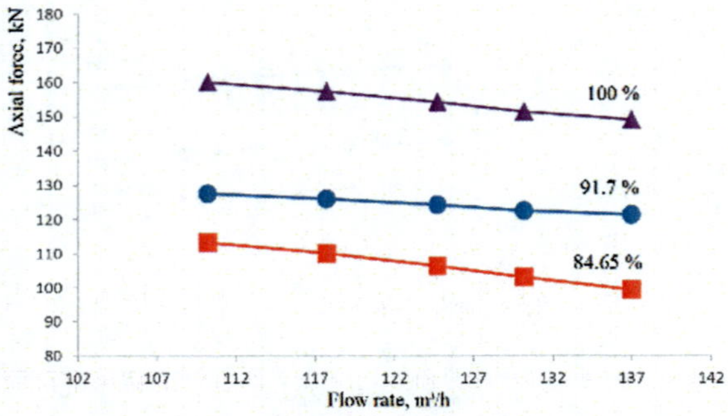

Fig. 14. Axial force versus flow rate [17].

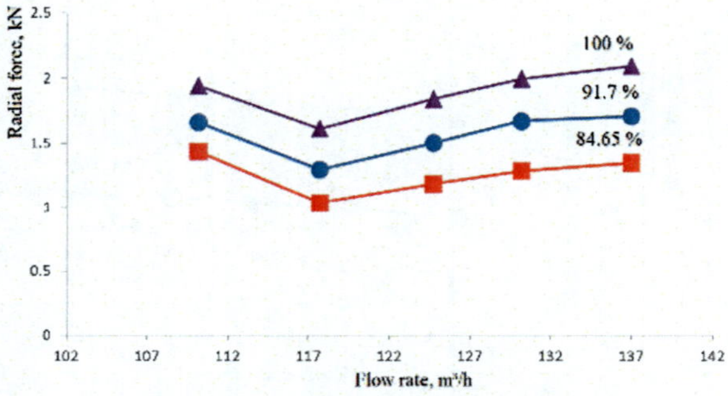

Fig. 15. Radial force versus flow rate [17].

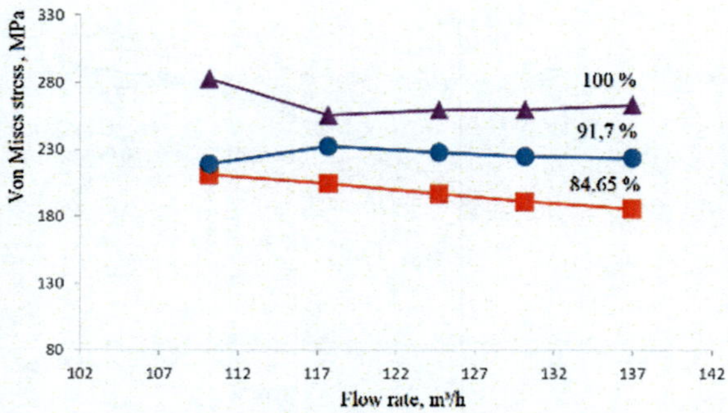

Fig. 16. Stress versus flow rate [17].

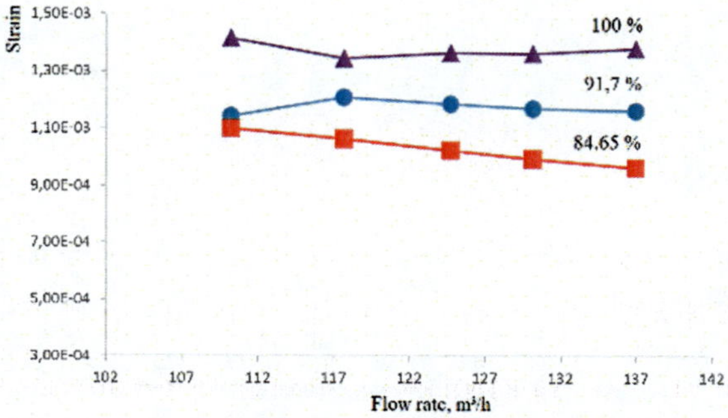

Fig. 17. Strain versus flow rate.

Effect of Stage Number. To examine the effect of the stage number on the pump shaft behavior, three pumps with 3, 5 and 6 stages are selected. Figures 18, 19, 20, 21, 22, 23, 24 provide some relevant information on the influence of the number of the pump stages on the pump performances, the axial and the radial forces, the stress and the strain. More the number of the stages is raised, the more the pump head is important as shown in Fig. 18. The brake horsepower increases with decreasing number of the pump stages as indicated in Fig. 19, whereas the efficiency is unchanged despite the number of the pump stages as shown in Fig. 20. In addition, it can be seen in Figs. 21, 22, 23 and 24 that the axial and the radial forces, the stress and the strain diminish with decreasing stage number.

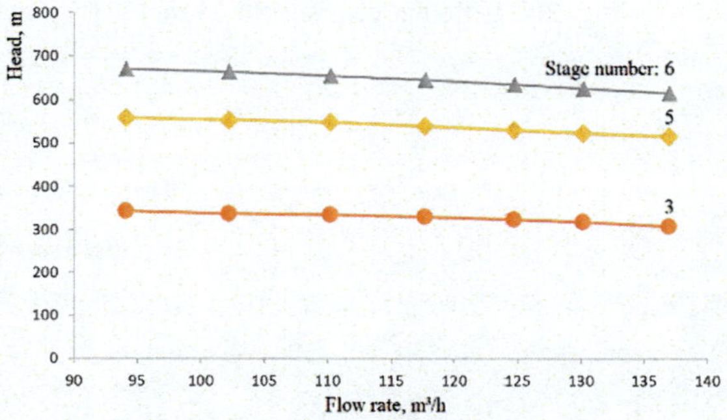

Fig. 18. Head versus flow rate [17].

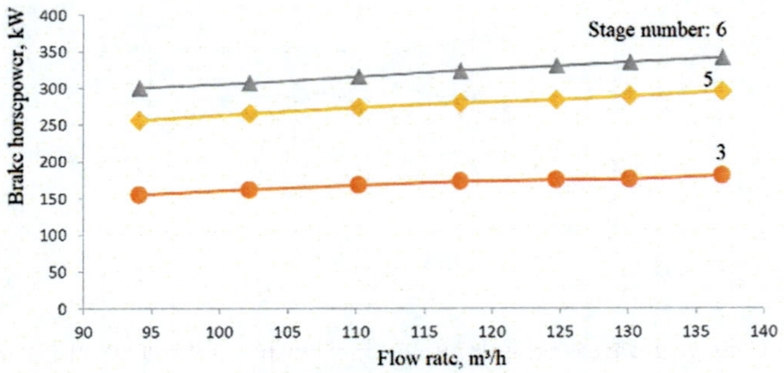

Fig. 19. Brake horsepower versus flow rate [17].

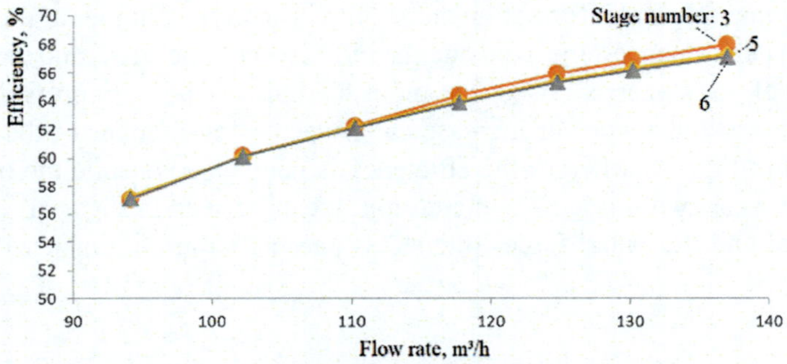

Fig. 20. Efficiency versus flow rate [17].

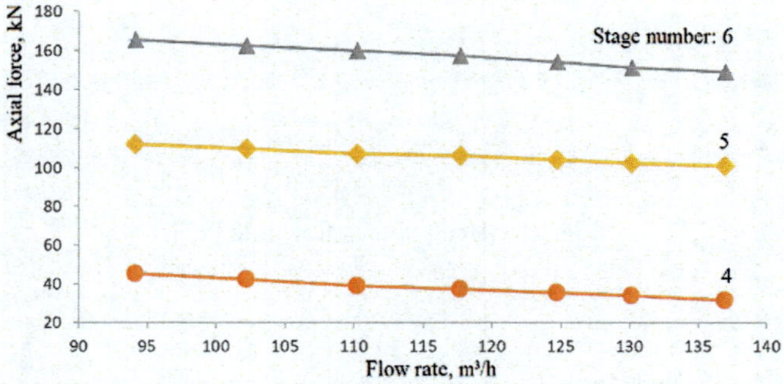

Fig. 21. Axial force versus flow rate [17].

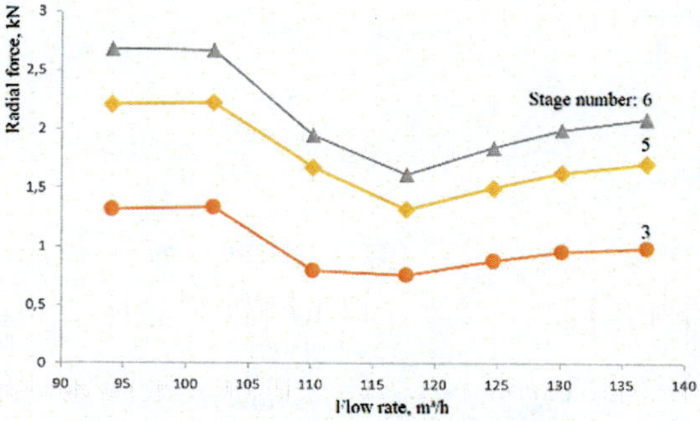

Fig. 22. Radial force versus flow rate [17].

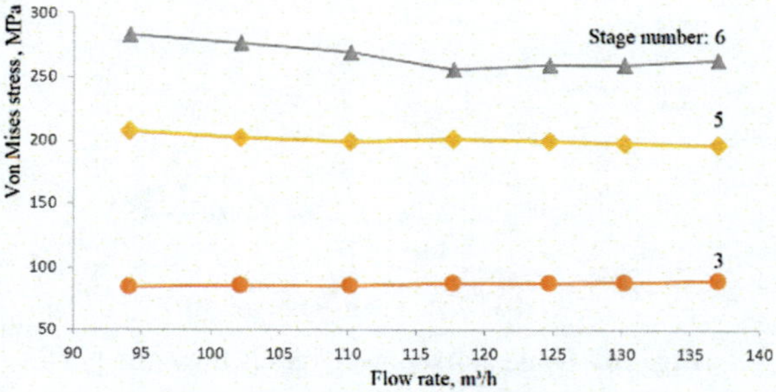

Fig. 23. Stress versus flow rate [17].

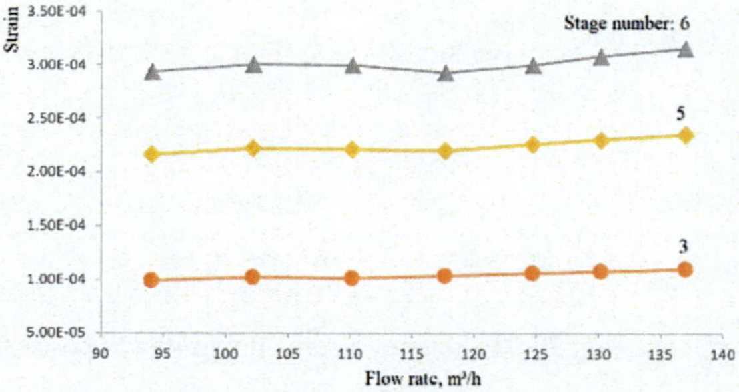

Fig. 24. Strain versus flow rate.

Effect of the Shaft Rotating Speed. This analysis concentrates on the effects of the shaft rotation speed on the performances of the six-stage vertical submersible pump accounting for the induced forces and the stresses on the shaft. Selecting the rotating speeds of 1800 rpm and 3600 rpm, Figs. 25, 26 and 27 show that the pump head and the brake horsepower grow with increasing the rotation speed, whereas the efficiency increases and decreases with the rotation speed in the considered flow rate range. Moreover, Fig. 28 and 29 indicate that respectively the axial and the radial forces increase with the augmentation of the rotating speed. In regard to the stress and the strain on the pump shaft, they raise with growing rotating speed as depicted in Figs. 30 and 31.

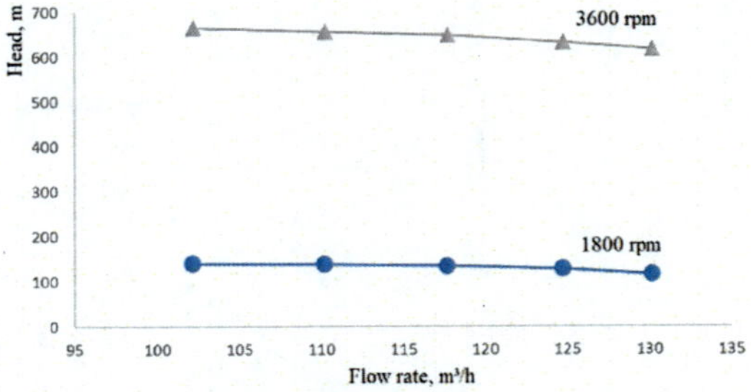

Fig. 25. Head versus flow rate [17].

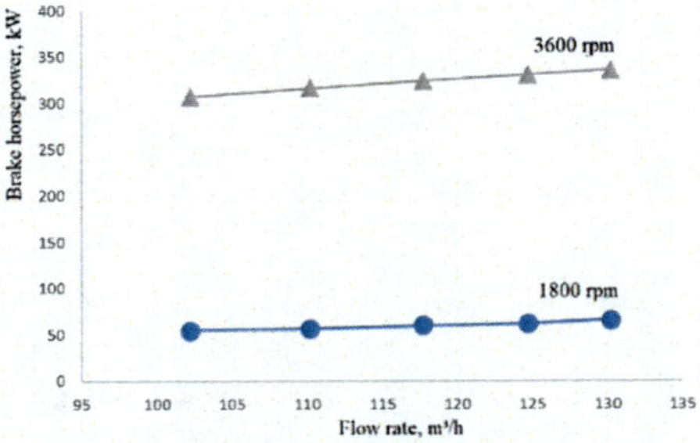

Fig. 26. Brake horsepower versus flow rate [17].

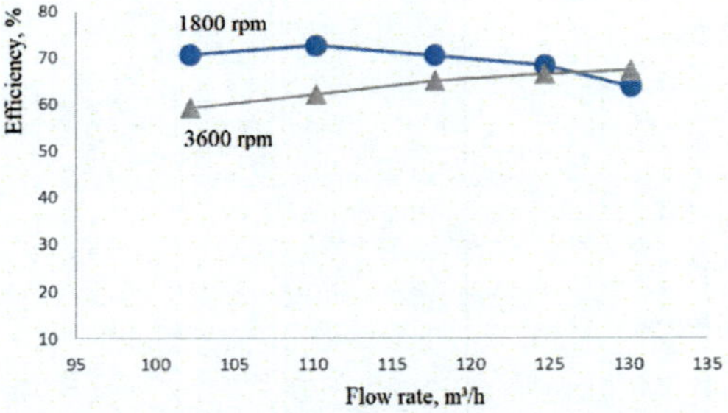

Fig. 27. Efficiency versus flow rate [17].

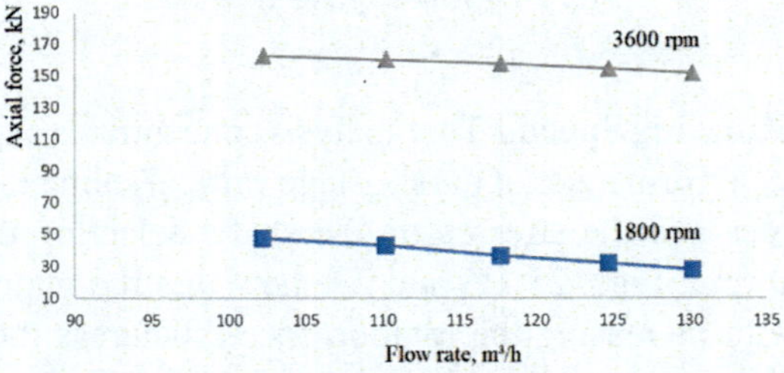

Fig. 28. Axial force versus flow rate [17].

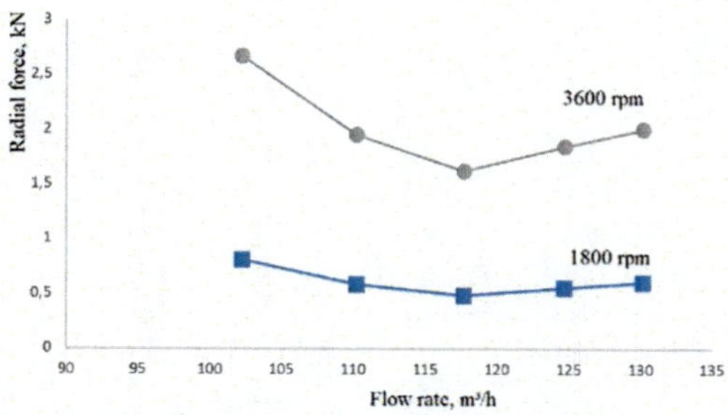

Fig. 29. Radial force versus flow rate [17].

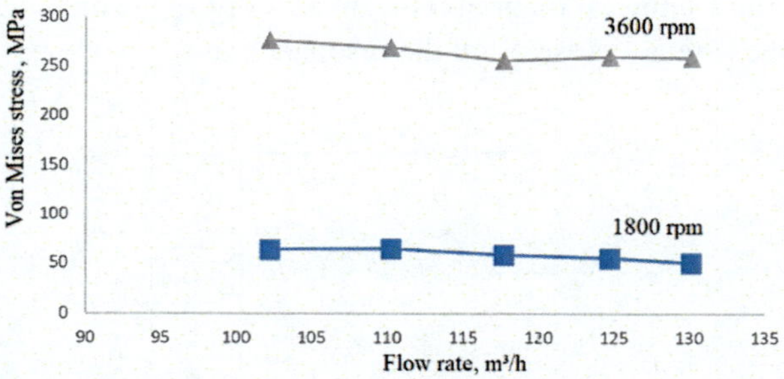

Fig. 30. Stress versus flow rate [17].

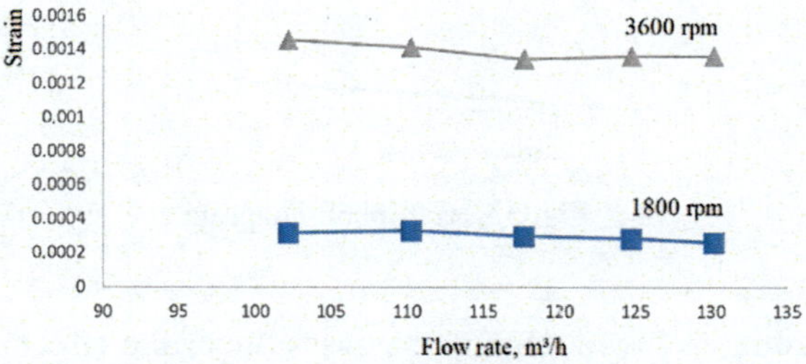

Fig. 31. Strain versus flow rate.

Effect of the Amplitudes and the Frequencies of Vibration on a Six-Stage Vertical Submersible Pump. After having determined the stresses and the strains induced in the pump, the dynamic behavior must be studied in the case of such a problem. A modal analysis and an examination of the harmonic response are done respectively with modal and harmonic response by means of the ANSYS software. The amortization and the hydraulic effect due to the liquid flow are not taken in account in these analyses. A modal analysis is necessary to know the intrinsic dynamic features of the structure. The results of the modal analysis consist of the natural frequencies and the vibration modes of the structure. Figure 32 illustrates the first six natural frequencies and the vibration modes corresponding to the six-stage vertical submersible pump. The natural frequencies allow to determine the degree of fragility of a structure. Besides, the structure is all the more fragile for the low frequency.

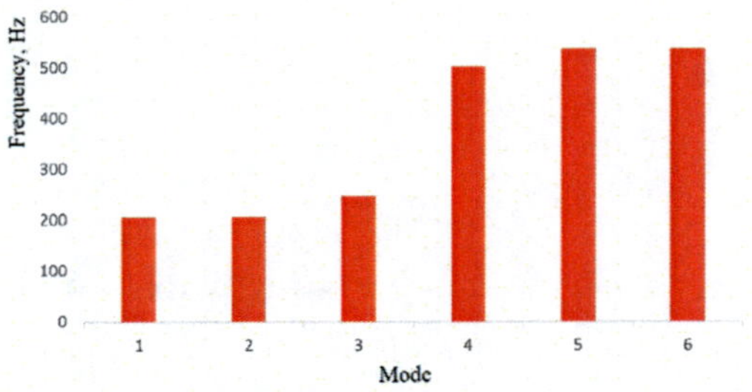

Fig. 32. Natural Frequency versus vibration mode.

In addition, the Campbell diagram is shown in Fig. 33. This diagram indicates the natural frequencies are not close or on the resonance line.

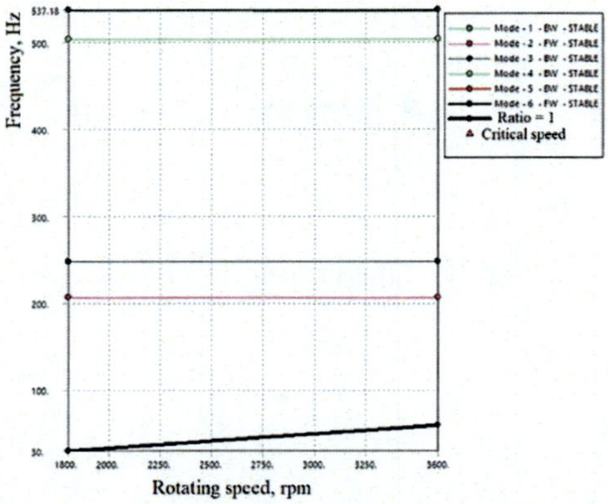

Fig. 33. Campbell Diagram

Besides, regarding to the harmonic responses under the effect of the displacement amplitude, Figs. 34 and 35 present the displacement amplitude and the phase angle respectively according to the frequency for the flow rates of 94.12 m^3/h; 110.22 m^3/h and 136.98 m^3/h. At the frequency of 247.5 Hz the structure is excited so that the maximal displacement in the direction of Z is 4.74×10^{-5} m for a flow rate of 136 m^3/h. So for the same frequency, Fig. 36 shows the displacement amplitude, the stress and the efficiency as a function of the flow rate. It is observed that the stress decreases until the flow rate of 117.7 m^3/h, then it rises, whereas the displacement amplitude and the efficiency grow with increasing flow rate.

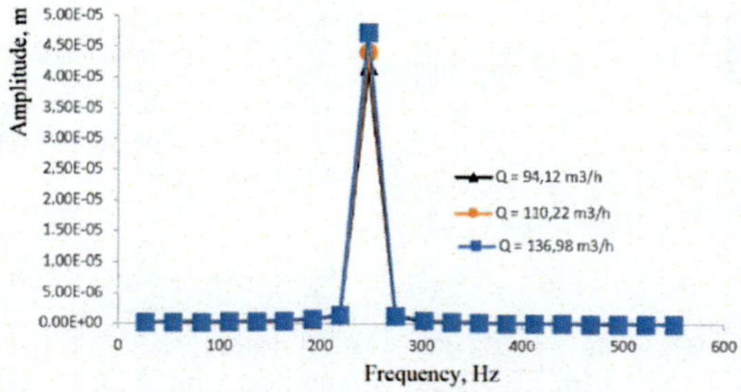

Fig. 34. Displacement amplitude versus natural frequency.

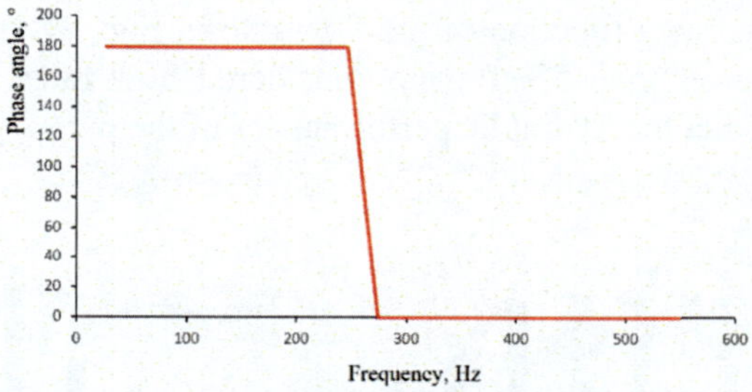

Fig. 35. Phase angle of displacement versus natural frequency.

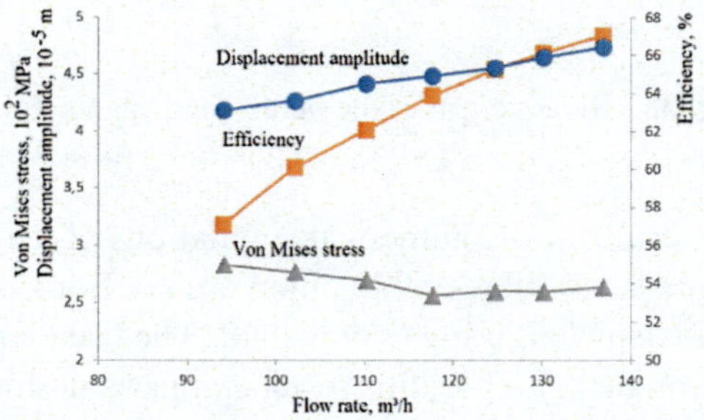

Fig. 36. Von Mises stress, displacement amplitude and efficiency versus flow rate.

Result Validation. The validation of the developed approach is realized comparing the results of numerical simulations with the experimental results provided by [17]. Figure 37 shows the numerical and the experimental curves of the head for a six-stage vertical submersible pump. A good agreement is observed between both curves.

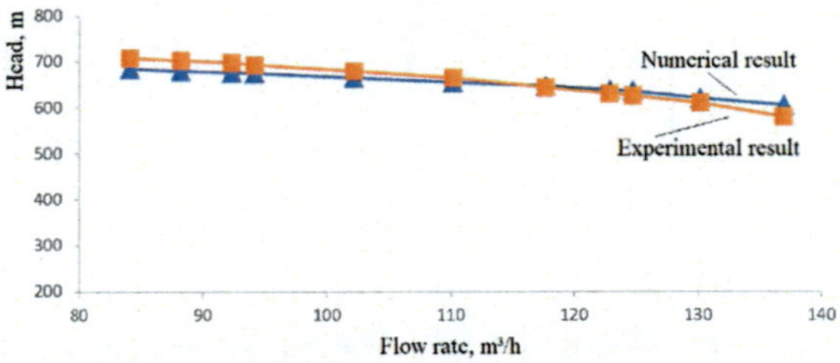

Fig. 37. Head versus flow rate [17].

The relative gaps in absolute value between the experimental and the numerical results are illustrated as a function of the flow rate in Fig. 38. It is remarked that the relative gaps are lower than 5% for the considered flow rates. Thus, the numerical simulations can predict the hydraulic performances of the pump.

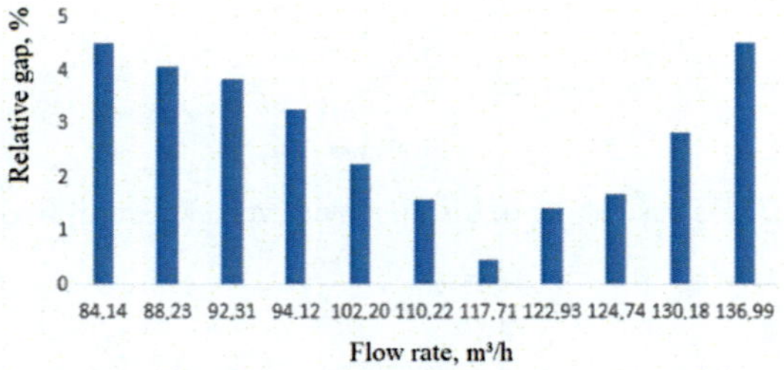

Fig. 38. Relative gap of the pump head versus flow rate.

Furthermore, the results of the numerical simulations of the stresses and the strains on the shaft of a single-stage submersible pump are confronted with the results of the classical equations, as depicted in Figs. 39 and 40. The used equations for the stresses and the strains are given by [9]. Thus, this result comparison shows good harmony. The relative gaps between the results can find their explanation in the hypotheses of classical and numerical approaches used.

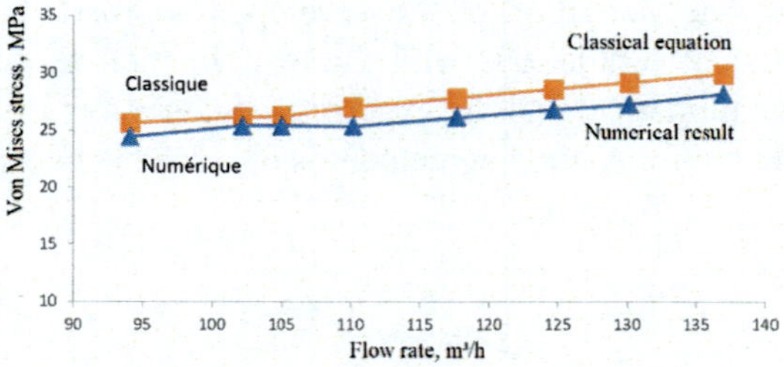

Fig. 39. Stress versus flow rate [17].

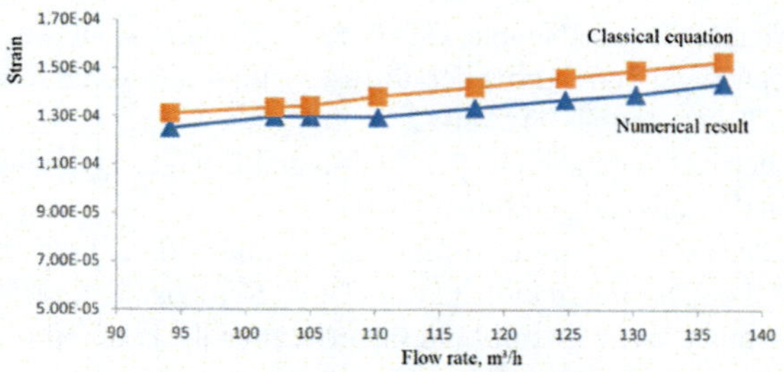

Fig. 40. Strain versus flow rate.

6 Conclusion

The purpose of this research was to design and to analyze a numerical model of the vertical submersible pump in the solid and the fluid domains while taking account of the radial and axial forces, the stresses, the strains, and the induced vibrations. The starting point was a design based on empirical methods. Following the determination of the key parameters of the components of a stage of the existing vertical submersible pump a fluid model was achieved and the results of numerical simulations permitted to observe different tendencies of the characteristic curves of the aforesaid pump. The validation of the numerical model was accomplished while confronting the results of the numerical simulations to the experimental results provided by the Technosub enterprise for a six-stage vertical submersible pump. This comparison demonstrated that the numerical curve follows the tendency of the experimental curve with a relative gap of less than 5% in terms of pump head. Thus, the torques, the axial and the radial forces generated by the impellers in rotation in the fluid domain allowed to determine the stresses, the strains and the vibrations induced in the solid domain of the pump while accounting for the operating conditions as for the flow rate and the rotating speed. The results of the stresses and the strains were validated using the classical equations. Thus, this study reveals, inter alia, that the impeller having a big outer diameter presents a stress and a strain more important considering a high torque developed and a great axial force induced in the pump shaft. Moreover, the stress and the strain become more pronounced as one increases the number of impellers. In addition, the stresses and the strains generated are highest for a pump running at 3600 rpm in comparison with a pump functioning at 1800 rpm. The torque also becomes considerable, so that the brake horsepower is more important at high speed. Finally, the displacement amplitude and the efficiency increase with the rise of the flow rate, whereas the stress decreases then grows slightly with increasing flow rate.

Acknowledgments. The authors are grateful to Technosub Inc., Industrial pumps manufacturing and distribution (Rouyn-Noranda, Quebec, Canada).

References

1. Gülich, J.F.: Centrifugal Pumps, 2nd edn. Springer, Heidelberg (2010). https://doi.org/10.1007/978-3-642-12824-0

2. La Roche-Carrier, N., Dituba Ngoma, G., Ghié, W.: Numerical investigation of a first stage of a multistage centrifugal pump: impeller, diffuser with return vanes, and casing. ISRN Mech. Eng. **2013**, 15 (2013). Article ID 578072
3. Takacs, G.: Electrical Submersible Pumps Manual-Design, Operations, and Maintenance. Gulf Professional Publishing, Oxford (2017)
4. Smith, D.R., Price, S.: Upthrust problems on multistage vertical turbine pumps. In: Proceedings of the 22nd International Pump Users Symposium (2005)
5. Wilk, A.: Laboratory investigations and theoretical analysis of axial thrust problem in high rotational speed pumps. WSEAS Trans. Fluid Mech **4**, 1–13 (2009)
6. Dong, W., Chu, W.-L.: Numerical investigation of the fluid flow characteristics in the hub plate crown of a centrifugal pump. Chin. J. Mech. Eng. **31**(1), 64 (2018). https://doi.org/10.1186/s10033-018-0264-z
7. TM.P. S.p.A. Termomeccanica Pompe: TERMOMECCANICA Centrifugal Pump Handbook, La Spezia, Italy (2003)
8. Badr, H.M., Ahmed, W.H.: Pumping Machinery Theory and Practice. Wiley, Hoboken (2015)
9. Budynas, R.G., Nisbett, J.K.: Mechanical Engineering Design, 10th edn. McGraw-Hill Education, New York (2014)
10. Wang, C., Shi, W., Si, Q., Zhou, L.: Numerical calculation and finite element calculation on impeller of stainless steel multistage centrifugal pump. J. Vibroengineering **16**, 1723–1734 (2014)
11. Matlakala, M., Kallon, D., Mogapi, K., Mabelane, I., Makgopa D.: Influence of impeller diameter on the performance of centrifugal pumps. IOP Conf. Ser.: Mater. Sci. Eng. **655**, 012009 (2019)
12. Bolade, P.S., Madki S.J.: Analysis of hydraulic thrusts in centrifugal pump to increase the bearing life. Int. J. Eng. Res. Technol. (IJERT), **4**(08), 760–763 (2015)
13. FLYGT: ITT Industries Engineered for life: Shaft and Bearings Calculations. 02.03.Eng. 0,5 M. 04.04; 892932 (2004). www.flygt.com
14. Lienau, W., Lagas, N.: Evaluation of rotordynamic criteria for multistage pump shafts. In: Proceedings of the 24th International Pump Users Symposium (2008)
15. Abdelouahab, M.-A., Ngoma, G.D., Erchiqui, F., Kabeya, P.: Investigation of induced loads, stresses, strains and vibrations in a high pressure multistage centrifugal pump. In: Obaidat, M.S., Oren, T., Rango, F.D. (eds.) SIMULTECH 2020. LNNS, vol. 306, pp. 184–208. Springer, Cham (2022). https://doi.org/10.1007/978-3-030-84811-8_9
16. He, H., Zhang, B., Pan, G.-P., Zhao, G.: Rotor dynamics of multistage centrifugal pump. In: 24th International Conference on Nuclear Engineering (2016)
17. Malonda, P.Z., Dituba Ngoma, G., Ghié, W., Erchiqui, F., Kabeya, P.: Characterization of a vertical submersible six-stage pump: accounting for the induced forces and stresses. In: Proceedings of the 11th International Conference on Simulation and Modeling Methodologies, Technologies and Applications (2021)
18. Technosub Inc. www.technosub.net
19. ANSYS inc. www.ansys.com
20. Karassik, I.J., McGuire, T.: Centrifugal Pumps. Springer US, Boston (1998). https://doi.org/10.1007/978-1-4615-6604-5
21. Karassik, I.J., Messina, J.P., Cooper, P., Heald, C.C.: Pump Handbook, 4th edn. McGraw-Hill, New York (2008)
22. Jino, T.: Hydraulic axial thrust in multistage centrifugal pumps. J. Fluids Eng. **102**(1), 6 (1980)
23. Wang, C., Shi, W., Zhang, L.: Calculation formula optimization and effect of ring clearance on axial force of multistage pump. Math. Probl. Eng. **2013**, Article ID 749375 (2013)
24. Gantar, M., Florjancic, D., Sirok, B.: Hydraulic axial thrust in multistage pumps - origins and solutions. J. Fluids Eng. **124**(2), 336–341 (2002)

25. Bolade, P.S., Madki, S.J.: Analysis of hydraulic thrusts in centrifugal pump to increase the bearing life. Int. J. Eng. Res. Technol. **4**(08) (2015). ISSN: 2278-0181
26. Popov, E.P.: Engineering Mechanics of Solids, 2nd edn. Prentice Hall, Hoboken (1999)

Comparison of Modelling Approaches and Solvers on Harmonic Studies for Renewable Energy Integration

Zhida Deng[1(✉)], Grazia Todeschini[1], and Kah Leong Koo[2]

[1] Department of Engineering, King's College London, London, UK
{zhida.deng,grazia.todeschini}@kcl.ac.uk

[2] Power Quality and Modelling Department, National Grid, London, UK
kahleong.koo@nationalgrid.com

Abstract. Renewable energy sources have been deployed in large amounts in the last decade, and this trend is expected to continue, and more likely to accelerate. These devices are interfaced to the power system by means of a power electronics interface, while other generating sources are directly connected to the grid. As a result, the development of dedicated models to study the integration of renewable energy sources is required. Once the models are developed, various types of system integration studies need to be performed to assess both steady-state and transient system operation. This paper focuses on a category of steady-state studies, referred to as harmonic analysis, that aims at assessing the impact of renewable energy sources on voltage and current distortion. Harmonic studies are gaining attention because under certain circumstances, the power electronics interface devices may generate undesirable harmonics, that may be further amplified under grid resonance conditions. This paper compares various cases studies where two different models, ideal and frequency-dependent impedance representation and various solvers are implemented. A nine-bus network model, including two wind plants and two photovoltaic installations, is used to demonstrate various case studies. Recommendations are proposed to appropriately model renewable energy sources depending on the specific application considered. The commercial software DIgSILENT PowerFactory is used to carry out the analysis, but the results can be generalized to other tools.

Keywords: Harmonics · Power quality · Power systems modelling · Renewable energy sources · Total harmonic distortion · Voltage unbalance · Impedance model

1 Introduction

Existing power systems were designed decades ago when fossil fuels (such as coal, gas, oil) and nuclear power plants were mostly employed to generate electricity. All these 'traditional' energy sources are characterized by being centralized, large and fully controllable – in other words, they can be adjusted almost in 'real-time' to meet demand.

G. Wagner et al. (Eds.): SIMULTECH 2021, LNNS 601, pp. 70–93, 2023.
https://doi.org/10.1007/978-3-031-23149-0_4

Growing population and industrial expansion have led to increasing electricity demands, pushing power systems close to their operational limits. This phenomenon has been accentuated by the difficulty to build new transmission lines, due to lack of space and environmental concerns. Simultaneously, the need to reduce greenhouse gas emissions from conventional power stations have raised additional concerns on the sustainability of fossil fuel-based electricity generation. These concerns have contributed to accelerating the deployment of Renewable Energy Sources (RESs), with the aim to meet future electricity demand and displace conventional generation. The most recent projections from the UK Transmission system operator (National Grid) estimate that the RESs (e.g. solar and wind) will make up to approximately 62.25% of total installed generation capacity by 2050 in the UK [1].

Conventional power systems have not been designed to operate neither with large number of RESs nor with increasing capacities, thus leading to concerns in terms of power quality. 'Power quality' is a specialist area within the general framework of power system analysis that aims at maintaining system operation close to the rated conditions in terms of frequency and voltage levels and to reduce waveform distortion due to harmonic components. Voltage and current are sinusoidal waveforms at 50 Hz or 60 Hz in an ideal power system. Harmonic components at frequencies up to a few kHz are generated by non-linear devices, such as power converters and motor drives [2], and contribute to detriment of both current and voltage waveforms. This phenomenon is not new, and small level of harmonics are tolerated by the system and the equipment. Power quality standards have been developed for harmonic control [2–4] these standards define acceptable harmonic limits and provide guidelines for harmonic measurement and assessment. This is because excessive harmonic levels may lead to various detrimental effects, including aging of dielectric insulation, damaging of electrical equipment, and false operation or failure of system relay [5].

Since RESs are connected to the power grid by means of power converters, they are a source of harmonics across a broad range of frequencies. Even if the contribution of individual units to overall harmonic distortion is negligible, with the growing penetration of these devices, the overall system harmonic levels are expected to increase in near future [6]. This phenomenon may be more visible in certain areas where large amounts of RESs are connected in proximity, due to favorable environmental conditions (for example, offshore wind farms in Scotland). Therefore, increasing harmonic level as one of the main concerns related to RESs integration has been raised by numerous transmission system and distribution network operators, and requires detailed investigations and needs to be properly managed, as reported in [7].

Computer simulations are used for the purpose of analyzing the impact of RESs on harmonic levels on the network. Two types of approaches are possible: time-domain and frequency-domain [8]. Time-domain methods characterize the system behavior using differential equations, and individual harmonic components can be obtained by using the Fourier transformation. This approach allows modelling the power converters and their controls in detail and studying time-varying conditions. However, it does not allow an easy calculation of frequency-dependent parameters and require a high computational power even for small networks [8, 9].

Frequency-domain methods are widely deployed in the engineering practice. These methods include frequency scans, harmonic penetration studies and harmonic load flows. A frequency scan consists of calculating the network impedance at various frequencies to determine the system frequency response and identify potential resonance conditions. For a thorough assessment, frequency scans are calculated at various points in the system and for varying grid conditions. A harmonic penetration study consists in carrying out nodal analysis for each harmonic order under the assumption that there is no interaction between fundamental and harmonic components. A harmonic load flow uses a Newton-Raphson or Gauss-Seidel based algorithm to solve unified fundamental and harmonic power flow equations [8, 10], thus iteratively assessing the interaction between various frequency components. A number of commercial software used for power system analysis allow to carry out harmonic studies, but the distinction between various solvers is not always clear to the user. However, various approaches may produce significantly different results and therefore, it is important to understand the assumptions underlying the solver under consideration. By performing harmonic load flows and harmonic penetration studies, harmonic current and voltage distortions on the network are calculated in frequency-domain. This process allows assessing compliance with the standards and, if required, informs on requirements for design of mitigating solutions, such as harmonic filters [11].

Frequency-domain analysis can be carried out for either balanced or unbalanced systems. System unbalance may be caused by asymmetry in the transmission system geometry or the unequal distribution of loads and generating sources across the three phases. In practical power system, the unbalance is normally very small, therefore a balanced frequency-domain analysis is commonly adopted. Unbalanced harmonic analysis provides more accurate results when studying asymmetrical systems, either in terms of network configuration [12], and/or loads. For example, power systems where RESs are single-phase connected, may experience unbalanced harmonic distribution and therefore an unbalanced harmonic study will lead to a more accurate assessment of voltage and current distortion.

In addition to unbalanced conditions, another factor to consider when carrying out harmonic studies is related to the summation of harmonic components due to different sources. When multiple harmonic sources are present in the system, they generate harmonic components with varying phase angle at each harmonic, and the collective distortion is then assessed at the point of common coupling (PCC) – this is referred as the point where one or more devices (generators or loads) are connected to the system. Each harmonic component is a complex number, consisting of magnitude and phase angle, but while the magnitude can be obtained from various sources, the phase angle of the harmonic source is generally not available. As a result, various approaches are possible to calculate harmonic summation. In the worst-case evaluation, the phase angle is ignored, thus effectively assuming that all harmonic components are in phase [13]. This approach however leads to very conservative results, where harmonic levels that are well above the ones observed in practical systems. Alternative methods to carry out harmonic summations include: (1) both magnitude and phase of each component are considered in the studies and a vectorial sum is calculated or (2) the summation rule described in standard IEC 61000-3-6 [4] is employed. In this case a summation exponent is considered to take into account that with higher harmonic orders harmonic phase angles tend to differentiate more. While the IEC summation rule is based on practical considerations, it may result in inaccurate assessment of harmonic levels at PCC [7, 14]. In [15], different modelling approaches are described to calculate the harmonic distortion at the PCC for an offshore wind farm, and the limitations of the IEC summation rule are highlighted.

It has been reported in [16] that harmonic components generated by power converters tend to be more variable in terms of relative phase angle. Under these conditions, it is more appropriate to carry out harmonic studies that not only consider the phase angle of the harmonic sources, but also include variability of the harmonic phase angles to provide a more accurate assessment. Although the topic of harmonic assessment for RESs is not new, to the knowledge of the Authors, no other paper has been published considering the simultaneous impact of three aspects: harmonic summation, harmonic models, and system unbalanced conditions.

Moreover, the network impedance characteristics can be affected when a large number of nonlinear RESs and their collector system is connected to the power grid. One of the main concerns is the introduction of unwanted resonances that may result in increased harmonic distortion on the network. Therefore, an appropriate representation of the RES equivalent impedance is required to carry out a reliable and accurate assessment. Traditionally, the ideal Norton/Thévenin equivalent representation is adopted, where the impedance is either assumed to be constant across the entire frequency range or neglected [17]. Nevertheless, it has been reported that the constant equivalent impedance is not able to capture the actual behavior of RES at varying harmonic orders [11]. Using frequency-dependent method are considered more appropriate as they allow modelling the effect of control topology and physical components (such as filters, cables and transformers) on the RES impedance at each frequency. However, it is very difficult to obtain the true impedance model due to the lack of manufacturing details and control strategy information, that are covered by NDAs. This work will look into adopting a frequency-dependent impedance model from [18], that has been chosen due to its generality. The impact of RESs impedance on the results will be assessed in comparison to the case of an ideal source.

The commercial software DIgSILENT PowerFactory (DPF) [19] is widely adopted in academia and in industry to carry out harmonic studies. It provides numerous options and settings, including various harmonic source models (named IEC source and Unbalance Phase Correct – UPC source, respectively), an option to include a constant or frequency-dependent equivalent impedance, choice of various summation rules and solvers. While these options provide flexibility, at the same time they may lead to significantly different results. Therefore, the user needs to be aware of the impact of different settings.

Previous work carried out in [20] aimed at filling in some of the gaps above by examining the use of different harmonic source models and harmonic load flow solvers available in DPF to calculate the harmonic impedance and harmonic levels in a typical power system. Additionally, it provided guidance to model unbalanced harmonic current sources using the UPC model, while at the same time providing comparable results as the ones provided by the IEC model. The advantage of the UPC model is that it provides more options to the user, while the IEC model is limited to inputting harmonic current magnitudes. This paper builds in the findings obtained in [20] by investigating the impact of the equivalent impedance of RES on harmonic assessments. It has become evident that RESs cannot be considered simply as current or voltage sources, but rather they present an impedance to the system. The impact of including the RESs impedance is assessed on both the network impedance characteristics and the harmonic levels. It is important to highlight that while the study is exemplified using DFP, the considerations and approaches presented in this paper are general. The methodology presented in this work will be exemplified by using a nine-bus network with seven harmonic sources, that was previously adopted in [19].

The paper is organized as follows: in Sect. 2 an introduction to power system symmetrical components and the harmonic source modelling approach are presented. In Sect. 3, the application of the proposed methodology is carried out on a nine-bus network and various case studies are presented using different solvers and models. For all cases, simulation results are presented in term of frequency scans and voltage distortion. For the first set of simulation results, the equivalent impedance of the harmonic sources is not included, to focus on the impact of other parameters on the results. Finally, the harmonic study is repeated by comparing the results with the cases where the equivalent impedance is included in the harmonic source model.

2 Modelling Approach

An introduction to symmetrical components is provided, followed by the description of the harmonic source models adopted in this paper.

2.1 Symmetrical Components

Transmissions and distribution systems are designed to operate in a balanced three-phase sinusoidal condition. Figure 1 shows two examples of phasor diagrams: the left one illustrates the case of a three-phase balanced systems, where voltage or current magnitudes are the same on the three phases and displaced by 120°. The right phasor diagram shows an example of unbalanced three-phase system. Unbalanced conditions may occur due to system faults, network structure, load conditions, among the most common reasons. To analyze unbalanced condition, symmetrical components are widely used. By applying the Fortescue transformation, unbalanced three-phase voltage or current phasors are represented by three sets of balanced three-phase sequence components, and each phasor (A, B and C) can be calculated as shown in Fig. 2 [21].

Although symmetrical components have been introduced to study unbalanced system at fundamental frequency, they find application within harmonic analysis too. It can be shown [22] that balanced integer harmonic components h follow the sequence order as shown below:

- Positive-sequence for $3h + 1$ orders
- Negative-sequence for $3h - 1$ orders.
- Zero-sequence for $3h$ (also known as triplen harmonics).

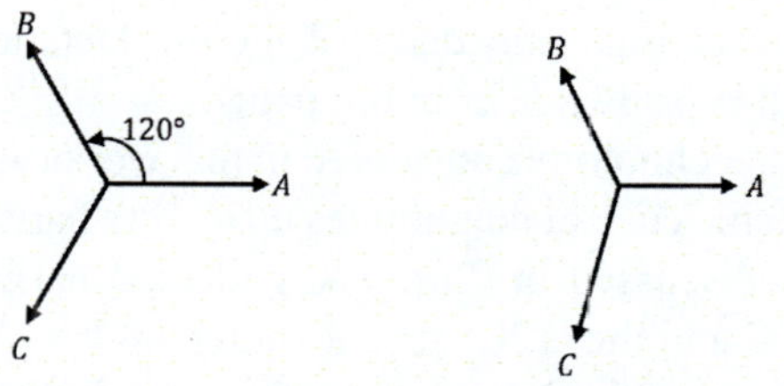

Fig. 1. Balanced (left) and unbalanced (unbalanced) three-phase phasor diagram.

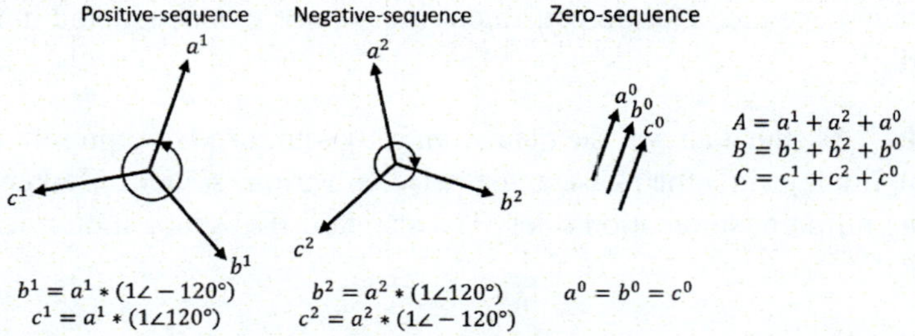

Fig. 2. Balanced three-phase sequence components for unbalanced phasors.

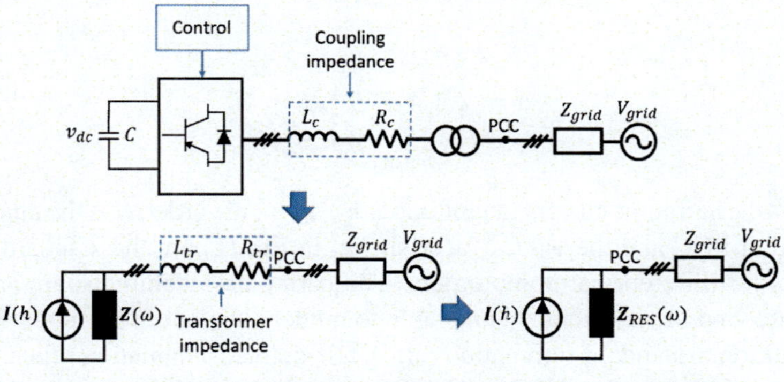

Fig. 3. Norton equivalent for RES harmonic source modelling.

Under unbalanced conditions, the same behavior as fundamental current and voltage applied and each harmonic set can be expressed into positive-, negative- and zero-sequence components.

2.2 Harmonic Source Modelling

In this work, each RES is modelled as a current source at both fundamental and harmonic frequency. A Norton equivalent representation is adopted, as shown in Fig. 3: this representation makes use of an equivalent impedance $\boldsymbol{Z}(\omega)$ in parallel with the harmonic current source. The equivalent impedance $\boldsymbol{Z}(\omega)$ can be either omitted, or a constant, or a frequency-dependent value. In general terms, a detailed impedance representation leads to a more accurate model. A further step consists in including the transformer equivalent impedance to calculate the total impedance $\boldsymbol{Z_{RES}}(\omega)$. Detailed steps to calculate both $\boldsymbol{Z}(\omega)$ and $\boldsymbol{Z_{RES}}(\omega)$ will be shown later in the paper.

For each generation technology considered in this work (wind and photovoltaic) and for loads, typical harmonic current magnitudes up to 50th harmonic order were retrieved from the literature, as discussed in [20]. The collected harmonic current magnitudes were applied to the IEC and the UPC current source model. The reference current was chosen as the rated current in both models: in this way, both models would lead to the amount of harmonic current injection.

In the following sections, the IEC and UPC harmonic current source model, and the frequency-depended Norton equivalent impedance model adopted in this work are introduced.

IEC Model. As stated above, the data from [20] were used to represent the harmonic current amplitude, while the phase angle between various sources is taken into account by applying different summation rules. The standard IEC summation rule is defined as follow [4]:

$$I_h = \sqrt[\alpha]{\sum\nolimits_{i=1}^{N} \left(I_h^i\right)^{\alpha}} \tag{1}$$

$$\alpha = \begin{cases} 1 & \text{if } h < 5 \\ 1.4 & \text{if } 5 \leq h \leq 10 \\ 2 & \text{if } h > 10 \end{cases} \tag{2}$$

where I_h is the harmonic current magnitude at h^{th} harmonic order, N is the number devices connected at PCC and the values for α shown in Eq. (2) are the standard summation exponents for different harmonic orders. The harmonic magnitudes are applied to all three-phase, and the IEC model is not able to model unbalanced harmonic injections.

In addition to standard summation rule, a self-defined summation rule is proposed to allow comparison with the UPC model. For the self-defined summation rule, the value of α is set to 1 for all harmonic orders [20]. By using the self-defined summation rule, the harmonic cancellation effect is not considered in the IEC harmonic current source model.

UPC Model. The UPC model is more flexible than the IEC model as the harmonic current magnitudes for the three phases can be set to different values. However, in this work, the three-phase harmonic current magnitudes set to be identical since only balanced conditions are studied. The three-phase harmonic phase angles φ_h^A, φ_h^B and φ_h^C are calculated as follows [19]:

$$\varphi_h^A = \Delta\varphi_h^A + h\varphi_1^A \tag{3}$$

$$\varphi_h^B = \Delta\varphi_h^B + h\varphi_1^B \tag{4}$$

Table 1. Parameters for building different RES impedance model.

RES type	S_{base} (MVA)	R_c (ohm)	L_c (mH)	R_{tr} (ohm)	L_{tr} (mH)
PV	0.1	10.9	1,040	0.19	500.0
WTG	6.5	1.46	31.0	0.03	67.22

$$\varphi_h^C = \Delta\varphi_h^C + h\varphi_1^C \tag{5}$$

where h is the harmonic order, and φ_1^A, φ_1^B and φ_1^C are the fundament current phase angles of phase A, B and C, respectively. The phase parameters $\Delta\varphi_h^A$, $\Delta\varphi_h^B$ and $\Delta\varphi_h^C$ can be set by the users in different ways:

- $\Delta\varphi_h^A$, $\Delta\varphi_h^B$ and $\Delta\varphi_h^C$ are 0° for positive- and negative-sequence orders. For triplen harmonics, $\Delta\varphi_h^A$, $\Delta\varphi_h^B$ and $\Delta\varphi_h^C$ are set to 0°, −120° and 120°, respectively. In this way, the UPC model matches the IEC model where the triplen harmonic are represented as positive-sequence.
- $\Delta\varphi_h^A$, $\Delta\varphi_h^B$ and $\Delta\varphi_h^C$ are calculated for each device, with the intention to obtain φ_h^A, φ_h^B and φ_h^C equal to 0°, −120° and 120° for positive- and zero-sequence harmonic orders, and 0°, 120° and −120° for negative-sequence harmonic orders. In this way, all UPC harmonic sources are in phase, similarly to the IEC harmonic source using the self-defined summation rule. This approach is labelled 'in-phase UPC model' and it is proposed to allow a straightforward comparison with the IEC harmonic source.

Frequency-Dependent Impedance Model. This work adopts a generic linear frequency-dependent impedance model [18], which is expressed by a transfer function in the Laplace s-domain as follows:

$$\mathbf{Z}(\omega) = \frac{R_c + sL_c + j\omega_1 L_c + D(F_{PI} - j\omega_1 L_c)}{1 - DH} \tag{6}$$

$$F_{PI} = k_p + \frac{k_i}{s}, \quad H = \frac{\alpha_f}{s + \alpha_f}; \quad D = e^{-sT} \tag{7}$$

where $\mathbf{Z}(\omega)$ is the frequency-dependent impedance of the device under consideration, ω_1 is the fundamental angular frequency, R_c and L_c are the coupling resistance and inductance of the power converter (as shown in Fig. 3), F_{PI} is the PI controller for the inner loop, H denotes the voltage low pass filter with bandwidth α_f, and D represents the voltage source converter delay function with time delay T. In general, the impedance and control parameters vary based on type of RESs, control strategy and manufacturer. In this study, the control parameters presented in [18] are applied, as they represent typical values. The parameters for Wind Turbine Generator (WTG) and photovoltaics (PV) are given in Table 1.

Expressions (6) and (7) only apply to individual RESs, however, in the practice, a RES power plant consists of multiple devices connecting to the PCC via step-up transformers. In this work, expressions (6) and (7) are extended as follows to express aggregation of multiple RESs and their step-up transformers:

$$\mathbf{Z}_{eq}(\omega) = \mathbf{Z}(\omega) + (R_{tr} + j\omega L_{tr}) \tag{8}$$

$$\mathbf{Z}_{RES}(\omega) = ceil\left[\frac{\mathbf{Z}_{eq}(\omega)}{n}\right] \tag{9}$$

where $\mathbf{Z}_{eq}(\omega)$ is the equivalent impedance of individual RES connecting with a step-up transformer, and R_{tr} and L_{tr} denote the transformer resistance and reactance referred to the PCC rated voltage (in this case 33 kV), included in Table 1 for PV and WTG. To model the impedance of a RES power plant, $\mathbf{Z}_{RES,i}(\omega)$ is applicable to n identical RES devices connected in parallel, where n is calculated as the rated power ratio between selected RES power plant (see Table 2 in Sect. 3.1) and individual RES (as given in Table 1).

Based on (6)–(9) and the parameters provided in Table 1 and 2, the frequency-depended impedance values (up to 2.5 kHz with 1 Hz increment) were calculated using MATLAB, and the obtained values are applied to each aggregated RES.

3 Simulation Results and Discussion

The modelling methodology described above was applied to a simple system to demonstrate the impact of using different models and solvers on the harmonic studies.

Section 3.1 will present the network under consideration.

Section 3.2 presents the impedance characteristics obtained from balanced and unbalanced frequency sweep analysis when the ideal IEC and the ideal UPC model (namely, with no equivalent impedance) are used. Then, harmonic load flow calculations are performed using different solvers, to compare the voltage THD values and examine the differences between the ideal IEC and the ideal UPC model. The voltage THD is defined as the root square summation of individual harmonic voltage components as follows [4]:

$$\text{THD} = \frac{\sqrt{\sum_{h}^{h_{max}} v_h^2}}{v_1} \tag{10}$$

where v_1 is the fundamental voltage magnitude, v_h are the voltage harmonic components and the maximum harmonic order h_{max} is set to 50 for this work. Next, an approach to match the ideal IEC model using an equivalent ideal UPC model is studied and discussed.

Section 3.3 compares the frequency scans and harmonic levels to analyze the differences between ideal and frequency-dependent harmonic source models.

To simplify the notation, in the following sections the names 'IEC' and 'UPC' model refers to the ideal model without considering equivalent impedance.

3.1 Network Description

The system under consideration is shown in Fig. 4. The three-phase network includes nine-busses, four transmission lines, two PV plants, two WTGs, three loads and five two-winding Wye grounded-delta (Yg-d) connected transformers with 30-degree phase shift. The external grid is represented by an equivalent impedance connected at bus B1 and acts as a slack bus for the load flow convergence. A distributed-parameter line model was adopted to consider the long-line effects, where the electrical distance of the line is increased at higher frequencies [11]. The colors indicate different voltage levels, and Table 2 provides the system component specifications and power flow parameters.

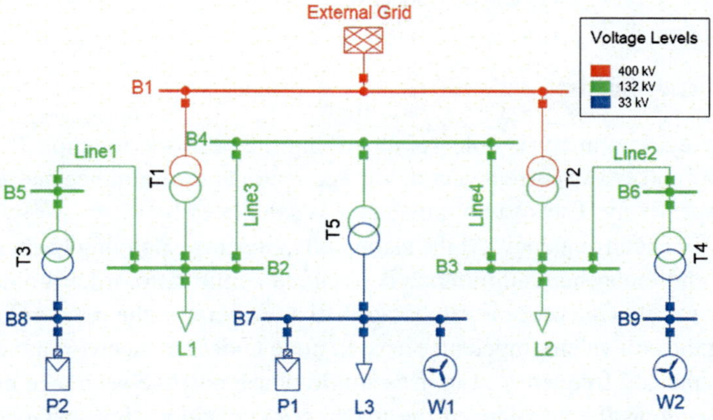

Fig. 4. Single-line diagram of the simulated network [20].

Table 2. System component parameters [20].

	L1	*L2*	*L3*	*P1*	*P2*	*W1*	*W2*
Sr(MVA)	20	30	100	12	21	60	30
P(MW)	19	28.5	90	8	18	40	20
Q(Mvar)	6.25	9.37	43.59	0	0	0	0
Transformer	*T1–T2*			*T3–T5*			
Voltage	400/132 kV			132/33 kV			
Sr(MVA)	255			90			
Impedance	16% short-circuit voltage with 1.8 MW losses			13% short circuit voltage with 0.25 MW losses			
	Line1	*Line2*		*Line3*		*Line4*	

(*continued*)

Table 2. (*continued*)

Length (km)	15	20	24	31
Impedance (Ω/km)	Positive/negative-sequence R and X: 0.0212 and 0.1162 zero-sequence R and X: 0.0848 and 0.4650			
	External grid			
Related parameters	Short-circuit power: 10000 MVA short circuit current: 14.43 kA; c-factor: 1.1 R/X ratio: 0.1, R: 1.75 Ω, X: 17.51 Ω			

The symbols S_r, P, Q, Z, R, X and L denote rated power, active power, reactive power, impedance resistance, reactance, and line length, respectively.

3.2 Frequency Scans

The frequency scan analysis consists in solving the network equation $\boldsymbol{I}^h = \boldsymbol{Y}^h\boldsymbol{V}^h$ [8], where $\boldsymbol{I}^h$, $\boldsymbol{V}^h$ and $\boldsymbol{Y}^h$ are current vector, voltage vector and admittance matrix at harmonic order h, respectively. The system impedance is calculated by the solver by injecting one *pu* current for each frequency. At the same node, the corresponding harmonic voltage is measured, and the system admittance is obtained as the ratio of harmonic voltage and harmonic current. This process can be applied anywhere on the network, thus allowing calculating the equivalent impedance at each node under balanced or unbalanced condition. The balanced frequency scan uses single-phase positive-sequence network representation, whereas the unbalanced frequency scan considers three-phase, and positive, negative and zero-sequence network representation.

When ideal harmonic source models are used, the network impedance is not affected by the magnitude harmonic injections, the type of harmonic source model and the number of harmonic sources. For the network under consideration, balanced and unbalanced frequency scans were calculated by injecting harmonic currents from 50 Hz up to 2.5 kHz with 1 Hz step size, under following study cases with seven harmonic sources (namely, three loads, two PVs and two WTGs):

- Case 1: IEC model applied to harmonic sources.
- Case 2: UPC model applied to harmonic sources.
- Case 3: no harmonic source is considered.

Both balanced single-phase and unbalanced three-phase impedance characteristics at the 400, 132 and 33 kV busbars (namely B1, B4 and B7) for Case 1, Case 2 and Case 3 are presented in Figs. 5, 6 and 7. The single- and three-phase impedance characteristics at the selected busbars for different cases presented same frequency responses independently on the type and number of harmonic sources, and the selection of balanced or unbalanced frequency scan. These findings were also verified and applied to other busbars. Additional frequency scan results (e.g. unbalanced sequence component frequency scans) at other busbars and using different settings can be retrieved in [20].

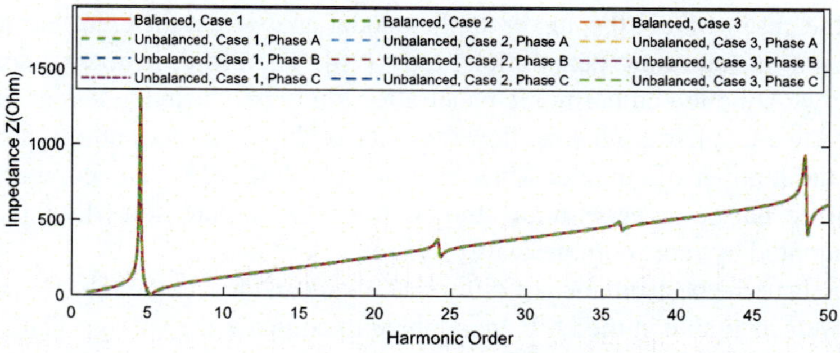

Fig. 5. Balanced and unbalanced impedance characteristics at 400 kV busbar B1 for various study cases.

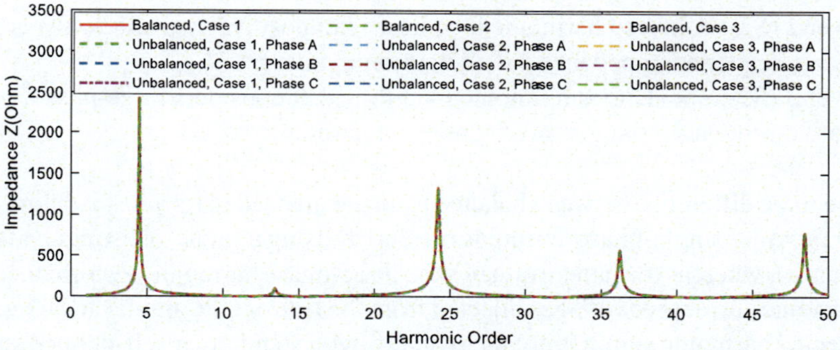

Fig. 6. Balanced and unbalanced impedance characteristics at 132 kV busbar B4 for various study cases.

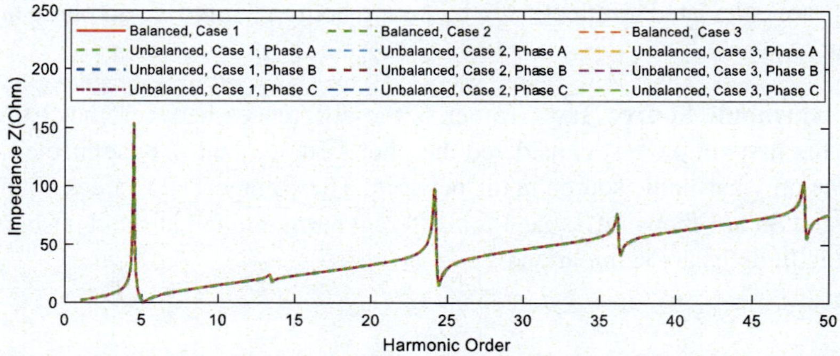

Fig. 7. Balanced and unbalanced impedance characteristics at 33 kV busbar B7 for various study cases.

3.3 Harmonic Load Flow

Different harmonic load flow techniques are available, and in some cases they may lead to distinct results. Examples included in [10] show that the discrepancies in harmonic voltage levels between different solvers can be up to 30%. Therefore, it is necessary to understand the load flow techniques available in order to carry out a correct assessment.

For the case of DPF, the solver applies nodal analysis at different harmonic orders to carry out an harmonic penetration study [20]. Therefore, it doesn't solve a 'true' power flow. Although an harmonic penetration study may provide less accurate results compared to a true harmonic load flow (as discussed in the introduction), it significantly reduces the number of equations involved in the solution [8]. The computational efficiency of the harmonic penetration study may be the main reason why this approach is widely adopted by most commercial software [11].

To analyze the performance of different solvers and harmonic models the following three solvers available in the DPF are adopted to compare the voltage THD when using the IEC and the UPC model:

- Solver 1 (S1): balanced harmonic load flow, considering single-phase, positive and negative-sequence equivalent impedance for relevant harmonic orders.
- Solver 2 (S2): balanced harmonic load flow with positive-sequence only, considering positive-sequence impedance for all harmonic orders.
- Solver 3 (S3): unbalanced harmonic load flow that considers three-phase, positive or negative-sequence components for relevant harmonic orders.

The main difference between balanced and unbalanced harmonic load flow is that the former provides single-phase harmonic results (by using a single-phase equivalent circuit impedance), whereas the latter evaluates the three-phase harmonic performance. In DPF, the harmonic currents or voltages injected from the IEC or UPC model are assessed using the selected harmonic summation rule (namely, with standard or self-defined summation exponents) when any IEC is included in the network. In other words, in the following sections, the UPC model is assessed by IEC summation rule if at least one harmonic source is using IEC model. When only the UPC model is used, the harmonic currents or voltages are calculated arithmetically by taking both magnitudes and phase angles into account.

Single Harmonic Source Test. To verify the differences between the IEC and UPC model, the first simple test considered the photovoltaic plant (P1) connected to busbar B7 as the only harmonic source in the network. The voltage THD values obtained from S1, S2 and S3 are shown in Table 3 for different harmonic models and busbars, and the following findings are summarized:

Table 3. Comparison of THDs between the standard IEC and the UPC models under different solvers (single harmonic source) [20].

	Standard IEC model		UPC model			
Busbar	S1 & S2 (%)	S3 3-Ph (%)	S1 & S2 (%)	S3 Ph-A (%)	S3 Ph-B (%)	S3 Ph-C (%)
B1	**0.007**	**0.007**	0.006	**0.007**	**0.007**	**0.007**
B2	**0.025**	**0.025**	0.023	**0.025**	**0.025**	**0.025**
B3	**0.031**	**0.031**	0.027	**0.031**	**0.031**	**0.031**
B4	**0.029**	**0.029**	0.026	**0.029**	**0.029**	**0.029**
B5	**0.035**	**0.035**	0.030	**0.035**	**0.035**	**0.035**
B6	**0.036**	**0.036**	0.032	**0.036**	**0.036**	**0.036**
B7	0.283	0.301	0.241	0.297	0.307	0.292
B8	0.035	0.037	0.030	0.035	0.038	0.037
B9	**0.036**	**0.036**	0.032	**0.036**	**0.036**	**0.036**

Bold values are all the same. 'Ph-A', 'Ph-B', 'Ph-C', and '3-Phase' refer to phase A, phase B and phase C and all three phases, respectively.

- For the standard IEC model, results from S1, S2 and S3 are identical. The only exception consists in the 33 kV busbars B7 and B8, because a Yg-d connection is used for the 132/33 kV transformers, thus resulting in slightly different fundamental power flow solutions for S3. The results shown in [20] indicate that the discrepancies at B7 and B8 were eliminated when a Yg-yg connection was used for the 132/33 kV transformers.
- For the UPC model, S1 and S2 provide different THDs compared to using the standard IEC model. This is caused by the neglect of triplen harmonics in the UPC model. When the triplen harmonics are excluded from the IEC model, the THD results from S1 and S2 were the same as for the UPC model. Moreover, the three-phase results (labelled as Ph-A, Ph-B and Ph-C) at the selected busbars (except at B7 and B8) produced by the S3 using the UPC model are identical as using the standard IEC model. The differences at B7 and B8 are caused by the transformer connection as discussed above. Previous work [20] shows that the same results are obtained when the Yg-yg connection was adopted for the transformers.
- For all study cases, transformer connection, solver and the harmonic model have no impact on the harmonic results at the high-voltage side (above 33 kV).
- The above findings were verified for cases when other loads, PVs or WTGs were considered as individual harmonic source.

Three Harmonic Sources at Same Busbar. To understand how multiple harmonic sources are processed under different solvers and models, a second test was carried out considering three harmonics sources (namely P1, L3 and W1) connected at the same busbar (B7).

Table 4 summarizes the voltage THDs when using the standard IEC model and UPC model at B1–B9 when different solvers are selected, and the following results are found:

Table 4. Comparison of THDs between IEC and UPC models for different solvers (three harmonic sources) [20].

	Standard IEC model		UPC model			
Busbar	S1 & S2 (%)	S3 3-Ph (%)	S1 & S2 (%)	S3 Ph-A (%)	S3 Ph-B (%)	S3 Ph-C (%)
B1	0.133	0.133	0.145	0.147	0.147	0.147
B2	0.520	0.520	0.564	0.570	0.570	0.570
B3	0.502	0.502	0.546	0.557	0.557	0.557
B4	0.598	0.598	0.640	0.648	0.648	0.648
B5	0.657	0.657	0.692	0.705	0.705	0.705
B6	0.638	0.638	0.677	0.689	0.689	0.689
B7	3.174	3.178	3.148	3.328	3.339	3.311
B8	0.657	0.658	0.692	0.705	0.706	0.705
B9	0.638	0.638	0.676	0.689	0.689	0.689

- For the standard IEC model, the THDs provided by different solvers present similar features as discussed in the previous test – three solvers produce identical THDs at busbar B1–B6 and B9, whereas small THD discrepancies between S1&S2 and S3 are found at B7 and B8.
- For the UPC model, because triplen harmonics are not considered by S1 and S2, the THDs for these two cases are slightly smaller than the results obtained with S3. When triplen harmonics were excluded from S3, the three solvers produced same results at most busbars (except for B7 and B8).
- Unlike in the single harmonic source test, the THDs results at the high-voltage busbars (B1–B6 and B9) when using the IEC and UPC for the various solvers are different even a Yg-yg connection is applied for all transformers included in the model. To explain these differences, the use of summation rule and the harmonic cancellation effect between harmonic sources are investigated below.

In [20], the self-defined IEC model (i.e. using the self-defined summation exponent introduced in Sect. 2.2) and the UPC model assessed by the standard IEC summation rule were investigated under different solvers. It was found that the self-defined IEC model under different solvers produced higher voltage THDs compared to the standard IEC model because the cancellation effect was ignored by the self-defined IEC summation rule (where all harmonics are assumed to be in-phase). The UPC model assessed by the standard IEC summation rule obtained smaller THD values compared to the UPC model results presented in Table 4 because it taken both summation rule and harmonic phase angles into account (over-emphasis on harmonic cancellation).

To better understand the reason of causing the discrepancies between the IEC and UPC model, harmonic current flows were analyzed in detail at various frequencies. For example, the 5th harmonic current flows obtained from using S1 for different models are shown in Fig. 8. It can be seen that the total harmonic current injection (labelled I_t and red circled) from source P1, L3 and W1 to the grid is not the same when different harmonic source models are applied. For each model, the total harmonic current injection from the three harmonic sources to the grid is calculated as follows:

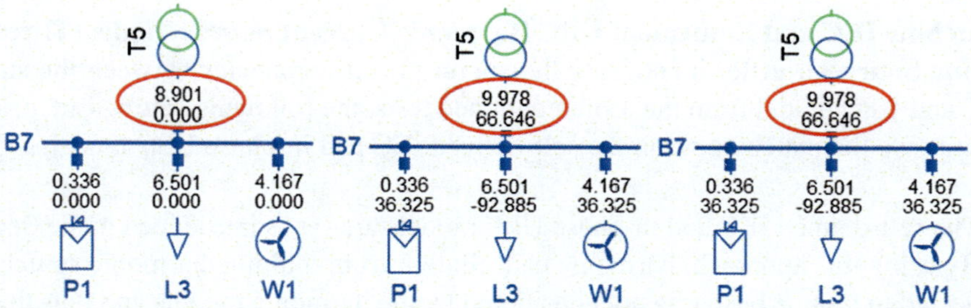

Fig. 8. 5th harmonic order balanced current flows obtained by using: the standard IEC model (left), the UPC model (middle), and the UPC model assessed by standard IEC summation rule (right) [20].

- Standard IEC model:

$$I_t = \sqrt[\alpha]{I_{P1}^{\alpha} + I_{L3}^{\alpha} + I_{W1}^{\alpha}} \tag{11}$$

where I_{P1}, I_{L3} and I_{W1} represent the harmonic current magnitudes for source P1, L3 and W1, respectively, and I_t is calculated using the standard summation rule introduced in the Sect. 2.2 (Eqs. (1) and (2)).
- UPC model:

$$T_t = |I_{P1}\angle\theta_{P1} - I_{L3}\angle\theta_{L3} + I_{W1}\angle\theta_{W1}| \tag{12}$$

The harmonic currents from using the UPC model are added arithmetically – considering both magnitudes and phase angles. Noticeably, the harmonic current injection from using the UPC model for the load component is considered in an opposite direction from using the IEC model in DPF.
- The UPC model assessed by the standard IEC rule:

$$\sqrt[\alpha]{|I_{P1}\angle\theta_{P1} + I_{W1}\angle\theta_{W1}|^{1.4} + I_{L3}^{\alpha}} \tag{13}$$

In this case, the source L3 used the IEC model and thus the UPC models used for P1 and W1 were also assessed by the standard IEC summation rule. Equation (13) shows that the harmonic currents from the UPC model based P1 and W1 takes both harmonic phase angles and summation rule into account. Note that current flow Eq. (11)–(13) were changed accordingly when the self-defined summation exponent was used.

Based on the results given in (11)–(13), one can conclude that the UPC model may be preferable under some circumstances (for example, when the harmonic injection phase angles are known), as the individual harmonic phase angles can be included in the UPC model. On the contrary, the IEC model always assumes fixed phase angles and the only variability can be introduced by defining the coefficients in the summation rule. Therefore, the IEC and UPC model will generally produce mismatching harmonic results. Nevertheless, the following sections proposed settings that allow the UPC model to provide comparable results with the IEC model.

Matching IEC and Equivalent UPC Harmonic Current Source Model (Three Harmonic Sources). After identifying the sources of discrepancies between the standard IEC and UPC model from the aforementioned tests, the following settings are proposed to improve the match between the self-defined IEC and in-phase UPC model:

- The self-defined IEC and in-phase UPC models are set as introduced in the Sect. 2.2. By using this approach, harmonic cancellation from multiple harmonic sources connected to a same busbar is not considered by the harmonic models and thus the same amount of harmonic current injection results from the self-defined IEC and in-phase UPC model.
- The harmonic current magnitudes for the UPC model are set to negative values when modelling a load component (as discussed in the three harmonic sources test).
- In order to maintain the same THD results at the low-voltage side, the Yg-yg connection is adopted for the 132/33 kV transformers.

Applying above settings to the three harmonic sources, the self-defined IEC and the in-phase UPC models result in the injection of identical harmonic currents to the network. As verified in [20], the THDs at the selected busbars obtained by using the self-defined IEC model under different solvers were the same as results of using the in-phase UPC model under S3. The differences observed in the UPC model of S1 and S2 are because these solvers ignore the triplen harmonics in the UPC model. Morover, it was noticed that the THD values of using the self-defined IEC and in-phase IEC model were the same under different solvers when triplen harmonics were ignored in both harmonic models.

Matching IEC and Equivalent UPC Harmonic Current Source Model (Seven Harmonic Sources). Further tests of the settings above were carried out by using seven harmonic sources (three loads, two PVs and two WTGs). In this case, the THD values of the in-phase UPC model produced by different solvers were slightly smaller compared to the results from using the self-defined IEC summation model, as discussed in [20]. This is because the harmonic current phase-shift caused by the network impedance is not considered when the IEC models are used. When multiple UPC harmonic source models located at different busbars, harmonic currents propagating in the network are not always in-phase and thus harmonic cancellation is introduced.

It was verified that the results obtained from the self-defined IEC model can be matched when using the in-phase UPC model if the impedance of transmission lines and the phase-shift of the transformers are ignored. As presented in Table 5, under these conditions, the self-defined IEC model under different solvers provides identical results as the in-phase UPC model under S3 (the differences under S1 and S2 are due to the triplen harmonics). In addition, the harmonic results were also verified when the in-phase UPC model was assessed by the self-defined summation rule (namely, by setting source L3 to use the IEC model). The results in Table 5 shows that the in-phase UPC model* matches the results from the self-defined IEC and in-phase UPC model under S3. The discrepancies between the in-phase UPC model* and the in-phase UPC model under S1, S2 and S3* are caused by the source L3 using IEC model considering triplen harmonics.

Table 5. THDs of self-defined IEC and in-phase UPC models obtained from using different solvers when impedance of transmission lines and phase-shift of transformers are set to zero and using Yg-yg transformers (seven harmonic sources) [20].

	Self-defined IEC model	In-phase UPC model*		In-phase UPC model	
Busbar	S1, S2, S3 (%)	S1, S2, S3* (%)	S3 3-Ph (%)	S1, S2, S3* (%)	S3 3-Ph (%)
B1	0.387	0.368	0.387	0.366	0.387
B2	1.518	1.444	1.518	1.439	1.518
B3	1.518	1.444	1.518	1.439	1.518
B4	1.518	1.444	1.518	1.439	1.518
B5	1.518	1.444	1.518	1.439	1.518
B6	1.518	1.444	1.518	1.439	1.518
B7	4.974	4.748	4.974	4.708	4.974
B8	1.854	1.723	1.854	1.719	1.854
B9	1.992	1.813	1.992	1.808	1.992

'In-phase UPC model*' means the UPC model is assessed by self-defined summation rule, and the '3*' means that the harmonic model under Solver 3 is not considering triplen harmonics.

The above results show that the in-phase UPC model does not allow exact match of the IEC model when multiple harmonic sources are located at different busbars, unless the self-defined summation rule is used for the IEC model and the network impedance is ignored.

3.4 Frequency-Dependent Impedance Model and Harmonic Analysis

The sections above investigate the impact of selecting different harmonic source model and solvers on system impedance and voltage THD when the harmonic source impedance is not considered, also referred to as ideal Norton equivalent model. This section compares the network impedance characteristics and harmonic levels with and without frequency-dependent impedance harmonic source model. For consistency with the tests above, three loads, two PV and two WTGs are considered as harmonic sources.

Network Impedance Characteristics. Balanced frequency sweeps up to 2.5 kHz with 1 Hz increment were performed to compare the impedance characteristics at 33 kV busbars B7, B8 and B9 for two cases: ideal RES equivalent model, and RES equivalent model with impedance, according to the modelling approach introduced in Section 2.2. These busbars were selected because they are the PCC for the harmonic sources. Since the selection of harmonic source type does not affect the impedance characteristics (as discussed in Sect. 3.1), the comparison between IEC and UPC model is not presented here, but rather the focus is on the impact of the harmonic source impedance, that is the same for both.

Figures 9, 10 and 11 show that the impedance characteristics at the selected busbars are visibly smaller when the impedance model is applied. More specifically, the observed resonance peaks are largely damped, while the resonances slightly shift toward higher frequency. However, for other 132 kV busbars and 400 kV busbar (where RESs are not connected), the main difference is that the resonance frequencies shift toward higher frequencies, whereas the impedance magnitude at most frequencies remains almost the same. The main reason behind this behavior is due to the introduction of damping resistance in the impedance model applied to RESs, that leads to a smaller network impedance characteristics [23]. This behavior corresponds to practical implementations, as the control strategy of the RES associated with filter is designed to mitigate resonance phenomenon to meet industry standards, although it may be only designed for a particular resonance and not for all frequencies.

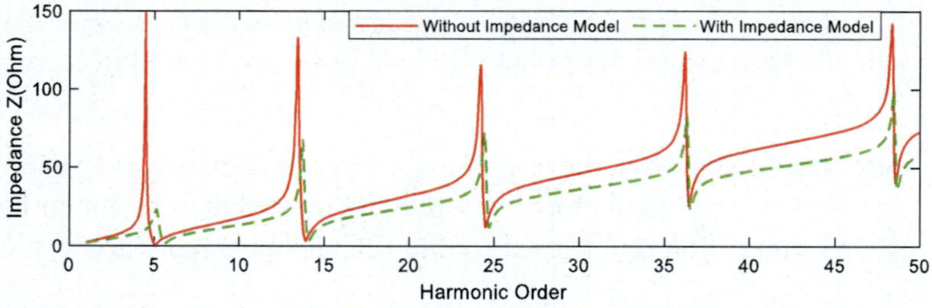

Fig. 9. Impedance characteristics with and without impedance model at B7.

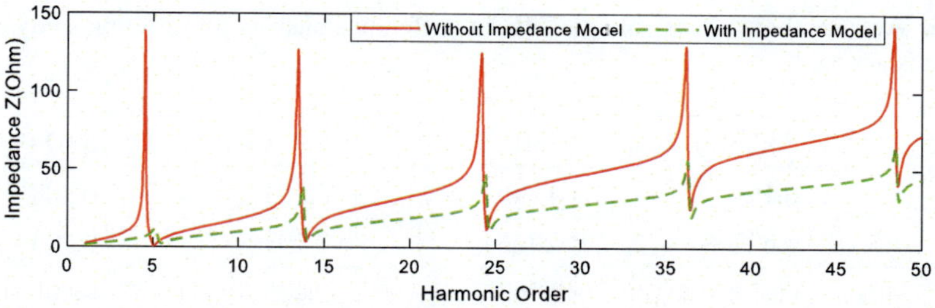

Fig. 10. Impedance characteristics with and without impedance model at B8.

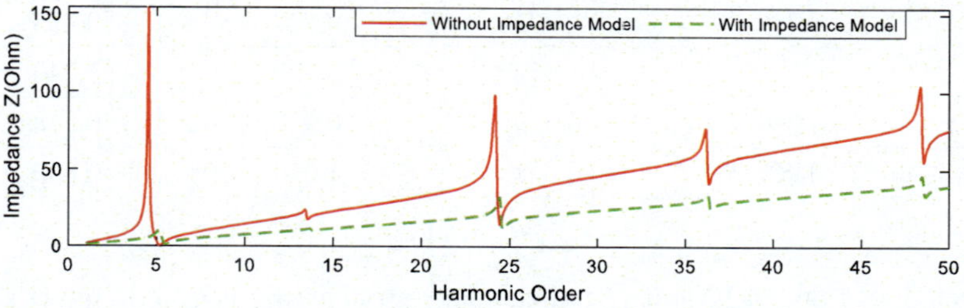

Fig. 11. Impedance characteristics with and without impedance model at B9.

Comparison of THD and Individual Harmonic Levels with and Without Impedance Model. In the previous section, it was noticed that the network impedance characteristics are significantly changed, especially at resonant frequencies, when the RES impedance model is applied. In order to analyze how the network harmonic levels are affected, unbalanced harmonic load flows were performed to compare the network harmonic levels when using IEC and UPC model with and without impedance model. The unbalanced harmonic load flow is selected because the balanced harmonic load flow does not consider the triplen harmonics when the UPC model is used. Since the unbalanced harmonic load flow corresponds to Solver 3, the cases studied in this section were labelled as follows:

- Solver 3A (S3A): Standard IEC harmonic source without impedance model
- Solver 3B (S3B): Standard IEC harmonic source with impedance model
- Solver 3C (S3C): UPC harmonic source without impedance model
- Solver 3D (S3D): UPC harmonic source with impedance model.

When modelling harmonic current injections of IEC and UPC model, the standard summation rule and phase angle settings introduced in Sect. 2.2 are adopted, respectively. Table 6 presents the average three-phase voltage THDs at different busbars for different study cases. The three-phase voltage THDs at 132 kV and 400 kV busbars are the same as discussed in Sect. 3.2, whereas the three-phase THDs at 33 kV busbars are slightly different in three phase due to the Yg-d transformer connection.

Comparing S3A and S3C with S3B and S3D, the voltage THDs of using IEC model are always slightly smaller than using UPC model due to its over-emphasis on harmonic cancellation effect. At the 33 kV busbars (B7–B9), the harmonic levels are significantly reduced when the RESs impedance model is applied, for either IEC or UPC model. This

Table 6. Average three-phase voltage THDs at difference busbars for different study cases when Solver 3 is adopted.

Busbar	S3A (%)	S3B (%)	S3C (%)	S3D (%)
B1	0.161	0.196	0.183	0.208
B2	0.627	0.726	0.703	0.773
B3	0.614	0.762	0.696	0.807
B4	0.704	0.796	0.783	0.848
B5	0.757	0.746	0.823	0.792
B6	0.739	0.801	0.827	0.855
B7	3.159	1.652	3.289	1.740
B8	0.884	0.519	0.937	0.548
B9	0.967	0.838	1.010	0.890

is expected, as Figs. 9, 10 and 11 indicated a significant reduction on impedance at the low-voltage busbars when the RESs impedance model is applied. More detailed analysis on individual harmonic voltage components up to the 50th harmonics are presented in Fig. 12 for busbar B8 (33 kV), where S3C and S3D are shown (similar results were found for S3A and S3B). Case S3D, when the RESs frequency-dependent impedance is introduced, presents significantly smaller harmonics than Case S3C.

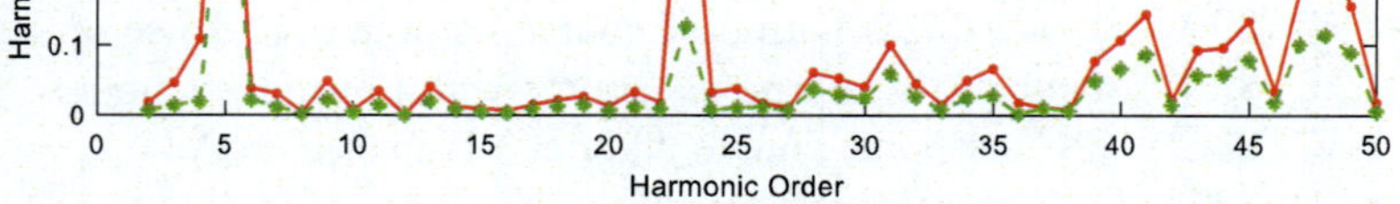

Fig. 12. Average individual three phase voltage harmonic level at busbar B8 for Case S3C and Case S3D.

Nevertheless, at the 400 kV busbar (i.e. B1) and 132 kV busbars (i.e. B2–B6), the THD levels of Case S3A and S3C (no RESs impedance model) are slightly lower than S3B and S3D (with RESs impedance model). This is because the impedances, except for resonance frequencies, at these busbars are almost not changed when the impedance model is applied.

Due to shift of the resonance frequencies, the system impedances at some frequencies are much higher when the RESs impedance model is applied. For example, at 400 kV busbar B1, the impedance around the 5th harmonic order is changed from 2.8 Ω to 310 Ω.

Figure 13 presents the individual harmonic voltage levels up to the 50th harmonic order for S3C and S3D at busbar B1. The harmonic voltage magnitude at the 5th harmonic order increases when the impedance model is applied, whereas the harmonic levels of the two cases are very similar at the other harmonic orders. These findings are also applicable to busbar B2–B6 and for S3A and S3B. Overall, harmonic levels at high-voltage busbars are slightly increased when the impedance model is applied due to the caused resonance shifts.

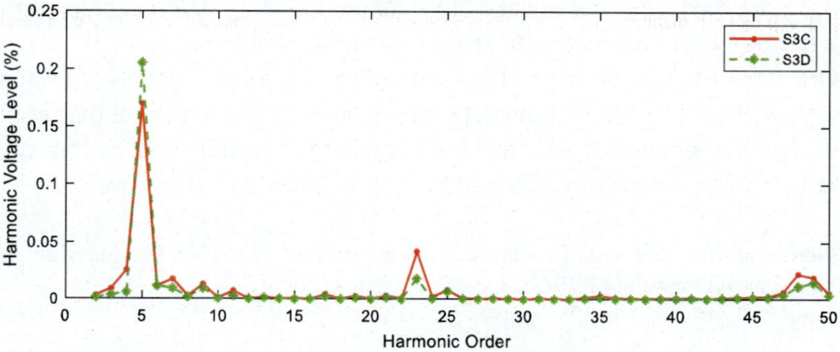

Fig. 13. Average individual three phase voltage harmonic level at busbar B1 for Case S3C and Case S3D.

4 Conclusion

This paper addressed different approaches for carrying out harmonic studies when RESs are connected to the power grids. Frequency scan analysis and harmonic penetration studies were carried out for a small network including up to four RESs. Nevertheless, the proposed approach can be extended to larger electrical systems.

The factors taken into consideration for these studies included: the solver, summation rule, harmonic source model and RES equivalent impedance.

When the RES equivalent impedance is included, the impact on the frequency scans may be significant. Therefore, it is important to ensure that this parameter is modelled as accurately as possible. When the impedance model is not included, the RESs model simplifies to a harmonic current source, and therefore does not affect the scans.

Numerous settings were tested to solve harmonic power flow using the two harmonic source models, named IEC and UPC model. For each harmonic current source model, both ideal and frequency-dependent equivalent impedances were considered. The discrepancies caused by the harmonic current source models, and the harmonic load flow solvers have been analyzed and explained by comparing different cases. In addition, an approach that allows modelling the IEC equivalent by using the UPC model has been described and verified.

It was concluded that the UPC model, the frequency-dependent equivalent impedance and the unbalanced harmonic load flow result in the most appropriate approach for harmonic analysis under certain operating conditions, for example (1) in stochastic harmonic analysis, (2) where it is deemed that the generic IEC summation rule may lead to underestimation or overestimation of harmonic levels, as it assumes a 'standard' cancellation of harmonic that may not take place in the practice, (3) when the network impedance changes significantly due to the presence of connected devices, such as loads an sources. More specifically, when power converter-based devices, such as RESs, are considered, it is recommended to adopt the UPC model and frequency-dependent equivalent impedance that accurately considers harmonic magnitude, phase angle and the output impedance. As a result, the harmonic cancellation effect and the impedance impact from RES control strategy and physical components are considered properly, and thus the harmonic assessment is more accurate and reliable.

Future work in this area includes the following topics: developing a non-linear frequency-dependent Norton equivalent impedance model; applying the proposed approach to a larger network including both transmission and distribution systems and studying increasing levels of RESs and their impact on harmonic levels.

Acknowledgements. The authors acknowledge the support of the UK Engineering and Physical Sciences Research Council (EPSRC); Project EP/T013206/1.

References

1. National Grid ESO: Future Energy Scenarios 2021 (2021). https://www.nationalgrideso.com/document/202851/download
2. IEEE Power and Energy Society: IEEE Recommended Practice and Requirements for Harmonic Control in Electric Power Systems. IEEE Std. 519-2014 (2014)
3. Energy Networks Association: Harmonic Voltage Distortion and the Connection of Non-Linear and Resonant Plant and Equipment to Transmission Systems and Distribution Networks in the United Kingdom. ENA Eng. Recomm. G5, no. 5 (2020)
4. IEC TR 61000-3-6: Electromagnetic Compatibility, Limits – Assessment of Emission Limits for the Connection of Distorting Installations to MV, HV and EHV Power Systems (2008)
5. Cherian, E., Bindu, G.R., Chandramohanan, P.S.: Pollution impact of residential loads on distribution system and prospects of DC distribution. Eng. Sci. Technol. Int. J. **19**(4), 1655–1660 (2016)
6. Koo, K.L., Emin, Z.: Comparative Evaluation of Power Quality Modelling Approaches for Offshore Wind Farms. In: 5th IET International Conference on Renewable Power Generation, pp. 1–7 (2016)
7. CIGRE Working Group JWG-C4/C6.29: Power Quality Aspects of Solar Power. CIGRE (2016). https://e-cigre.org/publication/672-power-quality-aspects-of-solar-power
8. Medina, A., et al.: Harmonic analysis in frequency and time domain. IEEE Trans. Power Deliv. **28**(3), 1813–1821 (2013)
9. Todeschini, G., Balasubramaniam, S., Igic, P.: Time-domain modeling of a distribution system to predict harmonic interaction between PV converters. IEEE Trans. Sustain. Energy **10**(3), 1450–1458 (2019)
10. Herraiz, S., Sainz, L., Clua, J.: Review of harmonic load flow formulations. IEEE Trans. Power Deliv. **18**(3), 1079–1087 (2003)
11. CIGRE Working Group JWG C4/B4.38: Technical Brochure 766: Network Modelling for Harmonic Studies. CIGRE (2019). https://e-cigre.org/publication/766-network-modelling-for-harmonic-studies

12. Jensen, C.F.: Harmonic background amplification in long asymmetrical high voltage cable systems. Electr. Power Syst. Res. **160**, 292–299 (2018)
13. Xiao, Y., Yang, X.: Harmonic summation and assessment based on probability distribution. IEEE Trans. Power Deliv. **27**(2), 1030–1032 (2012). https://doi.org/10.1109/TPWRD.2012.2187124
14. Eltouki, M., Rasmussen, T.W., Guest, E., Shuai, L., Kocewiak, Ł.: Analysis of harmonic summation in wind power plants based on harmonic phase modelling and measurements. In: 17th International Wind Integration workshop, pp. 1–7 (2018)
15. Ghassemi, F., Koo, K.: Equivalent network for wind farm harmonic assessments. IEEE Trans. Power Deliv. **25**(3), 1808–1815 (2010)
16. Strunz, K., Matvoz, D., Elvisa, B., Milanović, J.V., Hellmuth, S.: Deliverable 5.3: propagation of PQ disturbances through the power networks. MIGRATE – Massive InteGRATion of power Electronic devices (2018). https://www.h2020-migrate.eu/downloads.html
17. Technical Report IEC TR 61400-21-3:2019 - Wind energy generation systems - Part 21-3: Measurement and assessment of electrical characteristics - wind turbine harmonic model and its application (2019)
18. Cheah-Mane, M., Sainz, L., Prieto-Araujo, E., Gomis-Bellmunt, O.: Impedance-based analysis of harmonic instabilities in HVDC-connected offshore wind power plants. Int. J. Electr. Power Energy Syst. **106**, 420–431 (2019)
19. DIgSILENT: DIgSILENT PowerFactory 2020 (2020). https://www.digsilent.de/en/powerfactory.html
20. Deng, Z., Todeschini, G., Koo, K., Mulimakwenda, M.: Modelling renewable energy sources for harmonic assessments in DIgSILENT PowerFactory: comparison of different approaches. In: Proceedings of the 11th International Conference on Simulation and Modeling Methodologies, Technologies and Applications, pp. 130–140 (2021)
21. Grainger, J.J., Stevenson, W.D.: Power System Analysis. McGraw Hill Education, New York (1994)
22. Santoso, S.: Fundamentals of Electric Power Quality. CreateSpace, Scotts Valley (2012)
23. Arrillaga, J., Watson, N.R.: Power System Harmonics. Wiley, Hoboken (2003)

Stretching Simulation of Viscoelastic Fluid with Spring Connection

Nobuhiko Mukai[1,2,3(✉)], Asahi Onodera[1], Takuya Natsume[2,3], and Youngha Chang[1,2]

[1] Knowledge Engineering, Tokyo City University, 1-28-1 Tamazutsumi, Setagaya, Tokyo 158-8557, Japan
{mukai,onodera,chang}@vgl.cs.tcu.ac.jp

[2] Graduate School of Integrative Science and Engineering, Tokyo City University, 1-28-1, Setagaya, Tokyo 158-8557, Japan
natsume@vgl.cs.tcu.ac.jp

[3] Institute of Industrial Science, The University of Tokyo, 4-6-1 Komaba, Meguro, Tokyo 153-8505, Japan

Abstract. Liquid simulation is very difficult since it has a clear boundary and a large amount of dynamic deformation, while solid has a clear boundary with a small amount of deformation and gas has an unclear boundary with a large amount of dynamic deformation. There are two types of liquid: Newtonian and non-Newtonian fluid. Viscoelastic fluid is one of the non-newtonian fluids and the behavior has a unique feature called "spinnability", which shows long stretch and sudden shrink like a rubber since it has both characteristics of viscosity and elasticity. To simulate the behavior of the viscoelastic fluid, some constitutive equations were proposed; however, there is no established one. Then, we have been researching the method to simulate the behavior of spinnability by constructing the constitutive equation with the combination of the viscous and the elastic stresses; however, the summation of the stress coefficients did not equal 1.0. In addition, the stretch length was short because the middle area of the viscoelastic fluid became very thin and there was no particle in the area. On the other hand, real viscoelastic fluid is composed of so many small molecules even when it becomes very thin. Therefore, we propose a method to make the summation of the stress coefficients 1.0 and to insert a spring in the middle thin area of the viscoelastic fluid and interpolate the empty space with some virtual particles for visualization of spinnability until the spring is broken.

Keywords: Particle method · Viscoelastic fluid · Spinnability · Cauchy's equation of motion · Constitutive equation · Spring model

1 Introduction

One of the most challenging simulations is a visualization of liquid behavior because the liquid has a clear boundary and deforms dynamically. There are two types of fluids:

A. Onodera, T. Natsume, Y. Chang—These authors contributed equally to this work.

G. Wagner et al. (Eds.): SIMULTECH 2021, LNNS 601, pp. 94–105, 2023.
https://doi.org/10.1007/978-3-031-23149-0_5

Newtonian and non-Newtonian fluids. The simulation of Newtonian fluid is comparatively simple compared to that of non-Newtonian's because the shear stress of Newtonian fluid is proportional to the velocity gradient. However, it is not for non-Newtonian fluid. Viscoelastic fluid is one of non-Newtonian's, which has the characteristics of both viscosity and elasticity, and also has a unique feature called "spinnability". When the fluid is pulled, the middle area becomes very thin and it stretches so long. When it is broken, it shrinks very fast like a rubber. The basic equation for the analysis of viscoelastic fluid is Cauchy's equation of motion which is the fundamental equation for continuous material. On the other hand, the constitutive equation, which defines the relation between the stress and the distortion is unknown. There are some proposals, but, there is no established one. Some studies use the FEM (Finite Element Method), and the others employ the SM (Spring Mass) model or particle methods.

Then we have been researching the method of how to simulate and visualize the behavior of spinnability with a particle model since the topology of the fluid changes after it is broken. The particle method is very useful for the simulation where the topology changes. For the constitutive equation, we thought that the deviatoric stress should be composed of the combination of the viscous and the elastic stresses, and performed the simulation of spinnability. In our previous simulations, we tried the two types of combinations. One is the case where the summation of the coefficients of the viscous and the elastic stresses was 1.0, and the other is the case where the summation was not 1.0, and the coefficients were decided empirically because the characteristic of the fluid changes suddenly after it is broken. In this paper, we define the summation of the coefficients of the viscous and the elastic stresses as 1.0.

In addition, the stretched length was too short to visualize the behavior of spinnability since there was no particle in the thin middle area of the fluid when it was stretched. Real viscoelastic fluid, however, is composed of many small molecules, and there should be some particles in the middle area even when it becomes very thin. Then, we propose a method to insert a spring in the thin middle area and to interpolate the empty space with some virtual particles for the visualization.

2 Previous Studies

There are two kinds of fluid: Newtonian and non-Newtonian fluids. Water is the most representative Newtonian fluid so that there are a lot of studies for modeling, rendering, simulation and animation of water [4,9,12,19,27]. The main analytical methods are the Eulerian (grid based) method [8,10,17], the Lagrangian (particle based) method [22,28,35], and the hybrid method of them, which examples are an optimized grid construction for GPU (Graphics Processing Unit) [6], freezing ice with air bubbles [33], bubbles with Voronoi diagram [3], bubbles in water [18], and large flowing river [6]. In addition, some works used level-set algorithms [13,16,21,25].

On the other hand, there are also many types of researches on non-Newtonian fluid like viscous liquids [11,26]. Especially, viscoelastic fluid has so complex behavior that there are lots of approaches such as a spring-mass model [37], a finite element method [1,40]. A particle with spring method [7], a material point method [34], a position based dynamics method [2], a grid based method with level set [15], and a particle method [5].

There are varieties of methods mentioned above for the analysis of viscoelastic fluid since there is no established constitutive equation that defines the relationship between the stress and the distortion. However, there is the established governing equation for fluid called the Navier-Stokes equations, which is derivated from Cauchy's equation of motion that is the established governing equation for the continuous body. In addition, viscoelastic fluid has a unique characteristic called "spinnability", which shows the behavior of long stretch like a string, and fast shrink like a rubber. However, there was little research to represent spinnability.

With this research background, we have been performing the simulation of spinnability using Cauchy's equation of motion [29–32]. The basic idea is that the deviatoric stress, which constructs Cauchy's equation of motion, is composed of the linear combination of the viscous and the elastic stresses since the viscoelastic fluid has two characteristics of viscosity and elasticity. In the previous method [31], the summation of the two kinds of stresses coefficients did not equal 1.0 because the behavior of viscoelastic fluid changes suddenly after it is broken, and the coefficients were decided experimentally. Then, in this paper, we decide the coefficients to satisfy the condition that the summation equals 1.0.

In addition, viscoelastic fluid did not stretch so long because there is no particle in the middle area when it becomes thin. However, real viscoelastic fluid is composed of so many small molecules, and there are still some particles even when it becomes thin. Therefore, in this paper, we propose a method with spring for long stretch simulation, where spring is inserted into the middle area of viscoelastic fluid, and some particles are interpolated in the empty space for just the visualization until the spring is broken.

3 Simulation Method

3.1 Governing Equations

To perform the simulation of spinnability, which is a unique characteristic of the viscoelastic fluid, we have to treat the topology change between the before and the after of the break. In addition, viscoelastic fluid is non-Newtonian and incompressible fluid so we employ the MPS (Moving Particle Semi-implicit) method [23] in this research, which is a particle method developed for an incompressible fluid.

The governing equations of continuous body are the equation of continuity and Cauchy's equation of motion, which are described as Eq. (1) and Eq. (2), respectively.
Equation of continuity:

$$\frac{d\rho}{dt} = 0, \tag{1}$$

Cauchy's equation of motion:

$$\rho\frac{d\boldsymbol{v}}{dt} = \nabla\cdot\boldsymbol{\sigma} + \rho\boldsymbol{g} + \boldsymbol{f} = (-\nabla p\boldsymbol{I} + \nabla\cdot\boldsymbol{\tau}) + \rho\boldsymbol{g} + \boldsymbol{f}, \tag{2}$$

where, ρ is the density, $\boldsymbol{v}$ is the velocity, t is time, $\boldsymbol{\sigma}$ is the stress tensor, $\boldsymbol{g}$ is the gravity acceleration, $\boldsymbol{f}$ is the surface tension, p is the pressure, $\boldsymbol{I}$ is the unit matrix, and $\boldsymbol{\tau}$ is the deviatoric stress.

In the MPS method, a weight function is used to calculate the particle number density. The original weight function was defined as Eq. (3) [23]; however, the weight approaches infinity as the length between two particles of i and j, becomes 0. Then, this research employs the modified weight function shown in Eq. (4) [36].

$$w(\mid \boldsymbol{r}_j - \boldsymbol{r}_i \mid) = \begin{cases} \frac{r_e}{|\boldsymbol{r}_j - \boldsymbol{r}_i|} - 1 & (0 \leq \mid \boldsymbol{r}_j - \boldsymbol{r}_i \mid < r_e) \\ 0 & (r_e \leq \mid \boldsymbol{r}_j - \boldsymbol{r}_i \mid) \end{cases}, \tag{3}$$

$$w(\mid \boldsymbol{r}_j - \boldsymbol{r}_i \mid) = \begin{cases} (\frac{|\boldsymbol{r}_j - \boldsymbol{r}_i|}{r_e} - 1)^2 & (0 \leq \mid \boldsymbol{r}_j - \boldsymbol{r}_i \mid < r_e) \\ 0 & (r_e \leq \mid \boldsymbol{r}_j - \boldsymbol{r}_i \mid) \end{cases}, \tag{4}$$

where, $\boldsymbol{r}_i$ and $\boldsymbol{r}_j$ are the position vectors of particle i and j, respectively, and r_e is the radius of influence. With the weight function of Eq. (4), particle number of density i (n_i) is calculated as Eq. (5).

$$n_i = \Sigma_{j \neq i} w(\mid \boldsymbol{r}_j - \boldsymbol{r}_i \mid) \tag{5}$$

3.2 High Precision Calculation

There is an assumption that particles are regularly arranged in the original MPS method [23] so that the calculation result becomes unstable as the analysis proceeds because the particles are arranged irregularly. Then, this paper adopts two precise pressure calculation methods to stabilize the pressure. One is the method for the Poisson equation of pressure. The original Laplacian of pressure was calculated as in Eq. (6).

$$<\nabla^2 P>_i^{s+1} = \frac{\rho}{\Delta t^2} \frac{n^0 - n_i^s}{n^0} \tag{6}$$

where, $<\nabla^2 P>_i^{s+1}$ is the Laplacian of the pressure for a particle i at the time step $s+1$, Δt is the time step, n^0 is the initial particle number density, and n_i^s is the particle number density of a particle i at the time step s.

In this paper, we use the Laplacian of pressure with the velocity divergence term [39] as in Eq. (7) since the velocity divergence is actually not zero in the simulation calculation although it should be 0 for ideal incompressible fluid.

$$<\nabla^2 P>_i^{s+1} = \frac{\rho}{\Delta t} \nabla \cdot \boldsymbol{u}_i^* + \gamma \frac{\rho}{\Delta t^2} \frac{n^0 - n_i^k}{n^0} \tag{7}$$

where, $\boldsymbol{u}_i^*$ is the provisional velocity vector of a particle i, and γ is the relaxation coefficient, which is set as 0.02 according to the pre-calculation.

The other is the method for pressure gradient. The original pressure gradient was as follows [23].

$$<\nabla P>_i = \frac{d}{n^0} \sum_{j \neq i} \frac{P_j - \hat{P}_i}{\mid \boldsymbol{r}_j - \boldsymbol{r}_i \mid^2} (\boldsymbol{r}_j - \boldsymbol{r}_i) \omega(\boldsymbol{r}_j - \boldsymbol{r}_i), \tag{8}$$

where, d is the dimension, P_j and $\hat{P}_i$ are the pressure of a particle j and the minimum pressure in the radius of influence of a particle i, respectively. Then, $P_j - \hat{P}_i$ is always positive, and the repulsive force is generated between the particles of i and j. If two particles approach excessively, the repulsive force grows huge and the calculation becomes unstable. To prevent the excessive particle approach, this paper employs a high order gradient model [20] that considers an artificial repulsive force, which is written in the following.

$$<\nabla P>_i = \left[\frac{1}{n^0} \sum_{j \neq i} \frac{(\boldsymbol{r}_j - \boldsymbol{r}_i)}{\mid \boldsymbol{r}_j - \boldsymbol{r}_i \mid} \otimes \frac{(\boldsymbol{r}_j - \boldsymbol{r}_i)}{\mid \boldsymbol{r}_j - \boldsymbol{r}_i \mid} \omega(\mid \boldsymbol{r}_j - \boldsymbol{r}_i \mid) \right]^{-1} \left[\frac{1}{n^0} \sum_{j \neq i} \frac{P_j - \hat{P}_i}{\mid \boldsymbol{r}_j - \boldsymbol{r}_i \mid^2} (\boldsymbol{r}_j - \boldsymbol{r}_i) \omega(\mid \boldsymbol{r}_j - \boldsymbol{r}_i \mid), \right] \tag{9}$$

where, $\otimes$ is the tensor product. Equation (9) replaces d in Eq. (8) with the inverse matrix, and the pressure calculation becomes stable even when the arrangement of particles is unbalanced. In fact, the inverse matrix becomes a unit matrix when particles are regularly arranged, and Eq. (9) is equivalent to Eq. (8).

3.3 Constitutive Equation

In Eq. (2), τ is the deviatoric stress, and the constitutive equation that defines the relationship between the stress and the distortion has the key role to decide the behavior of the material. Some constitutive equations were proposed for viscoelastic fluid [14,24]; however, there are still no established equations and the proposed methods are very complex.

On the other hand, viscoelastic fluid has two characteristics of viscosity and elasticity so it should be considered that the deviatoric stress should be composed of the viscous and the elastic stresses. Thus we have been studying the method to represent the behavior of viscoelastic fluids by constructing the deviatoric stress using a linear combination of the viscous and the elastic stresses [29–32]. Then, the deviatoric stress ($\boldsymbol{\tau}$ in Eq. (2)) is written as in Eq. (10).

$$\boldsymbol{\tau} = C_v \boldsymbol{\tau}_v + C_e \boldsymbol{\tau}_e, \tag{10}$$

where, $\boldsymbol{\tau}_v$ and $\boldsymbol{\tau}_e$ are the viscous and the elastic terms of the deviatoric stress, respectively, and C_v and C_e are the linear combination coefficients for the viscous and the elastic stresses, respectively. The viscous term ($\boldsymbol{\tau}_v$) is written as follows.

$$\boldsymbol{\tau}_v = 2\eta_0 \boldsymbol{D}, \tag{11}$$

$$\boldsymbol{D} = \frac{1}{2}(\boldsymbol{L} + \boldsymbol{L}^t), \qquad \boldsymbol{L} = \nabla \boldsymbol{V}, \tag{12}$$

$$\boldsymbol{V} = (\boldsymbol{u}, \boldsymbol{v}, \boldsymbol{w}), \tag{13}$$

$$\nabla \boldsymbol{V} = \begin{bmatrix} \frac{\partial u}{\partial x} & \frac{\partial u}{\partial y} & \frac{\partial u}{\partial z} \\ \frac{\partial v}{\partial x} & \frac{\partial v}{\partial y} & \frac{\partial v}{\partial z} \\ \frac{\partial w}{\partial x} & \frac{\partial w}{\partial y} & \frac{\partial w}{\partial z} \end{bmatrix}, \tag{14}$$

where, η_0 is zero shear viscosity, $\boldsymbol{V}$ is the particle velocity.

On the other hand, the elastic term ($\boldsymbol{\tau}_e$) is written in the following.

$$\boldsymbol{\tau}_e = \lambda \boldsymbol{I} tr(\boldsymbol{\epsilon}) + 2\mu\boldsymbol{\epsilon}, \qquad \mu = \frac{E}{2(1+\nu)}, \tag{15}$$

where, λ and μ are Lame-parameters, $\boldsymbol{I}$ is the unit matrix, tr is the tensor trace, $\boldsymbol{\epsilon}$ is the distortion tensor, E is Young's modulus, and ν is Poisson's ratio. Here, the first and the second terms of the elastic stress ($\boldsymbol{\tau}_e$) are the components for the volumetric and the telescopic changes, respectively, and there is no volumetric change since the viscoelastic fluid is considered incompressible. Then, the elastic term ($\boldsymbol{\tau}_e$) is simplified as Eq. (16).

$$\boldsymbol{\tau}_e = 2\mu\boldsymbol{\epsilon} \tag{16}$$

Related to the calculation of the coefficients (C_v and C_e), the effect of viscosity is more significant than that of elasticity, although it becomes a little bit larger after it is broken because it shrinks very fast like a rubber. Then, C_v is larger than C_e all the time since the effect of viscosity is dominant. The value of C_e depends on the density of the narrowest area of the fluid because the fluid is broken at the narrowest area and the characteristic of elasticity appears. Then, C_e is calculated as in Eq. (17).

$$C_e = \left(1 - \frac{n^s}{n^0}\right) C_g, \tag{17}$$

where, n^0 is the initial particle number density, n^s is the particle number density of the narrowest area at the time s, and C_g is the global coefficient of elasticity, which is a parameter in this paper.

In the previous study [31], C_v and C_e were decided independently and $C_v + C_e \neq 1$. The deviatoric stress (τ), however, should be composed of the linear combination of the viscous stress (τ_v) and the elastic stress (τ_e) so that the both coefficients are decided to satisfy the condition of $C_v + C_e = 1$ in this paper. Then, C_v can be calculated as follows.

$$C_v = 1 - C_e = (1 - C_g) + \frac{n^k}{n^0} C_g \tag{18}$$

The coefficient value depends on the location of the fluid and is very different. Some areas have large values and the effect of viscosity is strong, while others have small values and the characteristic of elasticity is strong. Especially, the surface particles affect this influence. Then, the coefficient values are calculated as the average for all particles.

3.4 Spring Model

When viscoelastic fluid is stretched, the middle area becomes very thin and there are little particles there. Finally, it is broken before long stretching. Real viscoelastic fluid is, however, stretches very thin and long because it is composed of many small molecules. Then, we propose the method to insert a spring in the middle area when it becomes thin. If there is a particle that is judged as a free surface due to the small number of particles, two particles that has the maximum and the minimum vertical positions in the middle area are connected with a spring. The force that is generated by the spring is calculated with the following equation [38]. Some particles are pulled to the middle area by the force, which means that the upper particles are pulled down, and the lower particles are pulled up. As a result, some particles are kept in the middle area, and the viscoelastic fluid is stretched longer.

$$F_i = k(\mid \boldsymbol{r}_j - \boldsymbol{r}_i \mid - l^0)\frac{\boldsymbol{r}_j - \boldsymbol{r}_i}{\mid \boldsymbol{r}_j - \boldsymbol{r}_i \mid}, \tag{19}$$

$$F_i = -F_j, \tag{20}$$

where, F_i and F_j are the force working on the particles located in the maximum and the minimum vertical positions in the middle area, respectively, k is the spring constant, and l^0 is the initial spring length at the time when it is inserted. The middle part length, which is the distance between the maximum and the minimum vertical positions, is set as 5 layers in this paper, and k is a parameter that changes in the simulations.

4 Simulation and Results

Table 1 and Table 2 show the specification of the PC used for the simulation, and the parameters of the simulation, respectively. There were 29,791 particles for the viscoelastic fluid, 5,043 particles for a solid body that pulled up the viscoelastic fluid, and 51,483 particles for another solid body that was the ground so that 86,317 particles were used for the simulation in total. The simulation time was 720 ms per step.

Table 1. PC specification.

OS	Windows 10 Education 64 bit
CPU	Intel Core i5-4440 3.1 GHz
Main memory	8 GB
GPU	NVIDIA GeForce GTX 670 with 2 GB memory

Table 2. Simulation parameters.

Parameter		Value	Unit
Density	ρ	1.16	g/mm^3
Young's modulus	E	1.05	kPa
Poisson's ratio	ν	0.49	
Zero shear viscosity	η_0	0.36	Pa · s
Initial particle distance (= Particle radius)	l^0	0.3	mm
Pulling acceleration	α	1.8	mm/s^2
Time step	$\triangle t$	0.1	ms

Figure 1 shows the comparison between without and with a spring. The red spheres in Fig. 1(b) are the virtual particles that are interpolated by the spring connection between the maximum and the minimum positions in the middle area, which has become very thin when the viscoelastic fluid is stretched. The fluid is broken if there are no interpolated particles (Fig. 1(a)), while it is not broken and is still stretched longer if there are the interpolated ones (Fig. 1(b)). The fluid is broken when the number of particles in the middle area becomes less than two.

On the other hand, Fig. 2 and Fig. 3 show the simulation results that depend on the global coefficient of elasticity (C_g) for the spring constant $k = 5.0$ and $k = 500$, respectively. In general, the viscous feature appears more than the elastic one so that the global coefficient of elasticity (C_g) is set as small values for three cases: 0.01, 0.03, and 0.05. In all cases, the middle area of the viscoelastic fluid becomes very thin and the fluid shrinks after it is broken.

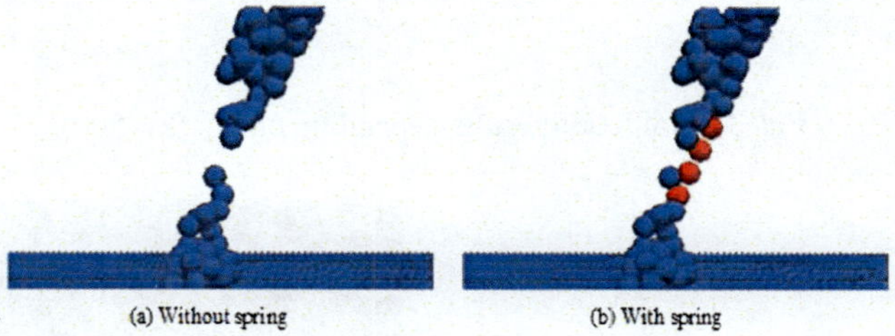

Fig. 1. Comparison of the stretching viscoelastic fluids without and with a spring.

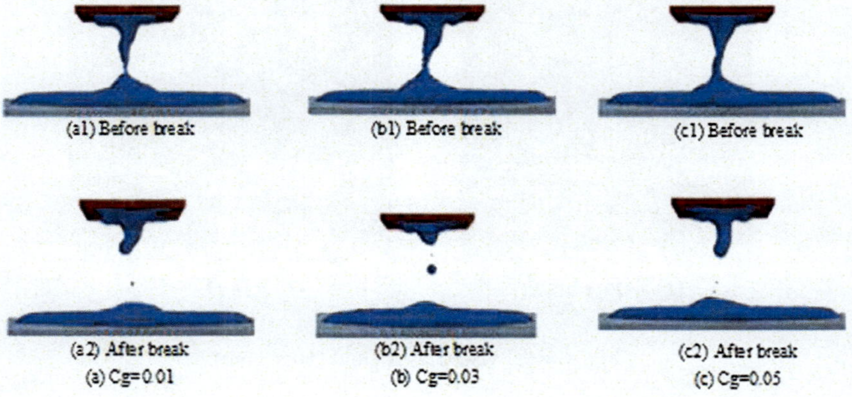

Fig. 2. Simulation results depending on C_g (k = 5.0).

In addition, Fig. 4 shows the state of the before and the after of the break using real viscoelastic fluid called "guar gum". With the comparison to Fig. 4, it seems that the case (c) in Fig. 2 and the case (a) in Fig. 3 are similar to Fig. 4 for $k = 5.0$ and $k = 500$, respectively, because the fluid has a thin, straight, and long part in the middle area like a string, and there is a small sphere-shaped part in the tip of the upper viscoelastic fluid. Here, C_g represents the global feature of elasticity and k shows the rigidness of the material. This indicates that if k is smaller, C_g should be larger (Fig. 2(c)), and if k is larger, C_g should be smaller (Fig. 3(a)). It means that there might be the best combination of the rigidness of the material (k) and the global feature of elasticity (C_g).

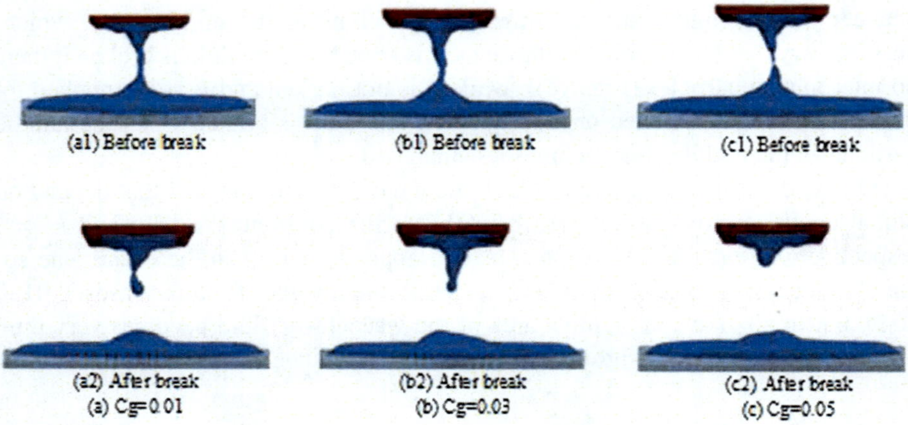

Fig. 3. Simulation results depending on C_g (k = 500).

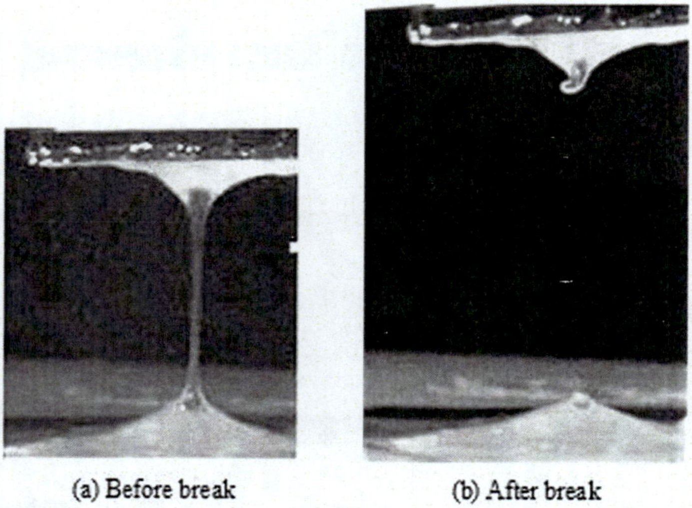

Fig. 4. Real viscoelastic fluid (guar gum).

5 Conclusions

It is a challenging issue to simulate the behavior of non-Newtonian fluid because its shear stress is not proportional to the velocity gradient. Viscoelastic fluid is one of the non-Newtonian fluids and has a unique characteristic called "spinnability". There have been many types of research related to viscoelastic fluid; however, there are few studies that focus on spinnability. Then, we have been trying to simulate and represent the spinnability of viscoelastic fluid. The constitutive equation, which decides the relationship between the stress and the distortion, has the key role for the simulation; however, there still has not been an established one and the equations that had been proposed are so complex. Then, we proposed a method to define that the deviatoric stress is composed of a linear combination of the viscous and the elastic stresses, and tried to simulate the long stretching; however, the summation of the linear combination coefficients was not 1.0. In addition, the stretched fluid was broken before it showed a long string since there was no particle in the thin middle area.

Therefore, we have proposed a new method to insert a spring between the maximum and the minimum positions in the thin area and to represent spinnability using virtual particles that interpolate the empty space in addition to that the summation of the linear combination coefficients equals 1.0.

As the result of the simulations, the viscoelastic fluid has stretched a little bit longer by inserting a spring and by interpolating the thin middle area with virtual particles. In addition, the shape before and after the break was similar to those of a real viscoelastic fluid called "guar gum". We have also found that if the rigidness of the material is larger, the global feature of elasticity should be smaller and vice versa. This indicates a possibility that there is the best combination of the rigidness of the material and the global feature of elasticity. In addition, the spring inserted in the middle area worked just for pulling some particles to the middle area, and the fluid was broken when the number of particles became less than two. However, if the fluid stretches until the force on the spring is over the elasticity limit, it can be stretched longer. Then, we have to solve these issues in the future.

References

1. Bargtei, A., Wojtan, C., Hodgins, J., Turk, G.: A finite element method for animating large viscoplastic flow. ACM Trans. Graph. **26**(3), 16 (2007)
2. Barreiro, H., García-Fermández, I., Alduán, I., Otaduy, M.: Conformation constraints for efficient viscoelastic fluid simulation. ACM Trans. Graph. **36**(6), 221 (2017)
3. Busaryev, O., Dey, T., Wang, H., Ren, Z.: Animating bubble interactions in a liquid foam. ACM Trans. Graph. **31**(4), 1–8 (2012). Article No. 63
4. Cao, R., Cao, R.: Computer simulation of water flow animation based on two-dimensional Navier-Stokes equations. Adv. Math. Phys. **2021**, 1–9 (2021). Article No. 5157197
5. Chang, Y., Bao, K., Liu, Y., Zhu, J., Wu, E.: A particle-based method for viscoelastic fluids animation. In: ACM Symposium on Virtual Reality Software and Technology, pp. 111–118 (2009)
6. Chentanez, N., Müller, M.: Real-time simulation of large bodies of water with small scale details. In: ACM SIGGRAPH/Eurographics Symposium on Computer Animation, pp. 197–206 (2010)
7. Clavet, S., Beaudoin, P., Poulin, P.: Particle-based viscoelastic fluid simulation. In: ACM SIGGRAPH/Eurographics Symposium on Computer Animation, pp. 219–228 (2005)
8. Cui, X., Jin, Y., Liu, X.: Real-time ocean wave in multi-channel marine simulator. In: ACM SIGGRAPH International Conference on Virtual Reality Continuum and Its Application in Industry, pp. 332–335 (2004)
9. Darles, E., Crespin, B., Ghazanfarpour, D., Gonzato, J.: A survey of ocean simulation and rendering techniques in computer graphics. In: Computer Graphics Forum, vol. 30, pp. 43–60 (2011)
10. Dupuy, J., Bruneton, E.: Real-time animation and rendering of ocean whitecaps. In: SIGGRAPH Asia, Technical Briefs, Article No. 15 (2012)
11. Foster, N., Fedkiw, R.: Practical animation of liquids. In: ACM SIGGRAPH, pp. 23–30 (2001)

12. Foster, N., Metaxas, D.: Modeling water for computer animation. Commun. ACM **43**(7), 61–67 (2000)
13. Geiger, W., Leo, M., Rasmussen, N., Losasso, F., Fedkiw, R.: So real it'll make you wet. In: ACM SIGGRAPH Sketches, Article No. 20 (2006)
14. Giesekus, H.: A simple constitutive equation for polymer fluids based on the concept of deformation-dependent tensorial mobility. J. Nonnewton. Fluid Mech. **11**, 69–109 (1982)
15. Goktekin, T., Bargteil, A., O'Brien, J.: A method for animating viscoelastic fluids. ACM Trans. Graph. **23**(3), 463–468 (2004)
16. Greenwood, S., House, D.: Better with bubbles: enhancing the visual realism of simulated fluid. In: ACM SIGGRAPH/Eurographics Symposium on Computer Animation, pp. 287–296 (2004)
17. Hinsinger, D., Neyret, F., Cani, M.: Interactive animation of ocean waves. In: ACM SIGGRAPH/Eurographics Symposium on Computer Animation, pp. 161–166 (2002)
18. Hong, J., Lee, H., Yoon, J., Kim, C.: Bubbles alive. ACM Trans. Graph. **27**(3), 1–8 (2008). Article No. 48
19. Iglesias, A.: Computer graphics for water modeling and rendering: a survey. Futur. Gener. Comput. Syst. **20**(8), 1355–1374 (2004)
20. Iribe, T., Nakaza, E.: An improvement of accuracy of the MPS method with a new gradient calculation model. J. Jpn. Soc. Civil Eng. (B2) **67**(1), 36–48 (2011). (in Japanese)
21. Kim, B., Liu, Y., Llamas, I., Jiao, X., Rossignac, J.: Simulation of bubbles in foam with the volume control method. ACM Trans. Graphi. **26**(3), 1–10 (2007). Article No. 98
22. Kipfer, P., Westermann, R.: Realistic and interactive simulation of rivers. In: Graphics Interface, pp. 41–48 (2006)
23. Koshizuka, S., Oka, Y.: Moving-particle semi-implicit method for fragmentation of incompressible fluid. Nucl. Sci. Eng. **123**, 421–434 (1996)
24. Larson, R.G.: Convection and diffusion of polymer network strands. J. Nonnewton. Fluid Mech. **13**, 279–308 (1983)
25. Losasso, F., Talton, J., Kwatra, N., Fedkiw, R.: Two-way coupled SPH and particle level set fluid simulation. IEEE Trans. Visual Comput. Graph. **14**(4), 797–804 (2008)
26. Miller, G., Pearce, A.: Globular dynamics: a connected particle system for animating viscous fluids. Comput. Graph. **13**(3), 305–309 (1989)
27. Mould, D., Yang, Y.: Modeling water for computer graphics. Comput. Graph. **21**(6), 801–814 (1997)
28. Müller, M., Charypsr, D., Gross, M.: Particle-based fluid simulation for interactive applications. In: ACM SIGGRAPH/Eurographics symposium on computer animation. pp. 154–159 (2003)
29. Mukai, N., Ito, K., Nakagawa, M., Kosugi, M.: Spinnability simulation of viscoelastic fluid. In: ACM SIGGRAPH Posters, Article No. 18 (2010)
30. Mukai, N., Matsui, E., Chang, Y.: Investigation on viscoelastic fluid behavior by modifying deviatoric stress tensor. In: SIMULTECH, pp. 216–222 (2019)
31. Mukai, N., Morooka, R., Natsume, T., Chang, Y.: Viscoelastic fluid simulation based on the combination of viscous and elastic stresses. In: SIMULTECH, pp. 172–178 (2021)
32. Mukai, N., Nishikawa, T., Chang, Y.: Evaluation of stretched thread lengths in spinnability. In: ACM SIGGRAPH Posters, Article No. 62 (2018)
33. Nishino, T., Iwasaki, K., Dobashi, Y., Nishita, T.: Visual simulation of freezing ice with air bubbles. In: SIGGRAPH Asia, Technical Briefs, Article No. 1 (2012)
34. Ram, D., et al.: A material point method for viscoelastic fluids, foams and sponges. In: ACM SIGGRAPH/Eurographics Symposium on Computer Animation, pp. 157–163 (2015)
35. Sims, K.: Particle animation and rendering using data parallel computation. In: ACM SIGGRAPH, vol. 24, pp. 405–413 (1990)

36. Tamai, T., Koshizuka, S.: Least squares moving particle semi-implicit method - an arbitrary high order accurate meshfree Lagrangian approach for incompressible flow with free surfaces. J. Comput. Particle Mech. **1**(3), 277–305 (2014)
37. Tamura, N., Tsumura, N., Nakaguchi, T., Miyake, Y.: Spring-bead animation of viscoelastic materials. In: ACM SIGGRAPH Sketches, Article No. 64 (2005)
38. Tanaka, M., Koshizuka, S.: Simulation of red blood cell deformation using a particle method. J. Jpn. Soc. Fluid Mech. **26**(1), 49–55 (2007). (in Japanese)
39. Tanaka, M., Masunaga, T.: Stabilization and smoothing of pressure in MPS method by quasi-compressibility. J. Comput. Phys. **229**, 4279–4290 (2010)
40. Wojtan, C., Turk, G.: Fast viscoelastic behavior with thin features. ACM Trans. Graph. **27**(3), 47 (2008)

SysML and Petri Nets Based Methodology for Analysis and Performance Evaluation in WSNs

Amel Berrachedi[1(✉)], Malika Ioualalen[2(✉)], and Ahmed Hammad[3(✉)]

[1] Faculty of Exact Sciences and Informatics, Hassiba Benbouali University of Chlef, Chlef, Algeria
a.berrachedi@univ-chlef.dz

[2] Department of Computing Science, MOVEP Laboratory, USTHB University, Algiers, Algeria
mioualalen@usthb.dz

[3] DISC/Femto-ST Department, Franche Comté University, Besançon, France
ahmed.hammad@femto-st.fr

Abstract. In order to satisfy their requirements, the verification and validation of the design of complex systems is a primary task. The main goal of the present work is to propose a model-based system engineering specification and verification methodology for the design and performance evaluation of an example of these systems, namely Wireless Sensor Networks (WSN). The proposed approach uses SysML language to describe the WSNs requirements, behaviors and performance parameters. Then, it translates the SysML elements to a Deterministic Stochastic Petri Net (DSPNs) and integrates them into an analytic model. The current paper improves the first part of this project; the transformation from SysML to a DSPN is done by integrating the performance parameters into the SysML activity diagram and not into the SysML parametric diagram. An other contribution is to refine the second part of our methodology by verifying the system behavior and evaluating its energy consumption.

Keywords: Model-based systems engineering · SysML activity diagram · Deterministic stochastic petri nets · Performance evaluation

1 Introduction

Recent advances in WSNs beget many challenges related to limited capacity of processing, storage and especially energy. So, when designing WSNs, it is important to consider these constraints for maximizing the network lifetime and to ensure the reliability required by these networks. To do so, it is attractive to reap the benefits of Model-Based Systems Engineering (MBSE) approaches [6]. This helps to produce easy and clear models, to reduce the time and the maintenance costs, and to increase the efficiency and the productivity.

Nowadays, Systems Modeling Language (SysML), which is a general-purpose graphical modeling language for the Systems-Engineering domain, is the most adopted modeling language because of its intuitive notations [8]. In addition, it provides several improvements, specifically, it considers the requirements modeling and takes into

G. Wagner et al. (Eds.): SIMULTECH 2021, LNNS 601, pp. 106–117, 2023.
https://doi.org/10.1007/978-3-031-23149-0_6

account the strong interaction between hardware and software system parts, which is an important condition for effective modeling. However, SysML is not formal. Accordingly, it does not provide the detailed execution semantics of models that allows the qualitative and quantitative analysis. Therefore, integrating SysML with other engineering analysis such as formal methods is necessary. Formal methods are well adapted for analyzing and validating complex systems that require rigorous verification. Different formalisms can be used to analyze WSNs and to evaluate their performances. Among these formalisms, Petri Nets (PN) [14] have many advantages, particularly, Deterministic and Stochastic Petri Nets (DSPNs) could be the most appropriate. In fact, they are very expressive and they represent a widely used high-level formalism for modeling concrete and discrete-event systems where events may occur either without consuming time, after a deterministic time, or after an exponentially distributed time [12].

The current work provides at first an overview of the proposed methodology of the specification and the verification of the non-functional properties in the WSN systems especially the energy dissipation. This methodology is based on SysML and DSPNs. We use SysML for describing the requirements, the behaviors and the parametric aspects of the WSN. Then, we establish the SysML behavioral specification, and thereafter, we construct the DSPN analytic model, which will be used to compute the elementary performance values of the designed WSN. In this step, TimeNET model-checker is used as an analysis and evaluation tool. The methodology might comprise a feedback, i.e. the results of the performance evaluation should be exploited by the SysML designers. These latter check if the requirements are satisfied regarding the computed performances values. After that, we improve the first part of this methodology has been proposed in our previous work. The amelioration consists of proposing mapping rules from the SysML activity diagram to its equivalent DSPN taking into account the performance parameters. Finally, we provide the second part of our approach which consists of analyzing the modeled system and evaluating its energy consumption.

The remainder of the paper is organized as follows. Section 2 presents the background with tools and languages used in our approach. The next section cites works related to methodologies dealing with the formalization of SysML diagrams, and then we compare them to our own. Section 4 explains the proposed methodology and the amelioration has been made for the mapping rules. After that, we run in the Sect. 5 these latter through an example of a hierarchical routing protocol in a WSN, and then, proceed to the functional properties verification of the analytical model and the evaluation of its performances, specifically, the energy consumption. At the end, we close the paper with some conclusions and possible improvements to the work.

2 Background

WSNs are so complex so that we can not design them without adequate high-level tools. This section presents the tools used during this work.

2.1 SysML

SysML is an OMG standard modeling language supported by leading organizations from the systems engineering industry, including the INternational Council On Systems

Engineering (INCOSE) [8]. SysML is an UML profile proposed to specify systems that include heterogeneous components. It reuses a subset of UML diagrams, extends others and defines new ones to provide specific systems engineering features. SysML diagrams cover four views of system modeling. The Requirement Diagram (RD) is used for better organizing requirements at different levels of abstraction and showing explicitly the various kinds of relationships between requirements and model elements. The PacKage Diagram (PKD), The Block Definition Diagram (BDD) and the Internal Block Diagram (IBD) are used to describe the structural aspects. The State Machine Diagram (SMD), the Sequence Diagram (SD) and the Activity Diagram (AD) are used to specify behavioral aspects. Finally, the Parametric Diagram (PD) is used to describe the mathematical relations between the system parameters.

One of the tools used to elaborate the SysML diagrams is Eclipse Modeling Framework (EMF). This project is a modeling framework and code generation facility for building tools and other applications based on a structured data model. From a model specification described in XMI, EMF provides tools and runtime support to produce a set of Java classes for the model, along with a set of adapter classes that enable viewing and command-based editing of the model, and a basic editor. Even if the models are described in XMI, they can be specified in UML/SysML documents. In order to create a working environment oriented to this specific area, toolkits can be added to EMF. This is the case of Papyrus which aims to provide an integrated environment for editing any kind of EMF model and particularly supporting UML and related modeling languages such as SysML.

2.2 Deterministic Stochastic Petri Nets

Deterministic and Stochastic Petri Nets (DSPNs) introduced by Ajmone Marsan and Chiola in [12] are a stochastic modeling formalism with graphical representation which include both exponentially distributed and deterministic delays. A DSPN is a 9-tuple $(P, T, I, O, V, W, \Pi, D, M_0)$, where:

- P is a finite set of places. $P = \{p_1, ..., p_n\}$;
- T is a finite set of transitions, disjoint from P, partitioned into three disjoint sets, T^I, T^E, and T^D, of immediate, exponential, and deterministic transitions respectively. $T = \{t_1, ..., t_m\}$;
- I is a set of the input arcs. $I \subseteq (P \times T)$;
- O is a set of the output arcs. $O \subseteq (T \times P)$;
- V is a set of the inhibitor arcs. $V \subseteq (P \times T)$, where $V \cap I = \emptyset$;
- W defines the weights of all arcs;
- Π is the priority function assigning a priority to each transition. $\Pi : T \rightarrow N^+$, N^+ the set of the positive natural numbers;
- D defines the firing times. $D : T \rightarrow \{0\} \cup R^+ \cup \Omega$, where R^+ is the set of positive real numbers and $\Omega = \{\lambda_1, ..., \lambda_l\}$ is the set of random variables with a given distribution;
- M_0 is the initial marking.

DSPNs have the same graphical notation of places and arcs in traditional PN. However, immediate transitions drawn as thin bars fire without delay, exponential transitions drawn as empty bars fire after an exponentially distributed delay, whereas deterministic ones drawn as black bars fire after a constant delay.

TimeNET (Timed Petri Net Evaluation Tool) [17] is a software package and an interactive toolkit for the modeling and evaluation of PNs in which the firing times of the transitions may be immediate, deterministic or more exponentially distributed. It has been developed at the Real-Time Systems and Robotics group of Technische University at Berlin, Germany. The project has been motivated by the need for powerful software for the efficient evaluation of Timed Petri Nets with arbitrary firing delays.

3 Related Work

A methodology is defined in [6] as a collection of related processes, methods, and tools. A MBSE methodology can be characterized as the collection of related processes, methods, and tools used to support the discipline of systems engineering in a model-based context. The author has provided a brief overview of MBSE methodologies. In this section, we relate existing research to our methodology, especially those which deal with the formalization of SysML.

In [16], the authors conduct a systematic mapping study to analyze SysML publications from 2005 to 2017. It has been found that this language is mostly used in the design or validation phase. In addition, there are approaches focusing on translation like transformation to PNs, Modelica, SystemC or Matlab/Simulink to build frameworks for the verification and validation of systems design. SysML is also used in combination with OCL, LTL or MARTE to support the implementation of an executable architecture that provides a feasible systems engineering solution. Furthermore, most of the publications deal with SysML profiles for facilitating the verification of functional and/or non-functional requirements and improving the application of SysML to complex systems.

A great number of methodologies deals with requirements that can be abstract when describing system objectives and can be more concrete when they relate to specific behaviors in the system or technical choices relating to its components. Specifying the structure and the behavior of a system is to describe it by conceptual models. The validation of the systems from the first design phases is necessary to assure the correction of the abstract models. In this context, works were proposed, in particular the method AVATAR [13] which is one surround including an equipped and adapted method to the real-time and distributed systems, and assisted by the tool Ttool. The language AVATAR is a profile of SysML. It extends SysML by proposing the language TEPE [11] for the expression of the properties. This methodology concerns only the verification of the properties by model checking. The traceability of the requirements and the validation of the not functional requirements are not taken into account in this environment. The approach OMEGA2 [1] includes a feasible profile UML/SysML dedicated for the specification and formal validation of Real-time systems. The model OMEGA2 uses the tool IFx or the simulation for properties verification.

All of these studies have in common that they do not consider PNs as a target formalism. Several initiatives have emerged such as [7] and [9]. These papers outline methodologies for modeling embedded systems. They transform SysML AD models into an equivalent PN, so that the analysis capability of PN can be applied. An other attractive work focusing on the verification of complex systems is that presented in [15]. The authors define a complete process to formalize and verify SysML functional requirements related to ADs. At first, they define a new language called AcTRL for the formalization of functional requirements at SysML level. Then, The verification is enabled by formalizing SysML activities with Hierarchical Coloured Petri Nets (HCPNs) and by automatically translating SysML requirements expressed on AcTRL into temporal logic.

According to the study that was done in [16], the non-functional properties did not have a big place compared to the functional ones. Among the works that propose methodologies dealing with quantitative properties, we cite [2] in which the authors associate SysML behavioral diagrams with the MARTE profile for the describing of real-time embedded systems behaviors, and then, transforming these extended diagrams into timed automata where time properties are expressed in temporal logic.

Among the formalisms which are very efficient in performance evaluation and which are different from the profiles extended to SysML, there are stochastic formalisms such as Markov chains and Stochastic Petri Nets (SPNs). According to the systematic mapping study cited above, there are only few methodologies in which the formalization of SysML were achieved by stochastic models. In [10], the authors present a model-based verification framework that supports the quantitative and qualitative analysis of SysML activity diagrams. To this end, they propose an algorithm that maps SysML ADs into Markov decision processes expressed by the language PRISM.

As non-functional parameters in SysML have not been widely used in the context of WSN, as well as the DSPNs despite their great utility, our main objective is to propose a methodology that combines these two tools. This is the idea we came up in our previous work [4], and started to explore it in [3].

4 Methodology and Mapping Rules

In this section, we will briefly explain the improvement has been made on the proposed methodology (see Fig. 1).

In [3], we proposed a methodology of the specification and verification of non-functional requirements of a WSN, focusing on SysML AD as a behavioral diagram. In addition, we focused on the RD and the PD diagrams that can be called during verification and performance evaluation after the transformation from the AD to a graphical formal model is performed. We realized that it would be wise to not use two types of SysML diagrams, namely the AD and the PD. So, we thought to express the performance parameters on the AD actions instead of expressing them through constraints on the PD. As a result, we'll have less work to do without losing information.

The way to do this is to attach a duration constraint to an action in order to specify a non-zero execution duration for it. This is a part of the mapping rules shown in the Fig. 2. Transition firing times are specified as delays for all transition types. Firing rates of exponential transitions have to be transformed into a delay by taking their reciprocal value: $delay_i = \frac{1}{\lambda_i}$. Also, we create a literal real for each action in order to specify the energy consumption of each sensor node task. The target element *Def* represents a

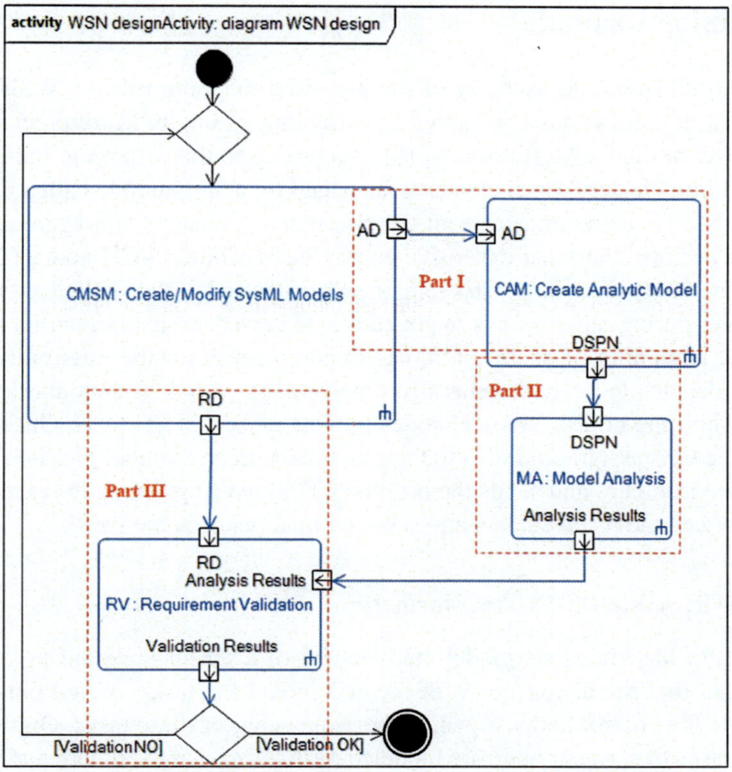

Fig. 1. The SysML model of the proposed approach (improved version of [3]).

Definition Measure, which will be used to compute the energy consumed by exponential and deterministic transitions in the resulting model (see Sect. 5.3).

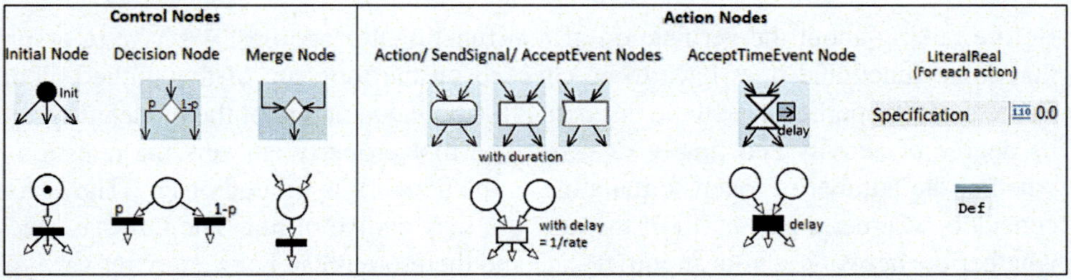

Fig. 2. Mapping rules from SysML-AD to DSPN.

Once the transformation is carried out, we proceed to the next step which allows us to analyze the modeled system and evaluate its performances. This is the subject of the rest of the paper which we'll elaborate through a running example.

5 Running Example

In order to illustrate the usability of the proposed mapping rules, a WSN based on a hierarchical topology was considered as a running example. More precisely, we take the example of the LEACH protocol [5]. According to this latter, the sensor nodes are organized into clusters. Each cluster is managed by a single node called Cluster Head (CH). Only CHs communicate with the BS, manage clusters and aggregate data. For that, they perform the most expensive energy tasks while no-CH nodes (or members) are dedicated only to sensing. CHs remain so for a period of time called round, then they switch roles during other rounds to get equitable power dissipation within the network.

At the beginning of each round, each node determines the possibility of being a CH. If it decides to be with generally a percentage of 5%, it announces its decision to all neighboring nodes. Non-CH nodes join the closest elected CH. Once the clusters formed, the CHs assign time slots to their members. Each member picks up information from its environment and sends them to the CH. Used as gateway to reach the BS, the CHs aggregate the received data and send the final result to the BS.

5.1 SysML AD to DSPN Transformation

At the left of the Fig. 3, we model the behavior of a sensor node during one round. It can be seen that the action nodes of the AD model the tasks carried out by a sensor node while the control nodes organize the sequencing of these tasks. On the one hand, the local tasks of a sensor node are modeled by Actions. On the other hand, the sending and receiving operations are modeled by SendSignalAction and AcceptEventAction nodes respectively. Concerning the tasks which consume a fixed period of time, they are modeled by AcceptTimeEventAction node. The DSPN model associated to the sensor node AD is given at the right.

5.2 Preliminary Verification

Before talking about the verification of non-functional properties, we have to assume that some functional ones have to be checked on the long run of the studied model. Three basic properties have to be discussed: the bounded nature of the modeled system, its degree of activity and finally its reset. The first property answers the question of whether the number of tokens circulating in the network is limited or not. The second considers whether a part or all of the network may evolve or not. The last one checks whether the network admits an initial state and therefore it can be reset. After verifying these properties, TimeNET displays analysis and performance results.

However, the initial node in the AD shall not have any incoming ActivityEdges. This implies that in the resulting DSPN, we cannot have an initial node with an input arc. In this case, we will have a DSPN which has a place that cannot be accessed after a certain time, which does not preserve a basic functionality called liveliness, so the second base property is not checked. However, this should not be a gap because we notice that there are unnecessary immediate transitions which will not influence the functioning of the DSPN model. They are useless since they do not express probabilities (in the case of parallelism) and they are not consuming either in terms of time or in terms of energy.

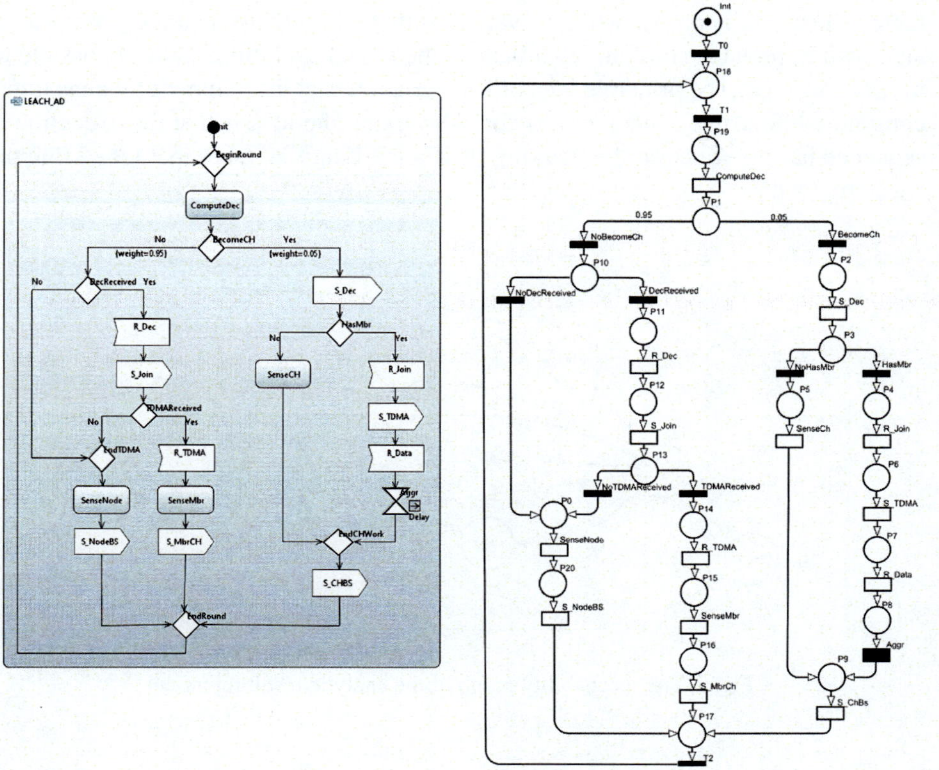

Fig. 3. Sensor node AD based on LEACH protocol and its associated DSPN [3].

We must therefore proceed to a reduction of the graph without losing the semantics of the resulting model. In this latter, T_0, T_1 and T_2 have to be omitted and therefore the source places too. The target places inherit the tokens and incoming arcs (if they exist) of the source places. Once the reduction is done, we lunch the structural and behavioral analysis. The results are given below.

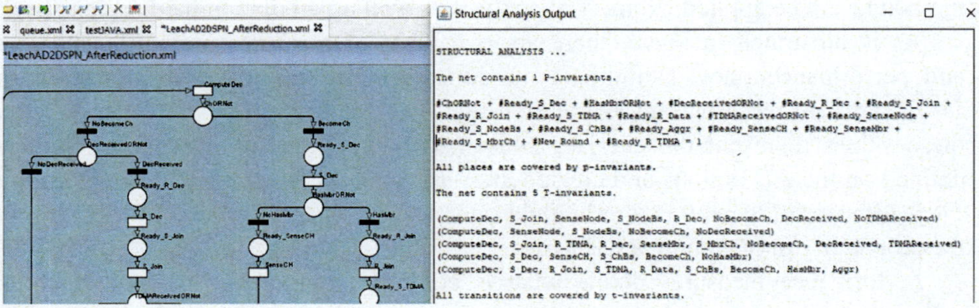

Fig. 4. The structural analysis after reduction: P-Invariants and T-Invariants.

Figure 4 shows the invariants generated through the resulting DSPN. This latter contains 1 P-Invariant and 5 T-Invariants, and, all places and transitions are covered by them. The essential property of an invariant is that the weighted count of tokens associated with it, remains constant regardless of the model evolution, and if a possible loop may be detected. The principle idea of P-Invariant is that the number of tokens remains constant while firing transitions. As for T-Invariant, the idea is that the execution of a sequence has no effect on the marking. It is a marking that takes us to the same point we started with.

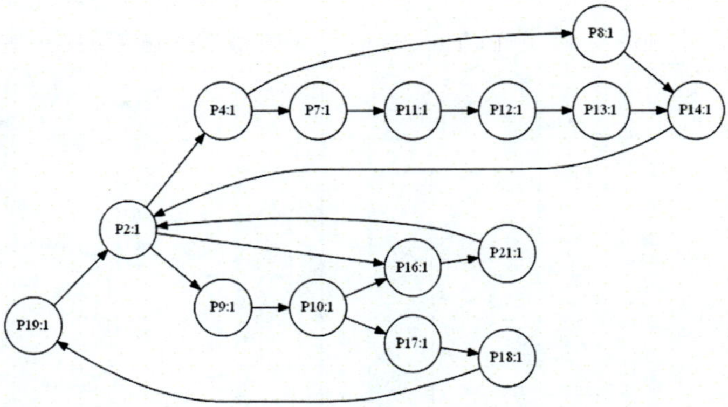

Fig. 5. The reachability graph of the analytic resulting model.

Figure 5 shows the reachability graph of the resulting DSPN. Once it is built, the behavioral properties can be evaluated. We notice that there is only one strongly connected component, so the previously mentioned properties are checked and the model admits a stationary state. Now, we can proceed to the verification of non-functional properties, the energy consumption in our case.

5.3 Energy Consumption Evaluation

In order to derive quantitative measures from a DSPN model, a performance evaluation method has to be applied. Numerical analysis as well as simulation methods exist.

As is illustrated in Fig. 6, there are two types of measures: definition measures and performance ones. Definitions measures can be created by using the button Def, and they represent any expression that may be as an input value. In our case, these values represent the energies consumed by each sensor node task. They are defined on the AD actions and created after applying the mapping rule. For example: $Def\_ComputeDec = 0.002$ milli joules indicates the energy necessary for calculating the probability of becoming a CH node.

Performance measures define what is computed during an analysis. A typical value would be the average number of transitions firings. For example, the average number of $ComputeDec$ firings defines the average of the energy consumption of this transition. We note: $(\#ComputeDec)$. So, the real energy consumption for calculating the probability of becoming a CH node during one round is equal to:

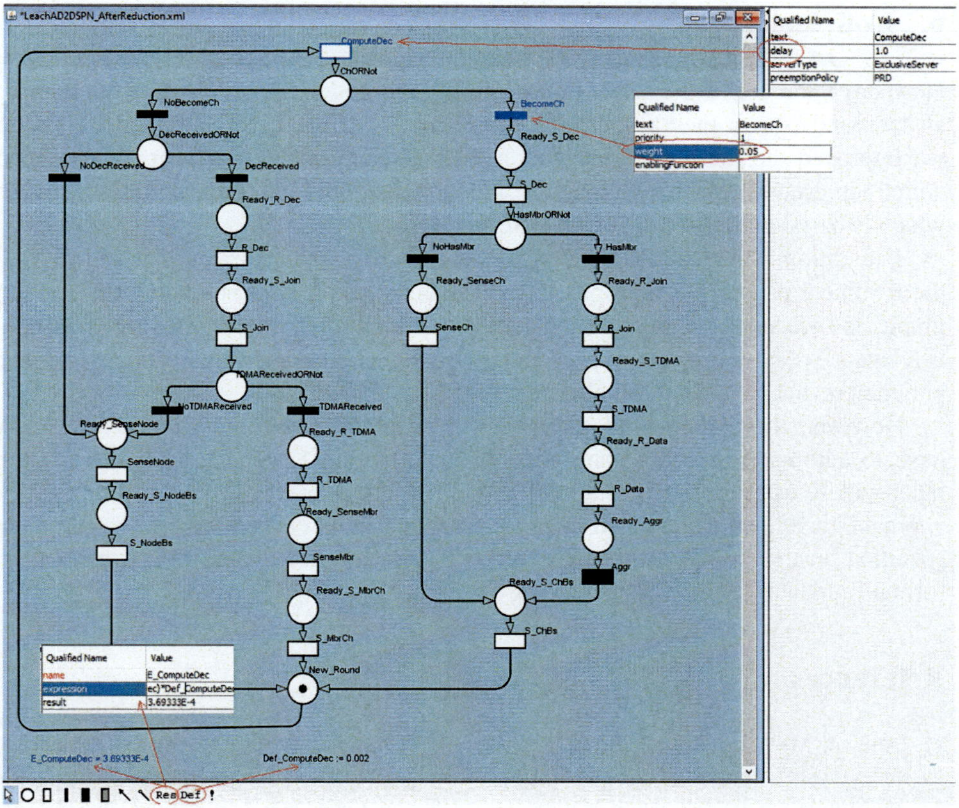

Fig. 6. Numerical analysis of the analytic resulting model.

$(\#ComputeDec) * Def\_ComputeDec$. By running the analysis at steady state, we get $3.69333E{-}4$.

By computing the average number of transitions firings during the steady-state simulation time, we can estimate the energy consumption of each node by multiplying each average by the consumed energy per the associated transition. Thus, the total energy consumed by a CH node or a member one is equal to the sum of the consumed energies of the transitions associated to each one.

According to this procedure, we can evaluate the lifetime of a sensor node, by setting the starting energy, and the execution times of each action. These time and energy parameters must be adjusted in order to meet the needs required by the application. This will be the subject of a future work where we will elaborate the third part of the proposed approach which allows us to check whether the results obtained following the analysis of the network and the evaluation of its performance do indeed meet the requirements specified in departure.

6 Conclusion

Nowadays, the use of the semi-formal and formal models, in order to design complex systems and express their properties, becomes a very active research topic. In this paper, we improved the first part and elaborated the second one of our proposed approach which represents a specification and verification methodology for designing WSNs and which is based on both SysML and DSPN models.

In addition, we got a model having unnecessary immediate transitions and in which the liveliness property is not done. We overcame these gaps by reducing the resulting graph, and we subsequently verified some basic functional properties. Once the model is relevant, we have proposed a procedure to follow in order to evaluate the performance parameters that have been established on SysML AD.

However, the mapping rules have not been performed in an automatic way. We have to automate them by using Atlas Transformation Language (ATL), one of the most widely used toolkits in the MBSE field. Moreover, we still have to finalize our approach by adjusting the performance parameters so that they verify the requirements specified on the SysML RD. Important work remains to be done to provide a better formal framework for WSNs technology.

References

1. Ahmad, M., Dragomir, I., Bruel, J., Ober, I., Belloir, N.: Early analysis of ambient systems SysML properties using Omega2-IFx (regular paper). In: International Conference on Simulation and Modeling Methodologies, Technologies and Applications, pp. 147–154. SciTePress, Septembre 2013. http://www.scitepress.org/
2. Baouya, A., Bennouar, D., Mohamed, O.A., Ouchani, S.: A probabilistic and timed verification approach of SysML state machine diagram. In: 2015 12th International Symposium on Programming and Systems (ISPS), pp. 1–9, April 2015. https://doi.org/10.1109/ISPS.2015.7245001
3. Berrachedi, A., Ioualalen, M., Hammad, A.: Towards the formal modeling methodology of WSN through the transformation of SysML into DSPNs. In: Proceedings of the 11th International Conference on Simulation and Modeling Methodologies, Technologies and Applications - SIMULTECH, pp. 83–91. INSTICC, SciTePress (2021). https://doi.org/10.5220/0010549200830091
4. Berrachedi, A., Rahim, M., Ioualalen, M., Hammad, A.: Validation of a SysML based design for wireless sensor networks. In: AIP Conference Proceedings, vol. 1863, no. 1, p. 330002 (2017)
5. Chandrakasan, A., Balakrishnan, H., Heinzelman, W.R.: Energy-efficient communication protocol for wireless microsensor networks. In: Proceedings of the 33rd Hawaii International Conference on System Sciences, vol. 2, pp. 1–10 (2000)
6. Estefan, J.: Survey of model-based systems engineering (MBSE) methodologies, Rev. B. INCOSE MBSE Focus Group **25**, 1–70 (2008)
7. Foures, D., Albert, V., Pascal, J., Nketsa, A.: Automation of SysML activity diagram simulation with model-driven engineering approach. In: Proceedings of the 2012 Symposium on Theory of Modeling and Simulation - DEVS Integrative M&S Symposium, pp. 1–6. No. 11 in TMS/DEVS 2012, Society for Computer Simulation International, San Diego, CA, USA (2012)

8. Friedenthal, S., Moore, A., Steiner, R.: OMG systems modeling language (OMG SysML^{TM}) tutorial. In: INCOSE International Symposium, vol. 18, pp. 1731–1862 (2008). https://doi.org/10.1002/j.2334-5837.2008.tb00914.x
9. Huang, E., McGinnis, L.F., Mitchell, S.W.: Verifying SysML activity diagrams using formal transformation to petri nets. J. Int. Council Syst. Eng. **23**(1), 118–135 (2020). https://doi.org/10.1002/sys.21524
10. Jarraya, Y., Debbabi, M.: Quantitative and qualitative analysis of SysML activity diagrams. Int. J. Softw. Tools Technol. Transf. **16**, 399–419 (2014). https://doi.org/10.1007/s10009-014-0305-6
11. Knorreck, D., Apvrille, L., de Saqui-Sannes, P.: TEPE: a SysML language for time-constrained property modeling and formal verification. ACM SIGSOFT Softw. Eng. Notes **36**(1), 1–8 (2011). https://doi.org/10.1145/1921532.1921556
12. Marsan, M.A., Chiola, G.: On Petri nets with deterministic and exponentially distributed firing times. In: Rozenberg, G. (ed.) APN 1986. LNCS, vol. 266, pp. 132–145. Springer, Heidelberg (1987). https://doi.org/10.1007/3-540-18086-9_23
13. Pedroza, G., Apvrille, L., Knorreck, D.: Avatar: a SysML environment for the formal verification of safety and security properties. In: Proceedings of the 2011 11th Annual International Conference on New Technologies of Distributed Systems, NOTERE 2011, pp. 1–10, May 2011. https://doi.org/10.1109/NOTERE.2011.5957992
14. Peterson, J.L.: Petri nets. J. ACM Comput. Surv. **9**, 223–252 (1977)
15. Rahim, M., Hammad, A., Ioualalen, M.: A methodology for verifying SysML requirements using activity diagrams. Innovations Syst. Softw. Eng. **13**(2), 1–14 (2017). https://doi.org/10.1007/s11334-016-0281-y
16. Wolny, S., Mazak, A., Carpella, C., Geist, V., Wimmer, M.: Thirteen years of SysML: a systematic mapping study. Softw. Syst. Model. **19**, 111–169 (2020). https://doi.org/10.1007/s10270-019-00735-y
17. Zimmermann, A.: Modeling and evaluation of stochastic petri nets with TimeNET4.1. In: 6th International Conference on Performance Evaluation Methodologies and Tools (VALUETOOLS), pp. 1–10 (2012)

Growing Bioinspired Synthetic Landscape Ecologies and the Adequacy of Object Oriented Programming

Jean Le Fur[1(✉)], Pape Adama Mboup[2], and Moussa Sall[3]

[1] Institut de Recherche pour le Développement (IRD), Centre de Biologie pour la Gestion des Populations (CBGP), CS 30016, 34988 Montferrier-sur-Lez, France
jean.lefur@ird.fr

[2] Lab. Informatique, Faculté des Sciences et Techniques, Univ. Cheikh Anta Diop (UCAD), Dakar, Senegal
papeadama.mboup@ucad.edu.sn

[3] Dépt. Informatique, Univ. G. Berger/Saint-Louis Sénégal and Lab. IRD-BIOPASS, Campus Bel-Air, Dakar, Senegal
sall.moussa@sanarsoft.com

Abstract. In this study we develop, using basic object-oriented paradigms, and in collaboration with biologists, a comprehensive model of landscapes and ecosystems dynamics based on bioinspiration principles. Faced with the issue of taking into consideration a variety of elements, processes, interactions, contexts, and scales simultaneously effective, we iteratively develop this model using successive aggregation of new components based on specific case studies. These were then generalized and consolidated to form a coherent platform. To address robustness, the model was continually reworked in search of the closest resemblance to the concrete workings of Nature.

We have arrived at a general architecture built from the bottom up that is both generic and as parsimonious as possible. The model emerging from this compilation is a shared class tree with three primary categories of variability: *(i)* cognitive living agents, *(ii)* containers of agents that can be nested at various functional scales, and *(iii)* particular genomes that instantiate attributes for each type of agent. The results of the iterative strategy to modeling synthetic ecology are discussed, as well as the suitability of object-oriented paradigms (composition, aggregation, inheritance, generalization…) for achieving the goal of bioinspired modeling.

Keywords: Bio-inspiration · Ecoinformatics · Natural Computation · Synthetic Ecology · Robustness · Object-oriented programming · Agent-based model · Rodent

Parts of this work have been presented on behalf of the Simultech (Internat. Conf. Simul. and Model. Method., Technol. and Applic.) conferences (see [1, 2])

G. Wagner et al. (Eds.): SIMULTECH 2021, LNNS 601, pp. 118–137, 2023.
https://doi.org/10.1007/978-3-031-23149-0_7

1 Introduction

In the field of bio-socio-ecological modelling, one's aim is to question natural processes using computer or formal engineering that are simplified abstraction of the processes under investigation. For example, differential equation sets can be used to model epidemic process such as in the classical susceptible-infected-recovered (SIR) model [*e.g.*, 3]. In some cases, bioinspired algorithms are used to formalize natural contexts. The so-called natural computing [4] or ecoinformatics [5] are approaches developed in that direction. These make use for example of cellular automata, genetic programming, swarm intelligence, artificial immune systems. However, and in this case also, these approaches may be not related to the biological processes explored. For instance, neural networks might be utilized to characterize a hierarchically structured ecosystem [6] or a bioinspired cell model (P-system) used to model a trophic web [7]. In any of these directions there is generally an assumed gap between the specific problem to solve and the model formalism (strategic models, [8, 9]). In the case of bio-ecological modeling, this gap may limit any model's application to the specific use case to which it is fitted to or trained on [10]. This eliminates the possibility of an explicative description that may be applied to other situations or yield useful forecasts in changing environments. The model's robustness may be called into question at this point.

Tactical models [8, 9] on the other hand are concerned with prediction and robustness. Their goal is to capture the processes that regulate real-world dynamics using virtual replicas that work in the same way [10]. Such approach can be found in synthetic biology, a discipline dominated by research at the microbiological level [11, 12]. At the broader scale of individuals, the so-called Functional-Structural Plant Modelling or FSPM [13] integrates biology and physiology to mimic plant growth [14]. At an even larger scale the emerging field of synthetic ecology proposes bioinspired solutions in a large range of academic domains, including city planning and sedimentation modelling [15]. In Ecology or natural landscape issues, DeAngelis and Mooij [16] also proposed a so-called "mechanistically rich" approach in which a maximum of the entangled components causing the natural dynamics investigated may be included.

The development of such promising techniques necessitates a strong emphasis on robustness. Indeed, it entails accounting for a wide range of entities with very varied distinct behaviors however at the cost of tough computation to manage this complexity and maintain coherence of the resulting model. One way to tackle coherence of such models is bioinspiration. Indeed, Nature appears to be the sole reliable 'model' in this regard, with universal drives, processes and components that work together and resist change. Bioinspired computation could be then one way to develop strong ecological or landscape models: as Nature is the only system that works reliably, the more similar processes are used for computation, the more robust analogies and hence simulations of the worlds studied can be produced.

We propose in this study to employ computer paradigms in partnership with bio-ecologists to replicate the known behavior of an application domain in Ecology, with the goal of producing *(i)* the most robust possible model and *(ii)* a comprehensive tool for biologists and modeling interaction.

Following, *e.g.*, [17], Object-Oriented Programming was selected as the privileged approach to develop a model as closed as possible to what is known of the functioning of Nature.

We first outline our strategy for creating a synthetic ecology that could be as bioinspired as possible. The structure of the resulting architecture is then described by showing the computational paradigms that have been found the most relevant for emulating the known functioning of ecosystems. From this perspective, the benefits and drawbacks of object-oriented programming are examined.

2 Material and Method

2.1 Model Purpose

The long-term goal of this project is to develop a general model of wild rodent ecology. The primary goal is to create dynamic simulations in which knowledge from multidisciplinary topics such as Bio-Eco-Sociology, Geography, and others may be articulated. The projected outcome would be a useful tool that experts might use to compare simulations to the majority of indicators and knowledge they are familiar with. Two specifications found the approach:

A) The model must be as comprehensive as possible to take into account the various sources of fluctuations at stake in the ecosystems or landscapes under investigation.
B) The model must hence be robust to multiple contexts. Robustness would permit also to instantiate a model that could be queried from multiple multidisciplinary points of view (*i.e.*, provide the variety of indicators with which bio-ecologists are used to).

We have kept as a premise that Nature is the only example of architecture and function that satisfies robustness and exhaustiveness in order to achieve (or rather, move in the direction of) this goal. As a result, the greater the fidelity to the Natural 'model,' the higher the assurance of robustness. As a result, we attempted to replicate existing bio-ecological knowledge using the most bioinspired modeling techniques feasible.

2.2 Context: Research in Rodents' Bio-Ecology and Epidemiology

This long-term experiment was carried out in a biology and ecology laboratory in collaboration with a group of scientists specialized in wild rodents, mostly in West Africa [18]. These populations are investigated as pests (crops or houses) or as pathogen reservoirs in a variety of epidemics.

Individual interviews with biologists affiliated with this laboratory were used to conduct a preliminary investigation. The interviews were conducted in order to determine the characteristics of the knowledge domain (rodents) that would be represented. The information of five interviews was then reified and combined to determine the many components, processes, variables, and indicators used in rodent ecology in a variety of settings.

The result of this work [19, 20] highlighted a rich field of knowledge with a great diversity of:

- constituents with, particularly, numerous species interacting,
- processes and scales, especially spatial ones, to be considered,
- approaches such as for example Agroecology, Biogeography, community Ecology, Eco-Immunology, landscape or population Genetics, Phylogenetics, Physiology,
- indicators used and the way they are obtained to observe the ecosystems concerned.

2.3 Approach: Growing the Model as a Heuristic

Given this complication, it proved impossible to create, from the ground up, a simulator capable of reproducing most components of a synthetic environment. As a heuristic, we therefore decided to 'grow' the model by formalizing concrete case studies one by one. In order to account for field knowledge, each new case study introduces new features and restrictions that must be resolved. It is therefore feasible to gradually evolve model structures, functions, and model parameters as well as advance in the construction of a synthetic model through the iterative integration of numerous scenarios. This method was also used to assess the model's robustness as it was being created (see below).

Each case study was chosen in the first place according to interest of biologists for a model of this type in their field. Beyond that, in the modelling project, we tried to select the most different studies possible, however still in the field of bio-ecology of wild rodents, so as to test the robustness of the model to various contexts: each particular situation or use case, once integrated, constituted in turn a constraint for the preceding ones that had to be accounted within the general code.

The model was created in Java using the agent-based Repast-Symphony platform [21] as well as Eclipse tools for refactoring operations. Out of the 11 use cases generated throughout the project, Table 1 and the text below describe six separate use cases.

Table 1. Main characteristics of the most distinctive case studies successively formalized to elaborate the model [2].

Use case		Agents	Objects Environments	Environmental forcing	cell/ domain size
1	 Common vole (*Microtus arvalis*) in a Changing Agricultural Landscape	Rodent, Burrow	crop, road, dwellings	Landscape structure, crop transition process	$56 m^2$/ $140 m^2$
2	 Hybridization of *Mastomys natalensis* & *erythroleucus* (cage experiment)	Rodent, 2 species	gene, chromosome, genome, cage	Simulated experimental protocol	$0.01 m^2$/ $21 m^2$
3	 CMR experiment on a *Mastomys* population in Bandia (Senegal) reserve	Rodent, Burrow	trap, savannah, road	Simulated trapping protocol	$1.73 m^2$ / 9 ha
4	 Black rat (*Rattus rattus*) colonization of Senegal over the past century	Rodent, Human, City	truck, train, boat, road, track, river, rail, region, city	Evolving trading areas, road taring, human population	$7.5 km^2$ / $315{,}000 km^2$
5	 Invasion of Senegal Sahel by nigerian gerbil (*Gerbillus nigeriae*)	Rodent, Owl, Burrow, Nest	crop, wild grass, tree, bush, landcover	Spatio-temporal rainfall	$1 km^2$ / 25 to $25{,}000 km^2$
6	 Zoonose propagation and transmission by house mouse (*Mus musculus*)	Rodent, Human, Tick, Bacteria, Cat, Nest	food source, room, yard, corridor, road, market, shop...	Landscape composition	$1 m^2$ / 5 ha

1. A preliminary investigation in a dynamic agricultural setting allowed the concept of nested spaces to be established, in which similar space cells can be grouped to form a higher level entity (a field, a road, etc.).
2. The second case study focused on simulating the genetic process of reproduction, which results in the creation of one being from two. The model depicts the genetic process, which includes gene, chromosome…, meiosis, chromosomal fusion and crossing, aberration rejection, and gene transfer. The model was validated by replicating the findings of an animal facility experiment [22] and it was then applied to all case studies.
3. These early stages of the model allowed us to carry out and enhance a rat trapping experiment in which we could formalize 'reality-inspired' rodent traps scientifically placed over the simulated area. Bio-inspiration allowed us to duplicate the trapping methodology and then produce ecological and genetic outputs we needed to reproduce and compare the specialized estimators that environmentalists are familiar with [23].
4. The following studies were interested in applying and adapting the model to much larger scales, allowing them to represent the historical evolution of commercial transport on the scale of a century and a country, as well as investigate the likelihood of rodents (black rats) boarding commercial vehicles (boats, trains, trucks) to gradually colonize the entire country. The general model was refined here by taking into account human agents as well as various adaptations such as graph algorithms to represent travels on roads or rail networks [24, 25].
5. Another implementation has been created to account for rodent behavior in an African Sahelian savannah. Using biologists' expertise, the hierarchy of behaviors has been rearranged. We took into account predation (owls that devour gerbils) and circadian activity in this study (cycles of animal activity different between night, day, dawn and dusk).
6. Finally, a recent study looked into the spread of zoonoses (epidemics) linked to urban mice and their proximity to humans (commensal animals). Biting ticks and pathogenic germs, cats, and human activity have all been integrated at high resolution to construct the epidemic chain. This problem has sparked research in specific phenomena, such as mice's ability to move around by walking along residential walls [26].

2.4 Use of Computer Paradigms

The model has been updated and changed in response to new features and constraints introduced with each new case study. This work mostly consisted of generalization, which entails factoring similar qualities or procedures across multiple studies and moving them up to a higher parent class (owl nest and rodent burrow generalized into animal home for example).

The repeated use of refactoring has also been a crucial element in this reworking. The 'natural' meaning of variables and procedures was improved as we progressed in our understanding of the emergent architecture, to conform as closely as feasible to biological knowledge. As a result of refactoring, the growing model's consistency, readability, and genericity increased with time, as did the model's maintenance and backward compatibility with all documented use cases. Finally, with the support of biologists, each specified behavior in the model was questioned as potentially generic. This resulted in the class hierarchy being enriched by the encapsulation of natural functioning at the exact level where it occurs.

Encapsulation, delegation, polymorphism, interface, and other paradigms of the sort have been used to replicate by analogy the structures known to occur in Nature and develop a clean architecture, depending on the unique needs. Apart from generalization, three specific object-oriented programming paradigms have been focused on to extend the model during this process: composition, aggregation, and inheritance (see results section).

At the conclusion of this collection of case studies, 1,143 minor or large model architectural and function reconfigurations (version commits) were completed. The final source code contains 22,248 lines (including data input, data retrieval, and display output), with 9,066 lines (41%) dedicated to the business model that underpins the synthetic ecosystem itself. There are 98 classes, 149 properties, 181 relations, and 761 methods in these 9,066 lines. There are 168 (22%) calls to super and 175 (23%) overridden methods in the latter.

3 Results

The model allows several simulations to be represented, each addressing a different component of dynamics, for a number of rodent species, at different geographical and temporal dimensions, and in different simulation scenarios. For example, only aspects relating to rodent cross-breeding were considered in Fig. 1a, whereas, in the example shown in Fig. 1b, studies of the diffusion of black rats over a century necessitated the simulation of a rich historical and geographical environment, including all forms of commercial transportation in the country concerned.

3.1 The Emerging Architecture

The main conclusion of this study is the model's structure, which evolved gradually from the collection of case studies. This approach provides three necessary and sufficient realms of diversity that are both essential and sufficient: simulated tangible entities (Fig. 2), many forms of substrate in which objects and agents can be found (Fig. 3), and genomes connected with living creatures (Fig. 4). When merging consecutive case studies, classes and procedures were aggregated, abstracted, relocated, or refined for each of these three domains.

Active Agents. The first domain constitutes the principal tree (Fig. 2). It describes agents that can intervene in the model.

At the base of this tree, any element of any system portrayed in the model is regarded a Nearly Decomposable System (NDS) [27]. This notion is used to illustrate hierarchical systems organized into a hierarchy [28]. It is functional from the perspective of

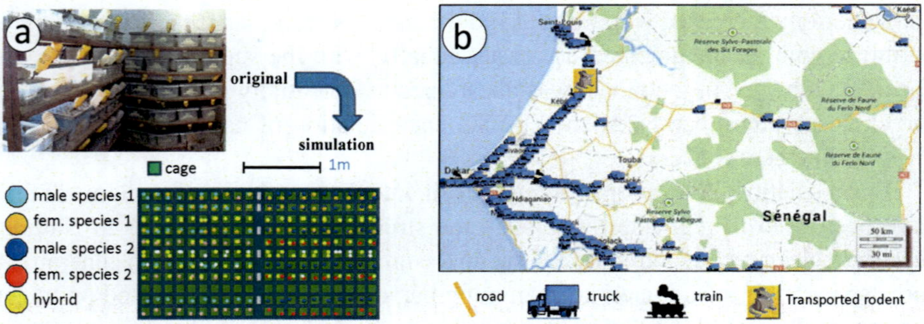

Fig. 1. Displays of simulations for two extreme case studies from the eight considered in this work (Table 1) [1]: *(a)* case study n°2: simulation of an animal room protocol for the crossing of rodents from sibling species in different configurations. The challenge here is to study the barriers to fertilization linked to chromosomal differences between the species. *(b)* case study n°4: simulation of the colonization of Senegal (West Africa) by the black rat, over a century. The biological, historical and geographical aspects relate to the likelihood of invasion by mean of carrying rodents in commercial vehicles (trucks, trains, boats).

bioinspired modeling in the sense that if a system can be defined as a developing entity resulting from the interaction of its components, it acquires uniqueness automatically. Once the creature's presence is exposed, it can be quickly identified (*i.e.*, named) and the beginning of its existence can be acknowledged. Finally, it must have a duration, which is codified by ageing, a method that can be recursively overloaded by the daughter classes. Indeed, agents belonging to the leaf classes accumulate 'skills' for 'getting older' all along the specialization chain. As a result, they develop sophisticated action skills or the ability to respond to their surroundings. These basic and minimalist principles outline the model's definition over time, allowing the dynamic component of the model to be encapsulated here.

The first specialization (A_VisibleAgent) concerns space and makes it possible to distinguish located objects within their context as well as instantiate the essential attributes associated with this characteristic (perception radius, birth place). With abstracted processes for localization, visible objects and agents implement the I_SituatedThing interface. The concept of environment takes into meaning in this paradigm as soon as an object is located. The faculty of perception is subsequently assigned to all agents. NDS becomes general beyond this class, affecting either the agents (organisms) or their containers (Fig. 3).

I_ReproducingThing is implemented in the third specialization (A_Organism), enabling basic reproduction functions (mate, give birth). Organisms have a genome as well, which gives unique features (see section Agents' biological properties below).

The model was gradually updated to include various agent attributes and behaviors from these three superclasses. Any function within a class was carefully matched to a biological reason for its qualifying within the hierarchy of life sciences. Animals (A_Animal) can move, burrowing rodents (C_RodentFossorial) dig burrows, colonial rodents (C_RodentFossorialColonial) interact socially, and so on. Agents follow a behavioral scheme that is a combination of PDE (perception-deliberation-execution) [*e.g.*, 29],

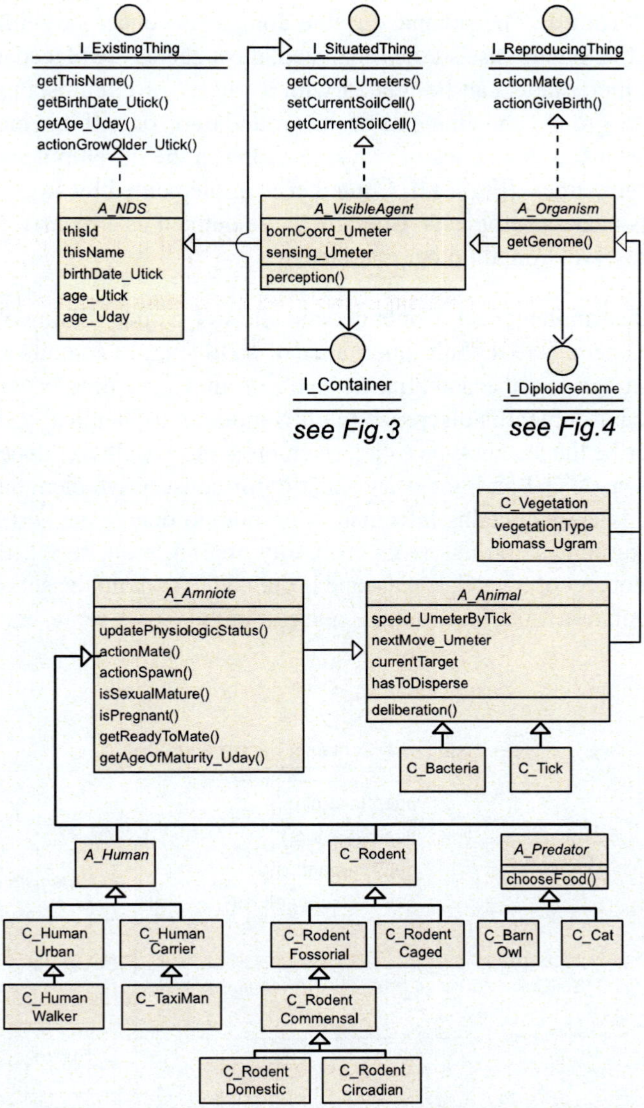

Fig. 2. UML-based class diagram describing agents within the model, as resulting from the case studies and engineered according to natural classifications. For the sake of clarity, only the relevant methods, properties or relationships are presented. Legend: A_: abstract class, I_: interface, C_: Class, _Uxxx: unit of the method or property, NDS: Nearly Decomposable System [27], see text.

and BDI (Belief-Desire-Intention) [*e.g.*, 30], and where deliberation is driven by a set of "desires" (foraging, reproducing, escape, dispersal, protection, suckling…).

The root classes were gradually discovered, and the methods and properties were polished by displacement or refactoring. Several rodent agent classes were added in stages between case studies, based on functionality criteria. The resulting classification (bottom of Fig. 2) was later found to correspond to distinct rodent social standings (commensal, fossorial, etc.).

Substratum/Ground. The second variable domain is concerned with the concept of space. Depending on the case studies analyzed, most agents are placed at a certain point in space, and their status can be related with a variety of circumstances. Objects that characterize the ground and elements of the ground upon or within which agents evolve have been rearranged in a series of ways, resulting in the development of the arbitrary concept of a 'container' (Fig. 3). I_Container is implemented by any structure that can contain agents or passive objects. Containers are defined as a recursive system, in that they may themselves contain other containers.

A_Container implements I_Container and allows I_SituatedThing objects to be contained in containers. Due to their inheritance of NDS (Fig. 2) containers are observable agents with an age and the potential to 'age', or alter over time according to specifications. The environment is discretized in this model into identically sized constituent cells that describe the terrain type (road, river, crop, etc.). Cell size depends on the level of detail chosen. A C_LandPlot is a collection of consecutive elementary cells of the same kind. This class contains information on how to demarcate and identify distinct zones inside simulated settings (*e.g.*, crop, city, region, road, river). Finally, for each case study, an object of type C_Landscape is defined to contain the set of visible objects within the simulation: this class contains one continuous space representing the topology

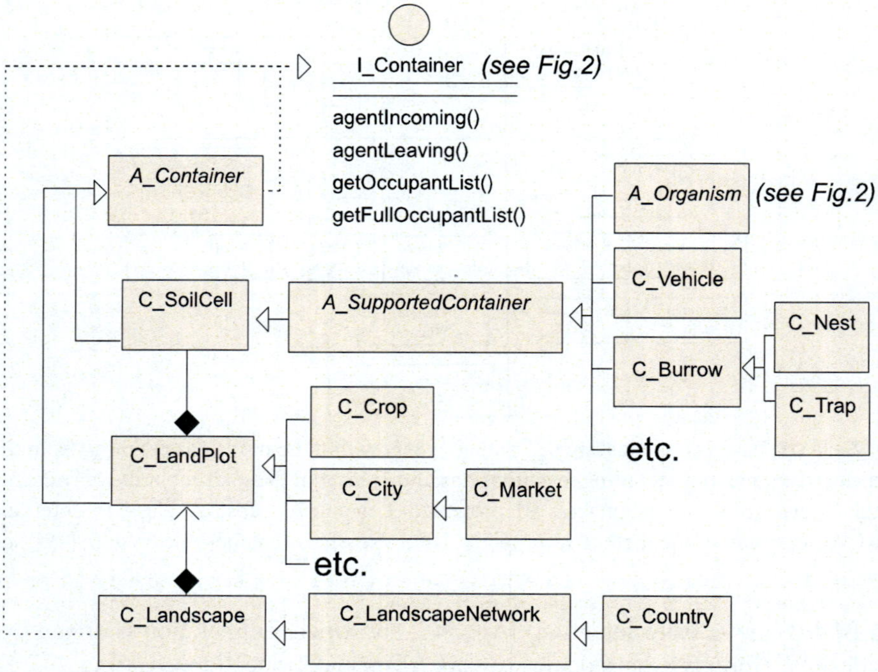

Fig. 3. UML-based class diagram of the types of ground organization, as determined from the aggregated case studies. Every instance in this hierarchy shares the I_Container features. Three successive levels of composition have been distinguished: landscape, land plot, soil cell. Each level of the composition is itself the root of a hierarchy that can be further extended (on the right) to describe various features of the worlds simulated. Legend identical to Fig. 2.

over which objects are located and agents move as well as a grid (matrix) of I_Container, the elements of which constitute a discretization of the environment and a topology used by the agents to perceive their neighborhood (Moore neighborhood).

As the model was improved, it became evident that nearly all relevant items in the represented systems (burrow, nest, vehicle, trap) could be assimilated as containers for agents, with a single abstract class (A_SupportedContainer) easing their integration into this scheme.

Agents' Biological Properties. The third hierarchy (Fig. 4) represents reproducing agents, which is a basic function of life. It is tied to an important feature of living creatures: the determination of the organism's traits by the genes it carries, which can be assigned values. Each agent in this class and its offspring has a genome that contains 'genes' that can be transcribed into attributes or parameters ('phenotypes' or 'life features' in Biology) that define how they function in their environment. This classification system organizes all aspects of living organism genetics: meiosis, segregation, fertilization, mutation, and recombination with the required mechanical elements (genes, alleles, and so on). The natural 'gene' - 'chromosome' - 'pair of chromosomes' - 'genome' composition is adapted from the work of [31] on locust genomes. In this sequence, any genome aggregates various pairs of 'chromosomes' that can be recombined and inherited, in part, during 'reproduction'.

Each agent (*viz.* Living organism) owns a genome (I_DiploidGenome) that corresponds to the species it belongs to. This branch uses the object paradigm's inheritance principle to heritably and cumulatively instantiate the properties (called 'traits') of living agents up to species-specific values. This part of the tree can also be considered an object-oriented knowledge base that exploit the analogy between the biological approach (Darwinian phylogeny) and object-oriented programming (inheritance). It allows for the valuation of movement speed, litter size, age at sexual maturity, and other genetically encoded characteristics for different species of agents. The mother classes are here also established using generalization: when a biological attribute matches to a natural classification, a class is generated. For example, amniotes, the branch of animals in which a fetus is created, have a property coding the length of gestation, while eukaryotes have a genome made up of pairs of chromosomes (illustration on Fig. 8).

3.2 Use of Object-Oriented Paradigms

This section focuses on programming paradigms that have been proven to be helpful in the development of a bioinspired synthetic ecology. Three programming paradigms were found to be particularly useful in developing and refining the code in this direction: composition, aggregation, and inheritance.

Composition. Composition was used in this model for two generic structures of ecosystems and landscapes.

The first is bio-inspired and straightforward to implement: genomes are known and described as composition and they were formalized in this way. Thus, during mating of any two reproducing agents, we obtain the following functional structure:

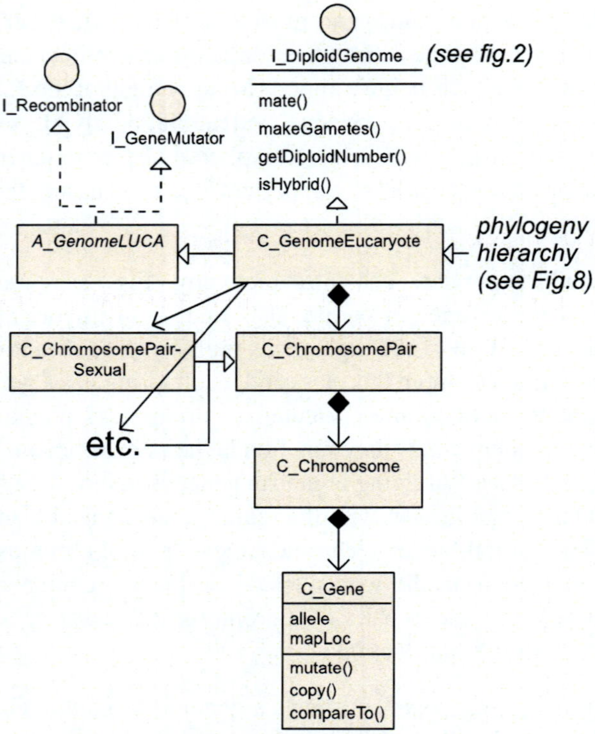

Fig. 4. UML-based class diagram of the genetic part of the model, as obtained from the case studies [1]. Left: Genetic structure allowing the transmission of the genetic heritage of a simulated agent to its 'progeny' following mating. The structure of the tree on the right-hand site resembles a knowledge base for instantiating, in the general scheme, any type of species (rodent, predator, man…, see Fig. 8) and its characteristics (life expectancy, age at maturity, perception range, speed,…). Legend identical to Fig. 2, LUCA: Last Universal Common Ancestor, the root of the species phylogeny [32].

genome ◆— diades (paired chromosomes) ◆— chromosomes ◆— genes.

The composition principle has made it here possible to computationally reproduce the functioning of a genome in a manner analogous to known reality.

The second use of composition concerns the formalization of space into successive entities such as:

cell —◆ region —◆ landscape

cell —◆ cage —◆ animal facility

However, this choice is here an arbitrary classification linked to observation and that does not correspond to a genuine architecture of Nature.

Aggregation and Recursion. In a bio-inspired perspective, the primary usage of aggregation was to account for a first fundamental organization of Nature: from the tiniest quantum atom to the entire Universe, nature is a nested system. Each level in this system emerges through interactions between components at the lower level: atom, molecule, cell, organ, organism, population… As a result, these several levels are entwined with one another. At the scale of an ecosystem or a landscape, each component is both a container for other components and a source of material in and of itself. All agents have gradually inherited from an abstract class (A_Container) that implements the same interface to reflect this architecture (Fig. 5).

interface Container
- methods returning booleans : *thingIncoming(thing)*, *thingLeaving* (thing)
- methods returning Collections : *getOccupantList ()*, *getFullOccupantList ()*
- optional methods : *setThisValue(value)*, *getThisValue ()*

Fig. 5. The minimal interface finally obtained for any object, included agent, sharing roles of container and content [2]. Methods underlined recursively run through the cascade of containers contained (see Fig. 7).

After the case studies integration was complete, the A_Container class became a foundation class, near to the business model's core, and all objects and agents inherit from it. As a result, a rat in a culture (in its diet), a prey in an owl's talons, a microbe in a tick, or an embryo within an animal might all be 'naturally' formalized (Fig. 6).

- Bacteria —◇ Tick —◇ Rodent —◇ Nest —◇ Wall —◇ Household —◇ Village
- Human carrier —◇ Truck —◇ Road —◇ Trade area —◇ Country
- Rodent —◇ Owl —◇ Nest —◇ Tree —◇ Bush —◇ Savannah
- Embryo —◇ Rodent —◇ Cage —◇ Laboratory

Fig. 6. Selected examples of embedded chains of containers currently observable in the model's simulations [2]. Each item is a Java class implementing the Container interface.

This implementation allowed us to use recursion to propagate global changes throughout the chain of containers and get in this way a parsimonious and robust management of the system (Fig. 7).

```java
/** Recursively set the value passed to self
    and contained containers field */
public void setAnyValue(double a) {
    this.myField = a;
    for (Container oneContainer : this.getContainerList())
            oneContainer.setAnyValue(a);
}
```

Fig. 7. Example (Java code) of recursion used for containers aggregation management [2]: propagation of a global change to all occupants within a Container.

Inheritance. The model resulting from the theory of evolution was structured using a second constitutive architecture of Nature [33]. Living organisms diversify over time by inheriting core features from their ancestors while developing unique functions or structures. As a result, organisms are classified according to a single phylogeny that is unique to the living world, and within which each individual of any species has genetically acquired all of its ancestors' unique characteristics. The object-oriented principle of inheritance is a phylogenetic paradigm. As a result, taxonomic hierarchy [34] was chosen as a reliable method for accounting for biodiversity as it occurs naturally. This method resulted in the creation of a genomic tree structure that was delegated to the modelled agents (see below). Each genome confers unique features on the agents who carry it.

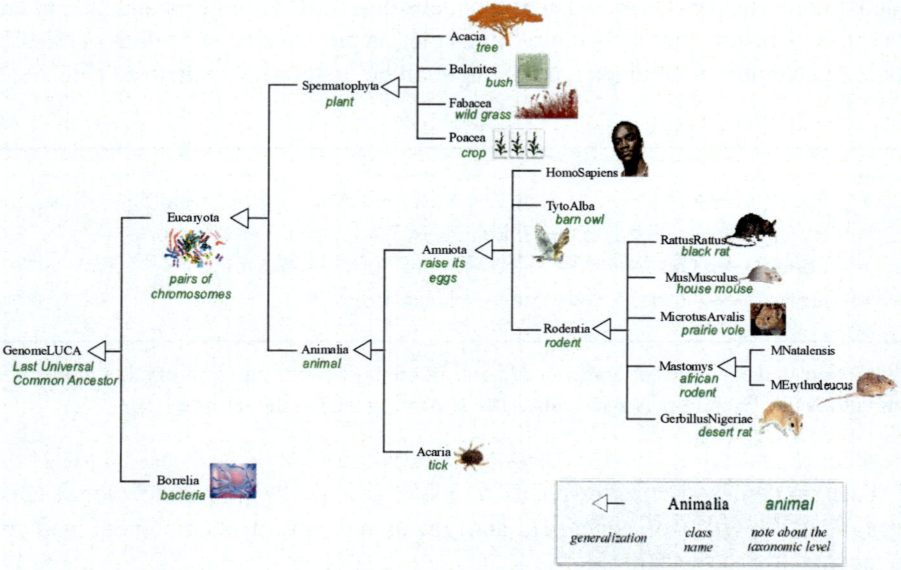

Fig. 8. The genome phylogeny model part that is arising from the sole compilation of case studies on wild rodents' ecology [2]. Genomes objects from the different classes are delegated to the corresponding type of agent in the functional hierarchy (see Fig. 9).

Each genome is thus equivalent to a knowledge base associated with every agent, allowing it to achieve its growth, reproduction, life cycle, and so on with the appropriate parameters (speed, litter size, weaning age, etc.) for its species. Without calling into doubt this design, it was possible to characterize plants, mammals, animals, birds, arthropods, and microbes (Fig. 8) during the course of the case studies. These qualities are passed down with each act of reproduction in a simulation, whether it's a mouse, a bird, or something else.

The inheritance paradigm is the basis of the functional model within which agents' behaviors and interactions are programmed, and it is linked to the previous genomic hierarchy as a founding principle in agent-based modeling. It was created in collaboration with biologists to differentiate biologically functional levels and assign their attributes based on current knowledge (Fig. 9). Each level was gradually distinguished and given features related to bio-ecological knowledge; for example, the class Organism manages the genome, the class Animal manages deliberation and movement, the class Amniote regulates reproduction, and so on. This arborescence grew over time as a result of use case modeling, model refactoring, and reorganization.

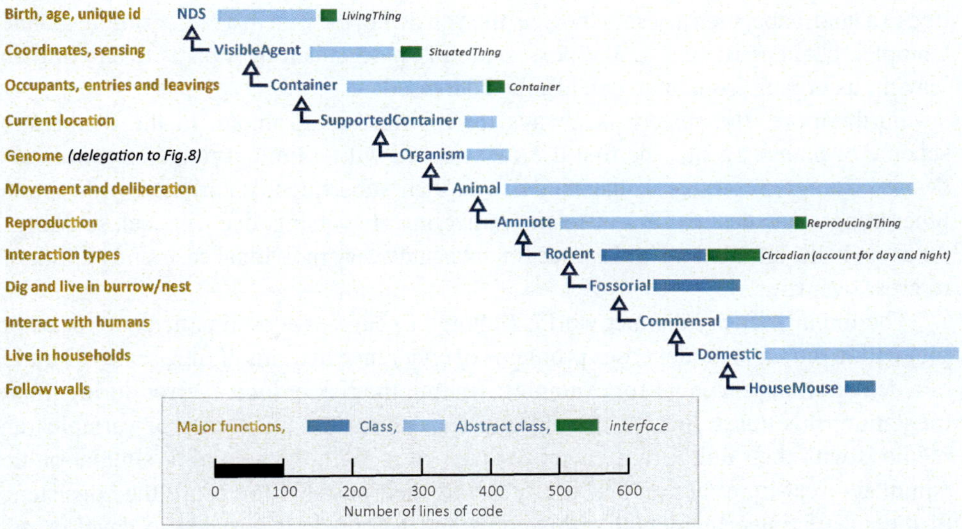

Fig. 9. Comprehensive functional hierarchy arising from 11 case studies [2]: Illustrative example of classes involved in a house mouse agent behavior (12 classes out of 98 in the business model). The number of lines of code (right) stands here as a proxy for the amount of behaviors and functions (left) dedicated to each level. A house mouse in this case owns the whole set of behaviors exposed. NDS: Nearly Decomposable System ([27], Fig. 2).

4 Discussion

4.1 Genericity, Robustness and Reusability

The model, which has been in development for more than ten years, can now reflect a laboratory experiment, historical processes at the country size, intracellular or ecological activities at the landscape, country, or city scales.

All studies remain active (a utility permits to switch from one case study simulation to another) and each enrichment of the platform's functionalities makes it possible to continuously improve the outcomes of each world formalized. This has led, for example, to the gradual addition of cellular processes, social behaviors, various types of movements or the sharing of indicators which can in many situations be reused from one case study to another within the global platform.

This technique achieves a robust modeling by attempting to systematically portray Nature as it is understood, and thanks to the generalization process, it is able to formalize and make interact in the same shared schema a large diversity of living organisms (humans, rodents, birds, plants, parasites, predators).

The architecture that results is both consistent and extensible. It allows you to set a tree in a landscape, then a nest in the tree, then an owl in the nest in a rather simple manner. Complex challenges, such as the possession of an organism by a parasite entering and leaving its organic container, can also be addressed.

Furthermore, the system is always improving. For example, in the deliberation scheme of mouse agents, the first use case started with a limited range of conceivable desires (forage, reproduce); the model was then subsequently enriched with, among other things, new desires such as resting, traveling, dispersing, flee, conceal, suckling... As a result, the simulations' relevance for each and every individual case study continues to grow over time.

The linking of very distinct worlds (laboratory cage or eco-climatic zone for example) within the same model poses problems of coherence in terms of time scales (duration of a deliberation procedure for example). To limit the risk of logical error during implementation, this constraint led to suffixing a large part of the names of variables and methods with their unit (gram, meter per day, etc.). With the help of a 'simulation tick /simulated real-time' conversion utility it was then possible to ensure the consistency of the calculations. Incidentally, this constraint has made it possible to develop, as a hoped-for but not expected result, a simulation environment with a configurable execution speed (continuous time scales from seconds to months) and hence multiply the points of view (coarse graining [35]) provided using the simulator. Such functionality has in particular made it possible to identify the ranges of simulation time steps within which the outputs of each model remain valid [36].

The incremental approach adopted assumes both continuous reworking of the code and checking for backward compatibility. Despite the security offered by the encapsulation of the object-oriented approach, the propagation of improvements throughout the tree structure often results in the need to review old parts of the code which thus turns out to be globally quite sensitive. The problems related to this technical sensitivity are a priori most often detectable and soluble.

Unresolved problems persist, however, which are mainly related to the spatial scales considered. For example, in a simulated world where spatial resolution is 8 m. (*e.g.*, case study 1, Table 1) a cultivated field will be the result of the aggregation of several unit cells of soil whereas in a world approached at a resolution of one kilometre (case study 4) a similar cultivated field will have the status of an object inscribed inside the unit cell.

4.2 Influence of Formalism on Results

This model, despite its generic nature, may not be a canonical depiction of these various universes. Indeed, each of the choices taken in computer modeling, such as the Perception-Deliberation (Desire)-Execution paradigm for agent actions or the discretization of space, are just one among several alternatives that could have been proposed. Other formalisms, such as 'Agent/Group/Role' [37] might have been used instead as a foundation for the model. Selecting one of the other options might have hence resulted in a different architecture.

The functional hierarchy that has been constructed is based only on the object-oriented approach's classic paradigms. However, with the help of biologists and modelers, this tree structure revealed itself to be biologically relevant. The model's robustness allows it to be expanded to include new sorts of creatures without putting the model in jeopardy. It is also possible that the model might be used to replicate real systems other than rodent bio-ecology, *i.e.,* if just the uppermost abstract classes and interfaces on Fig. 9) are retained.

The computational paradigms that have been discovered to be effective in replicating natural mechanisms (composition, recursion, aggregation, and so on) come from a variety of implementations that could be used for bioinspired modeling. Each is employed on a case-by-case basis, as needed and appropriate. They are thus a collection of '*ad hoc*' formalisms that can be used as a toolkit to represent specific parts of Nature in a bioinspired manner.

Certain aspects of Nature's operation, on the other hand, appear to be difficult to account for. This is true, for example, of processes that operate on multiple spatial and temporal scales at the same time. A study using the same model shown that algorithms formalizing this way of operation might be developed [38]. However, addressing this problem leads to unnecessarily complex algorithms that divert attention away from the bioinspired problem.

Finally, the question of whether or not using multiple inheritance to construct a synthetic ecology is relevant or necessary arises. For example, depending on the species, rodents can be colonial, fossorial, domestic, commensal (Fig. 9), in this study. An adapted computer formalism should correlate to this "logic" of Nature in a bio-inspired model. The interface paradigm, in the context of the Java language chosen at the outset of the project, is insufficient to account for multiple inheritance of behaviors because it necessitates code duplications. However, implementations using Java 8's so-called interface default methods [39] are available. Creating a synthetic ecosystem with a language devoted to multiple inheritance management [*e.g.*, 40] could thus be a promising way for developing bioinspired simulations of ecologies or landscapes. Multiple inheritance, on the other hand, can lead to inconsistencies like the diamond problem [41]. These issues may call into question the value of simulating the real-world functioning of Nature, which, fundamentally, is devoid of contradictions.

5 Conclusion

The overall model was hence grown iteratively by formalizing successive case studies. Each one articulates with the previous ones, takes advantage of them, and contributes to the overall development of the platform by feeding it with new facets. This leads to gradually broaden and deepen the class tree of the object model. A shared architecture thus emerges which can be considered as heading towards a generic, multi-purpose and knowledge-oriented, model of the general scientific domain addressed.

The approach adopted, based on the incremental articulation of contrasting case studies in the same bioinspired formal structure, seems to meet the requirements of robustness linked to long-term integration (consolidation or questioning by case studies at future).

It must however still be considered at an implementation stage given several unresolved questions, including issues of spatial scales. It does, however, make it possible to begin to discern which are the shared, shareable or irreconcilable components and, perhaps, the methods of a mutually beneficial integration of distinct disciplinary points of view.

Computer paradigms, particularly those connected to object-oriented formalism, represent a useful toolkit for progressing toward the building of bioinspired synthetic ecologies and landscapes. They enable us to mimic several forces or principles that Nature employs in order to create a functional and complete world. This toolkit of paradigms and formalisms is always being developed, for example, by making multiple inheritance more widely available and secure. Bioinspiration studies can then be used to direct further study in this way.

Acknowledgements. The authors would like to thank J.F. Cosson, J.P. Quéré, C. Berthier, B. Gauffre, J.M. Duplantier, L. Granjon, G. Ganem, J. Britton, O. Ninot, J. Lombard, P. Handschumacher, S. Piry, the scientists who kindly agreed to decipher their disciplinary expertise for the formalization of thematic case studies. This study owes much to the work done by Q. Baduel, A. Realini, J.E. Longueville, A. Comte and M. Diakhate, as part of their student internships. The study was supported by the Chancira (grant IRD-ANR-11-CEPL-0010), Cerise (grant IRD-FRB no. AAP-SCEN -20B III) projects, the French National Research Institute for Sustainable Development (IRD) and the 'Centre de Biologie pour la Gestion des Populations' (CBGP, UMR no. 22 INRAe/IRD/Cirad/Supagro). We also wish to thank the members of the BioPASS laboratory in Senegal for their decisive field support.

References

1. Le Fur, J., Mboup, P.A., Sall, M.: A simulation model for integrating multidisciplinary knowledge in natural sciences. heuristic and application to wild rodent studies. In: 7th International Conference on Simulation and Modeling Methodologies, Technologies and Applications, pp. 340–347. Scitepress (2017)

2. Le Fur, J.; Mboup, P.A., Sall, M.: Use and adequacy of computer paradigms to simulate bioinspired synthetic landscape ecologies. In: 11th International Conference on Simulation and Modeling Methodologies, Technologies and Applications, pp. 154–162. Scitepress (2021)
3. Chen, Y.C., Lu, P.E., Chang, C.S., Liu, T.H.: A time-dependent SIR model for COVID-19 with undetectable infected persons. IEEE Trans. Netw. Sci. Eng. **7**(4), 3279–3294 (2020)
4. Kari, L., Rozenberg, G.: The many facets of natural computing. Commun. ACM **51**(10), 72–83 (2008)
5. Yao, X., Liu, Y., Li, J., He, J., Frayn, C.: Current developments and future directions of bio-inspired computation and implications for ecoinformatics. Eco. Inform. **1**(1), 9–22 (2006)
6. Olden, J.D., Poff, N.L., Bledsoe, B.P.: Incorporating ecological knowledge into ecoinformatics: an example of modeling hierarchically structured aquatic communities with neural networks. Eco. Inform. **1**(1), 33–42 (2006)
7. Colomer, M.À., Margalida, A., Sanuy, D., Pérez-Jiménez, M.J.: A bio-inspired computing model as a new tool for modeling ecosystems: the avian scavengers as a case study. Ecol. Model. **222**(1), 33–47 (2011)
8. Holling, C.S.: The strategy of building models of complex ecological systems. In: Watt, K.E.F. (ed.) Systems Analysis in Ecology, pp. 195–214. Academic Press, Cambridge (1966)
9. Evans, M.R., et al.: Do simple models lead to generality in ecology? Trends Ecol. Evol. **28**(10), 578–583 (2013)
10. Svoray, T., Benenson, I.: Scale and adequacy of environmental microsimulation. Ecol. Complex. **6**(2), 77–79 (2009)
11. Dunham, M.J.: Synthetic ecology: a model system for cooperation. Proc. Natl. Acad. Sci. **104**(6), 1741–1742 (2007)
12. Shou, W., Ram, S., Vilar, J.M.: Synthetic cooperation in engineered yeast populations. Proc. Natl. Acad. Sci. **104**(6), 1877–1882 (2007)
13. Buck-Sorlin, G.: Functional-structural plant modeling. In: Dubitzky, W., Wolkenhauer, O., Cho, K.H., Yokota, H. (eds.) Encyclopedia of Systems Biology. Springer, New York, NY (2013)
14. Dejong, T., Da Silva, D., Vos, J., Escobar-Gutiérrez, A.: Using functional-structural plant models to study, understand and integrate plant development and ecophysiology. Ann. Bot. **108**, 987–989 (2011)
15. Cantrell, B., Holzman, J.: Protocols, simulation, and manipulation for indeterminate landscapes. Synthetic Ecologies (2014)
16. DeAngelis, D.L., Mooij, W.M.: In praise of mechanistically rich models. In: Canham, C.D., Cole, J.J., Lauenroth, W.K. (eds.) Models in Ecosystem Science, pp. 63–82. Princeton University Press, New Jersey (2003)
17. Clapham, W.M., Crosby, C.J.: Utility of object-oriented programming in complex system modeling. Math. Comput. Modell. **16**(617), 45–50 (1992)
18. Granjon, L., Duplantier, J.M.: Les rongeurs de l'Afrique sahélo-soudanienne. IRD Editions, Publications Scientifiques du Muséum - Collection Faune et Flore tropicales 43. (2009). ISBN IRD, 978-2-7099-1675-2
19. Le Fur, J.: A formal computer framework for linking multidisciplinary multiscale knowledge. a case study on rodent population dynamics and management. In: European Conference on Complex Systems (ECCS), Lisbon (2013a)
20. Le Fur, J.: Du foisonnement des disciplines à la recomposition d'une réalité partagée Élaboration d'une structure de modélisation dédiée à l'intégration de connaissances disciplinaires. In: XXIeme journées de Rochebrune "Multi-trans-interdisciplinarité". Rencontres interdisciplinaires sur les systèmes complexes naturels et artificiels (2014). https://hal.ird.fr/ird-03182988

21. North, M.J., et al.: Complex adaptive systems modeling with repast simphony. Complex Adaptive Systems Modeling, Springer, Heidelberg (2013). https://doi.org/10.1186/2194-3206-1-3
22. Comte, A.: Caractérisation des barrières à l'hybridation de deux espèces jumelles de rongeurs africains du genre Mastomys. Étude par simulation multi-agents à partir de deux expériences in situ. Rapp. M2 Écologie-Biodiversité Spécialité Biodiversité Évolution Parcours Génétique et Biodiversité, p. 44 (2012)
23. Diakhate, E.H.M., Diouf, N., Granjon, L., Konate, K., Mboup, P.A., Le Fur J.: Modélisation et simulation multi-agents d'un protocole de capture-marquage-recapture d'une population de rongeurs sauvages dans la réserve de Bandia (Sénégal). In: 12th African Conference on Research in Computer Science and Applied Mathematics (CARI), Saint-Louis Sénégal, pp.43–54 (2014)
24. Mboup, P.A., Konaté, K., Handschumacher, P., Le Fur, J.: Des Connaissances au Modèle Multi-agents par l'approche orientée événement : construction d'un simulateur de la colonisation du rat noir au Sénégal par les transports humains sur un siècle. Actes Colloque National sur la Recherche en Informatique et ses Applications, Thiès, Sénégal, p. 8 (2015a)
25. Mboup, P.A., Mboup, M.L., Konaté, K., Handschumacher, P., Le Fur, J.: Optimisation de l'utilisation de l'algorithme de Dijkstra pour un simulateur multi-agents spatialisé. In: World Congress on Information Technology and Computer Applications (WCITCA), Hammamet, Tunisia, pp. 1–6 (2015b). https://doi.org/10.1109/WCITCA.2015.7367061
26. Sall, M., Dembélé, J.M., Le Fur, J.: An hybrid algorithm to simulate mice following residential wall. In: Proceedings of the 8th International Conference on Simulation and Modeling Methodologies, Technologies and Application (Simultech). ScitePress Publication, pp. 368–375 (2019)
27. Simon, H A.: The architecture of complexity. Proc. Am. Philos. Soc. **106**(6), 467-482 (1962)
28. Le Fur, J.: Extending life concepts to complex systems Interdiscip. Descr. Complex Syst. **11**(1), 37-50 (2013b)
29. Macia-Perez, F., Lorenzo-Fonseca, I., VicenteBerna-Martinez, J.: A formal framework for modelling complex network management systems. J. Netw. Comput. Appl. **40**, 255–269 (2014)
30. Caillou, P., Gaudou, B., Grignard, A., Truong, C.Q., Taillandier, P.: A Simple-to-use BDI architecture for agent-based modeling and simulation. In: The 11th Conference of the European Social Simulation Association, Groningen, Netherlands hal-01216165 (2015)
31. Shaw, K.L., Wagner, K.: Cricketsim: a Genetic and Evolutionary Computer Simulation. J. Artificial Societies and Social Simulation, 11(3) (2008)
32. Forterre, P., Gribaldo, S., Brochier, C.: LUCA: à la recherche du plus proche ancêtre commun universel. Méd./Sci. **21**(10), 860–865 (2005)
33. Darwin, C.: On the Origin of Species by Means of Natural Selection, or the Preservation of Favoured Races in the Struggle for Life. John Murray, London (1859)
34. Marcos, E., Cavero, J.M.: Hierarchies in object oriented conceptual modeling. In: Bruel, J.-M., Bellahsene, Z. (eds.) OOIS 2002. LNCS, vol. 2426, pp. 24–33. Springer, Heidelberg (2002). https://doi.org/10.1007/3-540-46105-1_4
35. Gell-Man, M.: The Quark and the Jaguar: Adventures in the Simple and the Complex. H. W. Freeman, New York (1994)
36. Le Fur, J., Sall, M.: Using flexible time scale to explore the validity of agent-based models of ecosystem dynamics: application to simulation of a wild rodent population in a changing agricultural landscape. In: 8th International Conference on Simulation and Modeling Methodologies, Technologies and Application, pp. 297–304. Scitepress (2018)
37. Ferber, J., Gutknecht, O., Michel, F.: From agents to organizations: an organizational view of multi agent systems. In: Giorgini, P., Müller, J.P., Odell, J. (eds.) Agent-Oriented Software Engineering IV, LNCS, vol. 2935, pp. 214–230. Springer, Berlin, Heidelberg (2003)

38. Mboup, P.A., Konaté, K., Le Fur, J.: A multi-world agent-based model working at several spatial and temporal scales for simulating complex geographic systems. In: International Conference on Computational Science (ICCS), Zurich, Procedia Computer Science 108C, 968–977 (2017)
39. Mohnen, M.: Interfaces with default implementations in Java. In: Proceedings of the Workshop on Intermediate Representation Engineering for Virtual Machines, pp. 35–40 (2002)
40. Malayeri, D., Aldrich, J.: CZ: multiple inheritance without diamonds. ACM SIGPLAN Notices **44**(10), 21–40 (2009)
41. Truyen, E., Joosen, W., Nørregaard Jørgensen, B., Verbaeten, P.: A generalization and solution to the common ancestor dilemma problem in delegation-based object systems. In: Proceedings of the 2004 Dynamic Aspects Workshop, 103–119 (2004)

On Advanced Modeling of Compressors and Weighted Mix Iteration for Simulation of Gas Transport Networks

Anton Baldin[1], Kläre Cassirer[1], Tanja Clees[1,2], Bernhard Klaassen[1], Igor Nikitin[1(✉)], Lialia Nikitina[1], and Sabine Pott[1]

[1] Fraunhofer Institute for Algorithms and Scientific Computing, Schloss Birlinghoven, 53754 Sankt Augustin, Germany
{Anton.Baldin,Klaere.Cassirer,Tanja.Clees,Bernhard.Klaassen, Igor.Nikitin, Lialia.Nikitina,Sabine.Pott}@scai.fraunhofer.de

[2] Bonn-Rhine-Sieg University of Applied Sciences, Grantham-Allee 20, 53754 Sankt Augustin, Germany

Abstract. In this paper, the modeling of gas compressors with detailed representation of their calibrated characteristics is considered. A method is developed for transforming the characteristics from the space of calibration data to the space of transport variables in which the network problem is actually solved. This transformation satisfies the general stability conditions necessary for the convergence of Newtonian iterations in solving large scale network problems. In addition, a stabilization method of weighted relaxation for mixing iterations is presented, used to find the detailed gas composition. The tests on a large number of realistic gas networks demonstrate nearly 100% convergence of the methods.

Keywords: Complex systems modeling and simulation · Non-linear systems · Applications in energy transport

1 Introduction

This is an extended version of our conference paper [1], where a non-linear transformation algorithm for detailed modeling of gas compressors and drives has been described. In addition, we present an improved mix iteration for evaluation of detailed gas composition. The algorithms can be used for solution of large scale stationary network problems. Theoretical convergence issues have been addressed and extensive numerical tests have been performed.

The main elements of gas transport networks are pipes and control elements serving pressure drop and increase – regulators and compressors. The compressors are the most complex elements, possessing two levels of modeling: simplified 'free' model and sophisticated 'advanced' model. Our construction is built upon a theoretical result [2], which ensures that in a system composed of the Kirchhoff equations, describing flow conservation, and nonlinear element equations of a general form, under certain signature conditions on the derivatives, a nonsingular Jacobi matrix is obtained. This property ultimately guarantees convergence of the Newtonian algorithm with the Armijo

G. Wagner et al. (Eds.): SIMULTECH 2021, LNNS 601, pp. 138–152, 2023.
https://doi.org/10.1007/978-3-031-23149-0_8

line search stabilizer. These mathematical conditions can also be reformulated in terms of physical stability, namely, how the system reacts on small fluctuations of pressure and flows. Such fluctuations should be damped, not amplified by the system. It is also important that these signature conditions are met not only in the working region that is of interest to the user. They should also hold outside of this region, since the intermediate iterations of the solver algorithm can pass over the exterior regions as well. For this purpose, universal continuation formulas were developed in [2], that satisfy the signature rules globally.

By making the described continuations for all the main elements, we get a system that is guaranteed to have a solution when using the stabilized Newtonian method. The compressors are the most complex elements. The construction of their detailed modeling has been started in [3], continued in [4], and further improved in the present paper. Compressors possess individually calibrated characteristics, these data must be taken into account in the modeling. In this paper, we present an algorithm that allows to transfer the compressor calibration data from the initially taken variables to the transport variables, in which the problem is actually solved. At the same time, the fulfillment of the signature conditions necessary for the stability of the system solution can be controlled. The question of continuations in exterior domains arises for these elements as well, it can also be solved using universal formulas. In rare cases, folds occur in the system mapping, which should be reported to the user as possible instabilities appearing inside the working region. Technically, the 'free' model is modified by inserting a term corresponding to the calibration data to the element equation, forming the 'advanced' model. At the lowest level, the transformation is implemented using triangulated surfaces. We have tested the developed algorithm on a variety of compressors and problem settings, ranging from simple to really complex realistic networks.

Realistic modeling of gas networks also includes the calculation of detailed gas composition according to mixing equations. Prototypes of such equations were described in [5]. For large problems, these equations form a gigantic linear subsystem, much larger than the size of the main system describing pressures and flows. It would be nice to solve such a problem as a single coupled system, but this is not always possible. Sometimes the size and structure of the system makes it necessary to subdivide it into smaller parts and combine the solution of these parts using an external iterative loop. In this paper, we describe a method for stabilizing such external iterations using a weighted relaxation scheme. The developed algorithms extend our multi-physics network simulator MYNTS [6].

General aspects of gas network modeling have been described in [7,8]. These include the use of various friction laws for pipes: Nikuradse, Hofer, or Colebrook-White [9,10], various approximations for the equation of state: Papay, AGA8-DC92, GERG-2008 [11–13] and the Kirchhoff equations for the conservation of flows. The resulting non-linear system can be represented as a non-linear program (NLP) and solved by specialized NLP solvers such as IPOPT, SNOPT, MINOS [14–16]. We have also implemented our own solver based on the stabilized Newton method from [17].

This paper has the following structure. In Sect. 2 the general modeling of gas compressors is described. Section 3 presents the algorithm for the transformation from calibration parameters to the transport variables. Section 4 describes a weighted relaxation algorithm for mix iteration. Section 5 presents the numerical tests of the algorithms. Section 6 summarizes the obtained results.

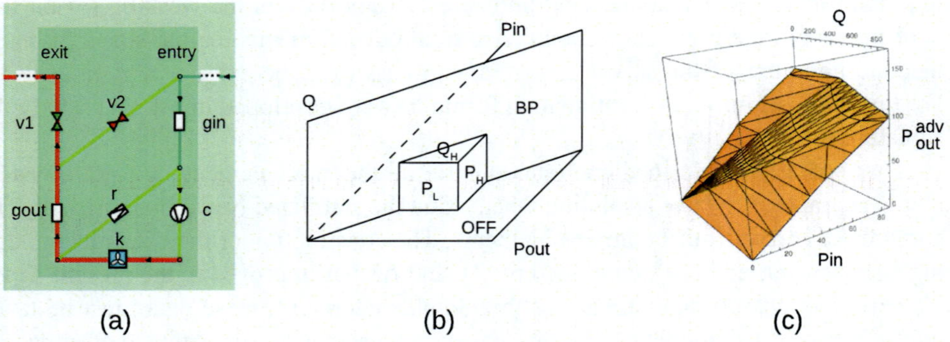

Fig. 1. Modeling of compressors: (a) structure of a compressor station; (b) compressor element equation in transport variables, 'free' model; (c) 'advanced' model. Images from [2–4].

2 Modeling of Gas Compressors

A strategy for stable representation of advanced compressors and drives is based on the following steps:

– eliminate all intermediate variables in the element equation;
– represent the equation in the space of transport variables;
– check monotonicity;
– use a monotone linear continuation outside of the working region.

The sets of variables and the transformation between them will be described further. The monotonicity condition is required for global convergence of the solver algorithm and is described in [2]. All element equations $f(P_{in}, P_{out}, Q) = 0$ should satisfy the following inequalities on their derivatives:

$$\partial f/\partial P_{in} > 0,\ \partial f/\partial P_{out} < 0,\ \partial f/\partial Q < 0, \tag{1}$$

meaning that the element equation function should monotonously increase w.r.t. P_{in} and monotonously decrease w.r.t. P_{out}, Q.

The basic continuation formula is also presented in [2]:

$$f(x_1, ..., x_n) = f(\hat{x}_1, ..., \hat{x}_n) + \sum_{k=1}^{n} (\min(x_k - a_k, 0) + \max(x_k - b_k, 0)), \tag{2}$$
$$\hat{x}_k = \min(\max(x_k, a_k), b_k).$$

It provides a continuation of the function of n arguments, monotonously increasing w.r.t. every argument, from a bounding box specified by $[a_k, b_k]$ limits to the whole space, preserving this monotonous property. For decreasing functions, coordinate reflections can be used.

The structure of the compressor station in its typical one-unit configuration is shown in Fig. 1a. It consists of the following elements: c – compressor, r – bypass regulator, gin/gout – input and output resistors, v1,2 – main and bypass valves, k – cooler. Entry/exit identify standard input and output nodes. In more complicated scenarios several units can be assembled together in parallel or/and serial connection.

Transport Variables. Figure 1b, c show the element equation of compressor in the space of transport variables. $P_{in,out}$ denote input and output pressures, normally measured in bars. Q is a throughput, standardly represented by mass flow Q_m in kg/s. With the known gas composition it can be converted to molar flow Q_ν in mol/s, or to normal flow Q_N in cubic meters of gas virtually decompressed to normal pressure and temperature, per second (often reexpressed to thousands of such normal cubic meters per hour). Special case is volumetric flow Q_{vol}. It is measured in m^3/s at current conditions, and due to compressibility of the gas depends on whether input or output conditions are meant (input conditions are taken by default).

Figure 1a presents the free model. The subscripts H, L indicate high and low control settings, defining upper and lower limits on the pressures and the flow; BP is open/bypass mode of compressor $P_{in} = P_{out}$; OFF is the closed mode $Q = 0$. This polyhedral surface, as all surfaces of this kind, can be represented by max-min formula [2]:

$$\begin{aligned} &\max(\min(\ P_{in} - P_L, -P_{out} + P_H, -Q + Q_H), \\ &P_{in} - P_{out}, -Q) + \epsilon(P_{in} - P_{out} - Q) = 0, \end{aligned} \tag{3}$$

where the last term with small positive constant ϵ serves regularization.

Figure 1c presents advanced compressor model. $P_{out}^{adv}(P_{in}, Q)$ is output pressure of compressor in the absence of control restrictions (also referred as compressor in MAX mode). It is considered as a function of the input pressure and the flow. This function represents the internal capability of compressor and its drive. It is combined with free diagram as follows [1]:

$$\begin{aligned} &\max(\min(\ P_{in} - P_L, -P_{out} + P_H, -Q + Q_H, \\ &-P_{out} + P_{out}^{adv}(\hat{P}_{in}, \hat{Q}) \\ &+ \min(P_{in} - P_{in,min}^{adv}, 0) + \max(P_{in} - P_{in,max}^{adv}, 0) \\ &+ \min(-Q + Q_{max}^{adv}, 0) + \max(-Q + Q_{min}^{adv}, 0)), \\ &P_{in} - P_{out}, -Q) + \epsilon(P_{in} - P_{out} - Q) = 0, \\ &\hat{P}_{in} = \min(\max(P_{in}, P_{in,min}^{adv}), P_{in,max}^{adv}), \\ &\hat{Q} = \min(\max(Q, Q_{min}^{adv}), Q_{max}^{adv}), \end{aligned} \tag{4}$$

here the second line represents the advanced surface, inserted into the free formula; the next two lines provide linear continuation of this surface outside of the bounding box; the last two lines define clamp functions to the bounding box.

The advanced surface is triangulated, every triangle is represented by own system of barycentric coordinates. For this purpose, on the plane $(P_{in}, Q) = (x, y)$ the vertices of triangle $\{v_1, v_2, v_3\}$ are defined. The point on triangle is then defined as

$\sum_i w_i v_i = (x, y)$, $\sum_i w_i = 1$. The system can be solved for the weights $w_i(x, y)$ by linear formulae $w_i(x, y) = c_{0i} + c_{xi}x + c_{yi}y$, with 3 constants (c_{0i}, c_{xi}, c_{yi}) per w_i precomputed. One formula can be spared using $w_3 = 1 - w_1 - w_2$. The point belongs to triangle, when all weights are non-negative $w_i \geq 0$.

The third coordinate $P_{out} = z$ is found by one more linear formula $z(x, y) = \sum_i w_i(x, y) z_i$. Altogether 9 coefficients (equivalent to 3 nodes x 3 coordinates) are precomputed. Explicit lengthy formulae for barycentric coordinates can be found in [4]. Finally, a function is implemented, searching for a triangle on xy-plane and evaluating z-coordinate and its xy-derivatives. The derivatives can be directly used to check monotonicity condition [1]:

$$-P_{out} + P_{out}^{adv}(P_{in}, Q) = 0,\ \partial P_{out}^{adv}/dP_{in} > 0,\ \partial P_{out}^{adv}/dQ < 0. \tag{5}$$

Equivalently, these conditions can be reformulated in terms of normals to triangles, which all should point to the octant $(P_{in}, P_{out}, Q) = (+, -, -)$.

Internal Variables. Density ρ is defined as monotonously increasing function of pressure P using equations of state (EOS), involving also the molar mass μ, the temperature T and the compressibility factor z. Different analytic or numerical EOS can be used, e.g., Papay [11], AGA8-DC92 [12], GERG-2008 [13]. The volumetric flow relative to input conditions is expressed via mass flow and density as $Q_{vol} = Q/\rho_{in}$. The revolution number rev and torque M_t describe the rotation of the engine. There are also energetic quantities characterizing compressors and drives: H_{ad} – increase of adiabatic enthalpy, η_{ad} – adiabatic efficiency, $Perf$ – performance power.

3 Warp Transformation

The transformation consists of a sequence of non-linear maps [1]:

$$(Q_{vol}, rev) \rightarrow (H_{ad}, \eta_{ad}, Perf_{max}) \rightarrow (\rho_{in}, H_{ad}, Q) \rightarrow (P_{in}, P_{out}, Q). \tag{6}$$

Step 1: standard 1D quadratic and 2D biquadratic models from [3]:

$$\begin{aligned} H_{ad} &= (1, rev, rev^2) \cdot A \cdot (1, Q_{vol}, Q_{vol}^2)^T, \\ \eta_{ad} &= (1, rev, rev^2) \cdot B \cdot (1, Q_{vol}, Q_{vol}^2)^T, \\ Perf_{max} &= (1, rev, rev^2) \cdot D^T, \end{aligned} \tag{7}$$

where A, B are constant 3×3 matrices and D is a constant 3-vector filled by calibration coefficients. $Perf_{max}$ is the maximal performance power provided by the drive at the given revolution number.

Step 2: temperature and gasmix independent models [3]:

$$Q = Perf_{max}\eta_{ad}/H_{ad},\ \rho_{in} = Q/Q_{vol}, \tag{8}$$

Step 3: temperature and gasmix dependence [3]:

$$\begin{aligned} &\alpha = (\kappa - 1)/\kappa,\ \gamma = RT_{in}/\mu,\ P_{in} = EOS_{inv}(\rho_{in}), \\ &z_{in} = P_{in}/(\gamma\rho_{in}),\ P_{out} = P_{in}(H_{ad}\alpha/(\gamma z_{in}) + 1)^{1/\alpha}, \end{aligned} \tag{9}$$

where κ is the adiabatic exponent, R is universal gas constant; the equation of state $\rho = EOS(P)$ is inverted to define P_{in}; the universal gas law $P = \rho RTz/\mu$ is resolved w.r.t. z; then H_{ad} definition from [3] is resolved w.r.t. P_{out}.

All equations are given in SI-units, practically conversion factors should be applied for the transformations W/kW/MW, bar/Pa etc.

Regions of the advanced surface are shown in Fig. 2a. The described transformations are used to construct the most important powmax region, where the performance of compressor is restricted solely by the power of its drive. It is bounded by $revmin/revmax$ lines and $Q_{vol,min}/\eta_{min}$ lines (also called surge line and choke line). Here $revmin$ and $revmax$ are given constants. Surge line is defined as [1]:

$$\begin{aligned} Q_{vol,min}(rev) &= \max(Q^{(1)}_{vol,min}, Q^{(2)}_{vol,min}, 0), \\ Q^{(1)}_{vol,min} &= (1, rev, rev^2) \cdot C^T, \\ Q^{(2)}_{vol,min} &= \arg\max_{Qvol} H_{ad}(Q_{vol}, rev), \end{aligned} \tag{10}$$

$Q^{(1)}_{vol,min}$ is given as 1D quadratic model with a calibration 3-vector C; the condition $Q_{vol} \geq Q^{(2)}_{vol,min}$ defines a physical decreasing branch of quadratic dependence of H_{ad} on Q_{vol} at fixed rev; the outflow condition $Q_{vol} \geq 0$ is also enforced. Choke line $\eta_{ad}(Q_{vol,max}, rev) = \eta_{min}$ with a given constant η_{min} is solved w.r.t. $Q_{vol,max}(rev)$. Generally it is a quadratic equation with two roots; the maximal root is taken. The region between $revmin/revmax$ and surge/choke lines is resampled to $N_{rev} \times N_\eta$ grid.

Revmax region: $rev = revmax$ side is taken and in (ρ_{in}, Q) projection proportionally scaled to the origin. Then it is mapped to the final (P_{in}, Q) coordinates by the inverse EOS transformation above.

Continuation regions 1 and 2 go downwards and upwards in the (ρ_{in}, Q) projection, respectively, till the limits of the bounding box. H_{ad} values in these continuations are kept constant.

P_{out}-coordinate in (9) lifts the whole construction to 3D space (P_{in}, P_{out}, Q), where the final surface is represented by triangulation. Orientation of normals allows to check monotonicity conditions for every triangle. Figure 2b indicates problems with monotonicity (blue triangles). These problems happen rarely and require a slight local adjustment of the diagram to satisfy the global convergence criterion.

In general, the revmin side of the powmax patch has a fold, shown in Fig. 2c. Physically, on the surge and revmin lines, a bypass regulator opens in compressor station (r in Fig. 1 top). It redirects a part of the flow to circulate through the compressor, preventing the compressor from going outside of the working region ($Q_{vol} < Q_{vol,min}$, $rev < revmin$). In our diagrams, total Q passing through the compressor and its bypass regulator is continued downwards from these lines. This continuation generally creates a fold, producing multiple solutions and degeneracy of the Jacobi matrix. Fortunately, for most of the cases, this fold is located beyond the physical domain of ρ_{in} or P_{in} and can be safely ignored. For extra safety, we define a $\rho_{in,max}$ value, cutting off the fold and restricting the patch by this value.

The other problematic case is displayed in Fig. 3a. It corresponds to the increasing torque dependence $M_t(rev) = Perf_{max}/rev$. If the drive is joined with a generic resistive load, $M_{t,sys}(rev) = c_0 + c_1 rev$ for dry and viscous friction, or other increasing $M_{t,sys}(rev)$ dependence, the stable intersection is ensured only when $M_{t,drive}(rev)$ is decreasing, Fig. 3b. Otherwise one can find such resistive system that none or multiple intersections exist, Fig. 3c. In this case multiple solutions or no solution exist for the whole network problem. Such problematic behavior is present in some electric engines (E-drives). The computations show that in this case the monotonicity conditions are violated in most of the diagram. The solution we have taken so far is to replace the actual $M_{t,drive}(rev)$ dependence with a weakly decreasing function that limits the real dependence from below (conservative), above (overestimation), or reproduces it on average. In practice, a constant $M_{t,drive}$ can be used here, since the regularization in the control equation removes all marginal degeneracies in the system.

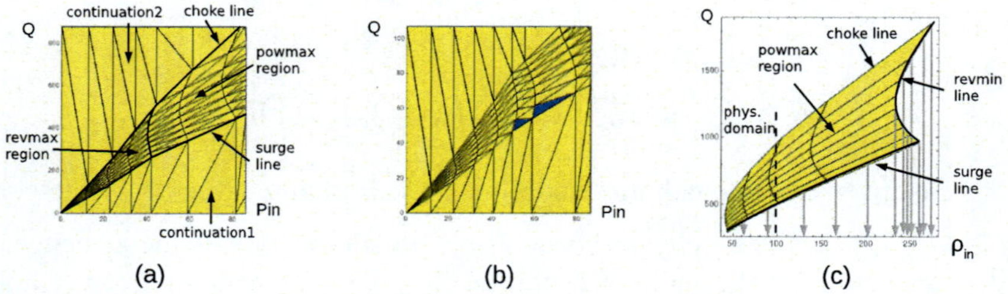

Fig. 2. Details of compressor diagrams: (a) regions of the advanced surface in (P_{in}, Q) projection: (b) problems with monotonicity detected (blue triangles); (c) the fold on $revmin$ line. Images from [1]. (Color figure online)

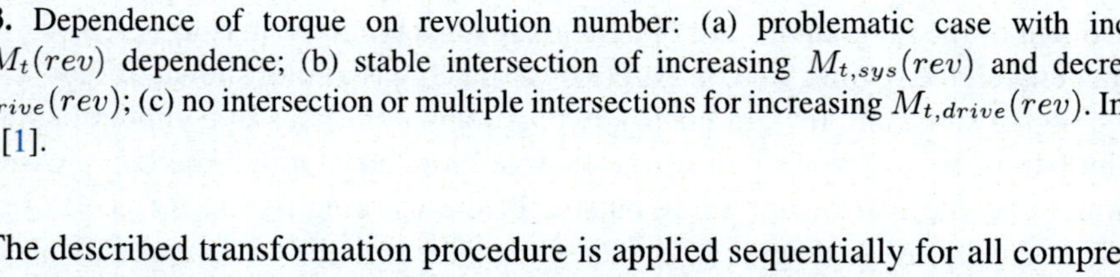

Fig. 3. Dependence of torque on revolution number: (a) problematic case with increasing $M_t(rev)$ dependence; (b) stable intersection of increasing $M_{t,sys}(rev)$ and decreasing $M_{t,drive}(rev)$; (c) no intersection or multiple intersections for increasing $M_{t,drive}(rev)$. Images from [1].

The described transformation procedure is applied sequentially for all compressor-drive pairs in the network, as shown in Fig. 4. Steps 1 and 2 of the transformation are performed once, in precomputation mode. The monotonous decrease of $M_t(rev)$, regularity of surge and choke lines, absence of folds on 2D diagrams is visually controlled. Step 3 of the transformation is applied repeatedly during the solution procedure, every time when the temperatures or/and gas composition changes. This step is just a monotonous remapping $\rho_{in} \rightarrow P_{in}$ according to the current EOS, not violating the verified monotonicity conditions.

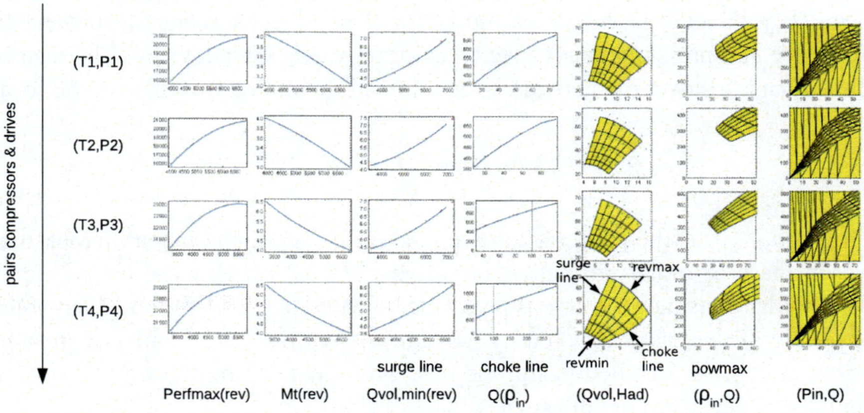

Fig. 4. Construction of advanced representation for all compressor-drive pairs in the network. Images from [1].

Station Resistors: shown by gin/gout in Fig. 1a, the resistors typically support constant pressure drops (REPD) on entry and exit of compressor station. They lead to trivial modification of the control Eq. (4), where in the first line $P_{in/out}$ represent the pressure in entry/exit nodes, while in the rest of the formula $P_{in/out}$ are replaced with the pressure at inlet/outlet of the compressor. These values differ by the given pressure drops on station resistors.

Ambient Temperature Dependence: compressor drives often possess an own dependence on ambient temperature, defined by biquadratic model [1]:

$$Perf_{max} = (1, rev, rev^2) \cdot D \cdot (1, T_{amb}, T_{amb}^2)^T, \tag{11}$$

used instead of (7). Here D is 3×3 calibration matrix and T_{amb} is absolute or relative temperature, with the corresponding recomputation. Note that the actual use of $Perf_{max}$ in step 2 of the transformation is a linear formula (8). As a result, the following linear algorithm can be used for precise account of T_{amb}-dependence. The step 2 precomputation is performed for three different temperature values $T_{amb,i}$, producing three 2-vectors $v_i = (\rho_{in}, Q)_i$. Then, in step 3, three weights are computed, defined by [1]:

$$w_1 = \frac{(T_{amb} - T_{amb,2})(T_{amb} - T_{amb,3})}{(T_{amb,1} - T_{amb,2})(T_{amb,1} - T_{amb,3})} \tag{12}$$

and the cyclic permutation of indices. Then the vector v is computed as the weighted average $v = \sum_i w_i v_i$ and the result is passed to step 3 of the generic computation. In this way, the variation of T_{amb} in particular scenario can be performed without repeating the steps1,2 in the chain.

4 Mix Iteration

In addition to the main (P, Q) variables in gas transport problems, many other variables are used that describe the detailed composition and thermodynamic properties of a substance. They include: molar mass, molar fractions of constituent substances, critical temperature and pressure, heat capacity, calorific value, enthalpy, etc. The distribution of these quantities over the network is determined by mixing formulas of the form

$$\sum_e prop_{e,in} Q_{e,in} = prop_{out} \sum_e Q_{e,out}, \tag{13}$$

where $prop_{in/out}$ is the property to be mixed; $Q_{in/out}$ are the incoming/outgoing gas flows, in mass or molar normalization, depending on whether the mixed values are mass or molar densities; the sum is performed over edges incident to a given node. The temperature of the gas is coupled by non-linear formulas to the enthalpy, resulting in a closed system for detailed gas properties. The low-level implementation of mixing formulas was described in our previous paper [5].

The problem here is that the number of additional variables significantly exceeds the number of the main ones, mainly due to the gas composition, which is usually described by 21 components. On the other hand, the mix equations are linear in additional variables (excluding temperature). Thus, a situation arises when the system being solved becomes more than an order of magnitude larger, but mainly consists of a stable linear subsystem. The remainder is represented by a non-linear and marginally stable PQ-system. Straightforward solution of such problems in a fully coupled form is still possible, for small networks.

For large problems, in order to optimize the solution process, it seems more convenient to separate the solution of the linear part into an independent subprocess. This subprocess is iterated together with the solution of the main PQ-system. As the first choice, a simple iteration can be taken: for a fixed gas composition, the distribution of pressures and flows is found, then, for fixed flows, the gas composition is found, and this procedure is repeated until convergence. As a convergence criterion, one can use the initial residual of the PQ-system before its solution. Indeed, if this residual is so small that the solution of the PQ-system can be skipped (since the answer has already been found with the required accuracy), then the flow distribution will not change, and the entire iterative process stops.

In a practical implementation, the mix iteration can include all complex calculations that can be moved from the PQ-iteration, such as adv-warp algorithm (described above), temperature coupling (or, better, its Newton-alike linearization), refinement corrections for sophisticated gas laws (AGA8-DC92, GERG2008), active set update for modeling of coolers and heaters, calculation of gas consumption in compressor drives, etc.

From a computational point of view, the following problem arises here. PQ and mix phases are combined in simple iteration sequence. Theoretically, it is not always convergent, and our numerical experiments presented in the next section confirm this fact. Special weighting relaxation scheme is implemented to overcome this problem.

We start with a theoretical discussion for a one-dimensional case. It is instructive and extendable to higher dimensions with contractive maps.

Let $x_{n+1} = f(y_n)$, $y_{n+1} = g(x_{n+1})$, where x – PQ variables, y – mix variables, f – PQ iteration, g – mix iteration. Substitution: $x_{n+1} = f(g(x_n)) = h(x_n)$. Weighted relaxation:

$$x_{n+1} = wh(x_n) + (1 - w)x_n. \tag{14}$$

Linear stability analysis: $h(x) = qx$, here constant is everywhere subtracted, only linear part remains. Simple iteration: $x_n \sim q_n x_0$, convergence range: $-1 < q < 1$. Weighted iteration: $x_{n+1} = (wq + 1 - w)x_n$, $x_n \sim (wq + 1 - w)^n x_0$, convergence range: $-1 < wq + 1 - w < 1$, or $-2 < w(q - 1) < 0$. Two cases: $q > 1 \Rightarrow -2/(q - 1) < w < 0$; $q < 1 \Rightarrow 0 < w < 2/(1 - q)$. If simple iteration converges, $-1 < q < 1 \Rightarrow 2/(1 - q) > 1$, therefore $w = 1$ can be taken, no stabilization needed. For $q < -1$ simple iteration has oscillations of exponentially increasing amplitude. Practically, the higher order terms can intercept this behavior and form a limit cycle. In this case $2/(1 - q) < 1$ and weighting scheme works with small positive w. When w is changed from $w_{max} = 2/(1-q)$ to 0, $q_{\text{eff}} = wq+1-w$ changes from -1 to 1, passing through $q_{\text{eff}} = 0$ (superconvergence) at $w = 1/(1 - q)$. For $q > 1$ simple iteration diverges monotonously (exponential explosion). This is not the case we usually observe for our iterations. However, even this case can be stabilized by the weighted relaxation with $-2/(q - 1) < w < 0$, e.g., small negative w.

Note: dynamical solver with small step in the vicinity of stable stationary point can behave like the weighted relaxation scheme with small positive w, i.e., weighted iteration in stationary solver and integration step in dynamical solver can possess similar stability criteria. Here $x_{n+1} - x_n = w(h(x_n) - x_n)$ is identical with explicit Euler integration scheme for ODE $dx/dt = h(x) - x$.

What normally happens with non-stabilized simple iteration, is usually demonstrated on the example of logistic map iteration: $x_{n+1} = rx_n(1 - x_n)$, where r is a constant. When r is increased, one observes a bifurcation of single root to 2-cycle, 4-cycle, and so on, to complete chaos. Weighted relaxation can help to bring such iterations to convergence. In the next section we describe the results of numerical simulation on realistic gas networks using weighted mix iteration algorithm.

Here we should emphasize that the coupled system solution looks more attractive from a numerical point of view. In this case, it is necessary to solve only one system and monitor only one sequence of iterations. These iterations are of the Newtonian type, they also have proven methods of stabilization.

Other factors speak against coupled system:

- in such system, mix subsystem is not linear anymore, since its coefficients w.r.t. mix variables depend on Q-variables;
- presence of active set conditions in temperature part makes it a linear complementarity problem (LCP), highly non-linear and non-smooth;
- the theory of convergence is available only for the PQ-phase;
- some parts of mix subsystem, especially AGA8-DC92 and GERG2008 gas laws, do not have all analytical derivatives, necessary for implementation of Newton iterations;
- the number of mix variables is much greater than that of the PQ-variables, mostly due to the gas composition variables, meaning drastic increase of the system size.

The analysis of iterative-vs-coupled tradeoff is the topic of our future work.

5 Results of Numerical Tests

The described algorithms have been tested on a number of real-life gas networks. Parameters of the test networks are given in Table 1. The number of elements is given before applying topological cleaning procedure, described in [3]. This procedure removes trivial elements, such as valves, shortcuts, short pipe segments, and can significantly reduce the size of the network in certain cases. While the transport networks mainly consist of pipes with a nearly quadratic friction law, their computational complexity is defined by the most non-linear elements, namely compressors and regulators. In the networks, there are two types of supply nodes, the ones with a pressure setpoint ($Pset$) and the ones with an inflow setpoint ($Qset < 0$). Many outflow nodes ($Qset > 0$) exist, representing the large number of gas consumers in the network.

Table 1. Parameters of test networks [1].

Network	Nodes	Edges	Pipes	Compressors	Regulators	Psets	Qsets
N1	100	111	34	4	4	2	3
ME	437	482	370	20	24	3	164
N85/L	3232–3886	3305–3974	2406–2835	1–7	59–77	6–7	625–843
N85/H	2914–3818	2989–3952	1498–1937	16–42	59–107	5–9	328–505

Table 2. Convergence results [1].

Test networks	Total num.	Converged	
		Old	New
N1	1	1	1
ME	1	1	1
N85/L	23	23	23
N85/H	62	31	62

The small and medium size networks are presented in Fig. 5. Network N1 has 100 nodes and 111 edges, while ME has 437 nodes and 482 edges and possesses a more complex topology. In addition, we use a set of 85 large networks received from our industrial partner for benchmarking. They are subdivided to L- and H-type denoting gas with low and high calorific value. Although the calorific value itself has no influence on the convergence properties, the L-networks contain considerably less compressors and are topologically more simple than their H-counterparts. As a result, L-networks typically possess better convergence then the H-ones.

The test networks were subjected to the solver procedures of two types. One used the 'old' type of the compressor modeling, where all intermediate variables were present and constrained by the corresponding equations. The other one used the 'new' type, with intermediate variables eliminated and the problem formulated completely in terms of the transport variables. The convergence results are shown in Table 2. While the old procedure was sufficiently stable to process simple N1 and ME networks, as well

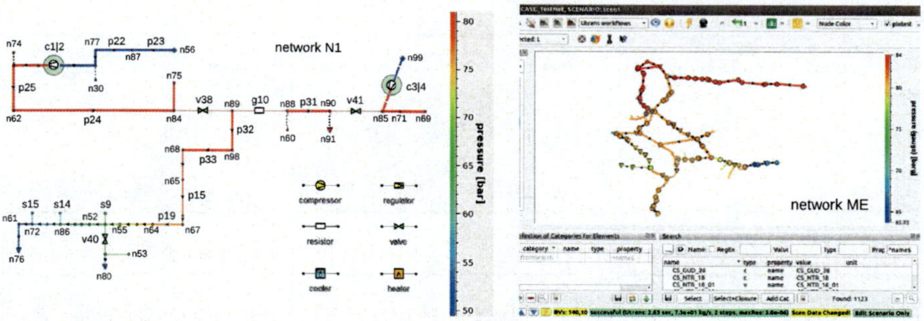

Fig. 5. Test networks N1 and ME. Images from [1].

as L-type N85 networks, it diverged in the half of H-type N85 networks. Our main achievement is that the new procedure converged in our tests in 100% of cases, also in the most complex N85/H-type ones.

It should be noted, however, that in spite of the theoretical guarantee for the convergence of the algorithm, the control Eqs. (3) and (4) contain problematic marginally degenerate terms. They correspond to the faces of the 'free' diagram (Fig. 1, center), with normals directed exactly along coordinate axes. On such faces, some derivatives in (1) vanish and the whole network problem degenerates. Regularizing the ϵ-term in the control equations formally removes this singularity. However, precise physical modeling requires ϵ to be as small as possible while for numerical stability larger values of ϵ are preferable. In our applications, a compromise value in the range $\epsilon = [10^{-6}, 10^{-3}]$ is selected.

As a result of the marginally singular problem statement, the solution procedure cannot be started from an arbitrary point, as it should be possible for the absolutely stable globally convergent algorithm. It still requires empirics in the definition of a starting point, for which we use a 'gradual sophistication' strategy. It starts from 'forced' goals of compressors and regulators and proceeds via 'free' to 'advanced' modeling. We have found that the solution procedure can randomly diverge under variations of the problem settings. It happens rarely, by our experience in $\sim$1% of cases. In these special cases the adjustment of ϵ value, global per network or local per problematic element, may help.

Weighted Relaxation algorithm has been applied on the same networks, with different setting for weighting parameter w. Precision is measured as PQ initial residual. Since the equations are standardly normalized to $\sim$100 value, it is equivalent to relative precision in percents.

Figure 6 shows the behavior of the algorithm for H-type networks:

$w = 1$ corresponds to unweighted algorithm, for most of cases it has no convergence, oscillates;
$w = 0.875$: already slight weighting improves convergence drastically;
$w = 0.75$: the slope of convergence increases, according to the theory;
$w = 0.5$: the slope starts to decrease, overall convergence improves;

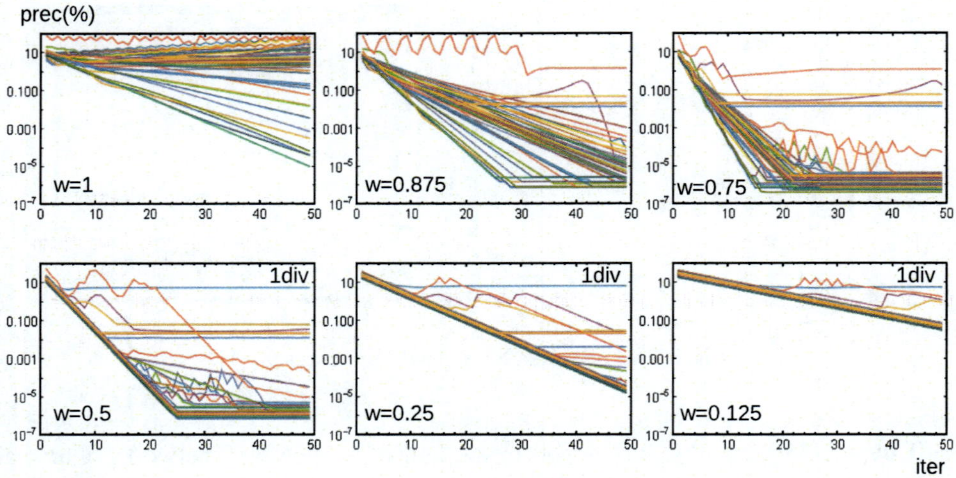

Fig. 6. Convergence of weighted mix iterations for H-type networks. Percentual precision vs iteration number at different w factors.

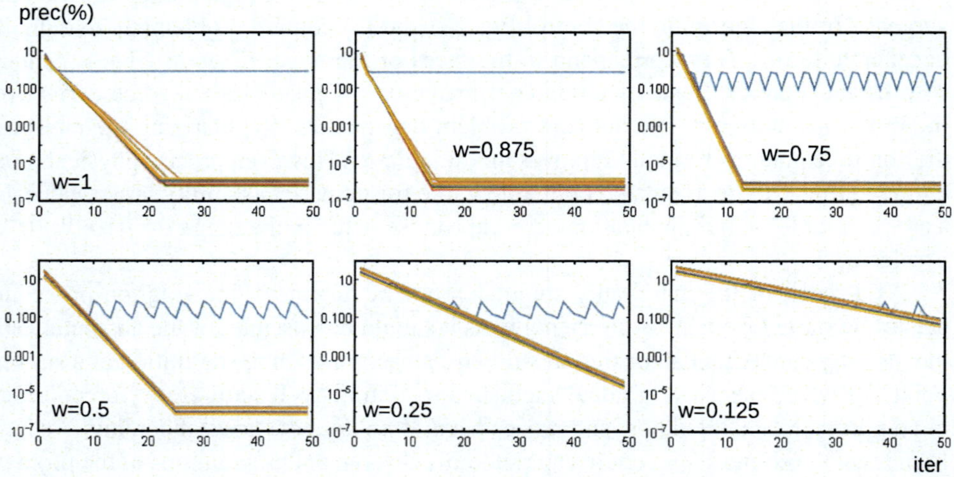

Fig. 7. Convergence of weighted mix iterations for L-type networks. Percentual precision vs iteration number at different w factors.

$w = 0.25, 0.125$: theoretical tendency continues, practically convergence becomes too slow.

Remarks: starting from $w = 0.5$, one network diverges in PQ-phase. As we have previously mentioned, such divergence is caused by marginal degeneracy of the problem and can happen at rare instances. Setting $w = 0.5$, nmixiter = 10 provides $\sim 0.1\%$ precision for almost all test networks, selected as default in our software.

Figure 7 shows the behavior of the algorithm for L-type networks. While these cases are expectedly more stable, one network oscillates for all w, on the level 0.1–1%, going to a stable limit cycle. The other networks show the same theoretical tendency, and $w = 0.5$, nmixiter=10 can be taken as default as well.

As a result, we have forced PQ-mix iterations to converge for most of the test cases. Weighted relaxation parameters $w = 0.5$, nmixiter = 10 present a good tradeoff between precision and convergence speed.

6 Conclusions

In this paper, we have considered the modeling of gas compressors with detailed representation of their calibrated characteristics. A method was developed for transforming the characteristics from the space of calibration data to the space of transport variables in which the network problem is actually solved. This transformation satisfies the general stability conditions necessary for the convergence of Newtonian iterations in solving large scale network problems. This algorithm should be applied every time when the temperature or gas composition change. In addition, in this paper, the stabilization method of weighted relaxation for mixing iterations was considered, used to find the detailed gas composition. These algorithms have been tested on 87 realistic gas networks up to 8K elements in size and give nearly 100% convergence.

The remaining problems include the marginal degeneracy of the system associated with the control equations of compressors and regulators. This degeneracy enhances the effects of divergence. In the future, we would like to investigate the influence of this instability on various solution algorithms, including dynamic methods for integrating differential-algebraic equations. In addition, we would like to explore the alternative between iterative and coupled methods for solving mixing equations.

Acknowledgments. We acknowledge the support of the German Federal Ministry for Economic Affairs and Energy, project BMWI-0324019A, MathEnergy: Mathematical Key Technologies for Evolving Energy Grids and by the project TransHyDE-Sys, grant 03HY201M. We are grateful to the organizers and participants of SIMULTECH 2021 conference for the opportunity to present our work and for the fruitful discussion.

References

1. Baldin, A., et al.: AdvWarp: a transformation algorithm for advanced modeling of gas compressors and drives. In: Proceedings of SIMULTECH 2021, International Conference on Simulation and Modeling Methodologies, Technologies and Applications, pp. 231–238. SciTePress (2021)
2. Clees, T., Nikitin, I., Nikitina, L.: Making network solvers globally convergent. In: Obaidat, M.S., Ören, T., Merkuryev, Y. (eds.) SIMULTECH 2016. AISC, vol. 676, pp. 140–153. Springer, Cham (2018). https://doi.org/10.1007/978-3-319-69832-8_9
3. Clees, T., Nikitin, I., Nikitina, L., Segiet, L.: Modeling of gas compressors and hierarchical reduction for globally convergent stationary network solvers. Int. J. Adv. Syst. Measur. **11**, 61–71 (2018)
4. Baldin, A., Clees, T., Klaassen, B., Nikitin, I., Nikitina, L.: Topological reduction of stationary network problems: example of gas transport. Int. J. Adv. Syst. Measur. **13**, 83–93 (2020)

5. Baldin, A., et al.: Universal translation algorithm for formulation of transport network problems. In: Proceedings of SIMULTECH 2018, International Conference on Simulation and Modeling Methodologies, Technologies and Applications, vol. 1, pp. 315–322. SciTePress (2018)
6. Clees, T., et al.: MYNTS: multi-phYsics NeTwork simulator. In: Proceedings of SIMULTECH 2016, International Conference on Simulation and Modeling Methodologies, Technologies and Applications, pp. 179–186. SciTePress (2016)
7. Mischner, J., Fasold, H.G., Kadner, K.: System-Planning Basics of Gas Supply. Oldenbourg Industrieverlag GmbH (2011). (in German)
8. Schmidt, M., Steinbach, M.C., Willert, B.M.: High detail stationary optimization models for gas networks. Optim. Eng. **16**(1), 131–164 (2015)
9. Nikuradse, J.: Laws of flow in rough pipes. NACA Technical Memorandum 1292, Washington (1950)
10. Colebrook, C.F., White, C.M.: Experiments with fluid friction in roughened pipes. Proc. R. Soc. Lond. Ser. A Math. Phys. Sci. **161**(906) 367–381 (1937)
11. Saleh, J. (ed.): Fluid Flow Handbook. McGraw-Hill Handbooks, McGraw-Hill (2002)
12. DIN EN ISO 12213-2: Natural Gas – Calculation of compression factor. European Committee for Standartization (2010)
13. Kunz, O., Wagner, W.: The GERG-2008 wide-range equation of state for natural gases and other mixtures: an expansion of GERG-2004. J. Chem. Eng. Data **57**, 3032–3091 (2012)
14. Wächter, A., Biegler, L.T.: On the implementation of an interior-point filter line-search algorithm for large-scale nonlinear programming. Math. Program. **106**(1), 25–57 (2006)
15. Gill, P.E., et al.: SNOPT: an SQP algorithm for large-scale constrained optimization. SIAM Rev. **47**, 99–131 (2005)
16. Murtagh, B., Saunders, M.: Large-scale linearly constrained optimization. Math. Program. **14**, 41–72 (1978)
17. Kelley, C.T.: Iterative Methods for Linear and Nonlinear Equations. SIAM, Philadelphia (1995)

Quasi-static Optimal Control Strategy of Lattice Boom Crane Based on Large-Scale Flexible Non-linear Dynamics

Lingchong Gao(✉), Xiaobing Dai, Michael Kleeberger, and Johannes Fottner

Chair of Materials Handling, Material Flow, Logistics, Technical University of Munich, Boltzmannstrasse 15, 85748 Garching, Germany

lingchong.gao@tum.de

Abstract. Lattice boom cranes are usually used to lift heavy loads with the optimized lattice structure of the boom structure. Considering the huge mass of the payload and the crane itself, the flexibility of the crane boom structure cannot be ignored. The elastic vibrations mainly accrue at the lattice boom, the luffing system, and the cables. In this paper, several flexible multibody dynamic models are established as the beam elements (spatial Timoshenko beam), the rod elements (strut tie model), and the rope elements (ideal cubic spline model). In addition, a super truss element formulation for the regular truss structure is proposed to reduce the number of degrees of freedom of the complex lattice boom components. For controlling the large-scale nonlinear dynamic system, a quasi-static optimal control strategy is designed to realize the controllable motions for the specified complex system. This method combines the static mapping relationship with the target optimal trajectory to generate the optimal trajectories of control inputs. Through the elementary motions, the dynamic calculation of the lattice boom crane is performed to simulate the lifting, the luffing, and the slewing stages. In the aspect of control, the static mapping relationship between the key state variables and the control variables is established. A specified lifting task is designed to verify the quasi-static optimal control strategy.

Keywords: Lattice boom crane · Non-linear dynamics · Model reduction · Optimal control

1 Introduction

In the industry, different types of cranes are widely used, such as tower cranes used in the construction of tall buildings, floating cranes used in bridge and port construction, etc. Most of these cranes are equipped with a boom system. Because of the optimized structure, lattice booms (truss booms) have higher load capacity compared with telescope booms, especially in the field of steel, building, and bridge construction [1]. Because of the large mass of the load and the crane itself, the flexibility of the crane elements should not be ignored. Besides the long boom system, the deformation of the rods and cables is also large and determined, which makes the dynamic modeling much more difficult. The

G. Wagner et al. (Eds.): SIMULTECH 2021, LNNS 601, pp. 153–177, 2023.
https://doi.org/10.1007/978-3-031-23149-0_9

commonly used dynamic models, either with FEM or with nonlinear dynamic models, will cause an explosion of the number of degrees of freedom (DoFs). Nonlinear dynamic models need relatively fewer DoFs, but much higher complexity and nonlinearity. Both high numbers of DoFs and nonlinearity make the control for lattice boom cranes very hard. In the following text, suitable dynamic modeling methods and implementable control strategies are studied.

Among these dynamic models, lattice boom is most difficult because of its uneven structure and plenty of beam elements. The most common way is to ignore the particularity of the truss structure and model it by using a continuous flexible beam, which loses some of the important characteristics. Another way is to model all the elements in the lattice boom [2], which needs hundreds of DoFs and needs more calculations, and is time-consuming. To simplify the model of truss boom, a static condensation method is proposed [3] but is not suitable for dynamic calculation. For dynamic condensation, the Craig-Bampton method converts the dynamic equation from time-domain to frequency-domain to get the key information [4]. But it is hard to be implemented to the large-scale nonlinear system, such as lattice boom.

The vibration of flexible large structures is a severe test for the stability and safety of large structures [5]. To reduce the impact of dynamic loads on the crane structure, a controller is needed which can reduce the vibration while meeting specific operation objects [6]. In addition, the movement of such large machinery is very energy-consuming, it is necessary to investigate the optimal trajectory to meet the requirements of moving time, mechanical stability, and energy consumption. Currently, general optimal control algorithms are only used for simpler models. The research on optimal control algorithms for large-scale dynamic models is very necessary.

In this paper, several dynamic models are established for beams, rods, and ropes. Besides a super truss element is proposed to create the dynamic model for lattice structure with the minimum number of DoFs and acceptable calculations. A detailed dynamic model for lattice boom crane is built with two lattice booms, luffing module, hoist cable, and sub-cable. In the aspect of control, a quasi-static optimal control strategy is proposed to apply the optimal control on the large-scale nonlinear system. Then through the simulation, the dynamic model and control strategy are verified.

2 Modeling of Lattice Boom Crane

2.1 Single Beam Elements

Timoshenko Beam Model. All the elements in the lattice boom, which have no connection in the middle of the elements, will be modeled as the beam element. Spatial Timoshenko beams take both shear deformation and bending effect into account and are suitable for modeling the beams with relative short in length. The calculation for spatial Timoshenko beam in this section is based on [7]. In [7], the virtual power is calculated by using triple integral. Here the equivalent properties will be calculated to convert the triple integral into the integral along the arc-coordinate.

The spatial Timoshenko beam is modeled based on the co-rotational coordinate $\boldsymbol{q}_{\mathrm{B}}$, which describes the undeformed state. The deformation is defined as the deviation from the undeformed state in the co-rotational coordinate. To avoid the shear lock, a cubic shape function with respect to the arc coordinate is established to calculate the deformation of any point $\boldsymbol{q}_{\mathrm{d,c}}$ based on the deformation at the end section $\boldsymbol{q}_{\mathrm{d,end}}$ [8].

$$\boldsymbol{q}_{\mathrm{d,c}} = \boldsymbol{N}_{\mathrm{c}}\boldsymbol{q}_{\mathrm{d,end}},\ \boldsymbol{q}_{\mathrm{d,c}}^{\mathrm{p}} = \partial^{\mathrm{p}}\left(\boldsymbol{q}_{\mathrm{d,c}}\right)/\partial s^{\mathrm{p}} = \boldsymbol{N}_{\mathrm{c}}^{\mathrm{p}}\boldsymbol{q}_{\mathrm{d,end}} \tag{1}$$

where

$$\boldsymbol{N}_{\mathrm{c}} = \begin{bmatrix} \boldsymbol{N}_{\mathrm{c,r}} \\ \boldsymbol{N}_{\mathrm{c},\varphi} \end{bmatrix} = \begin{bmatrix}
N_{\mathrm{u,u0}} & & & & & \\
 & N_{\mathrm{v,v0}} & & & & N_{\theta,\mathrm{v0}} \\
 & & N_{\mathrm{w,w0}} & & N_{\psi,\mathrm{w0}} & \\
 & & & N_{\varphi,\varphi0} & & \\
 & & N_{\mathrm{w},\psi0} & & N_{\psi,\psi0} & \\
 & N_{\mathrm{v},\theta0} & & & & N_{\theta,\theta0} \\
N_{\mathrm{u,ue}} & & & & & \\
 & N_{\mathrm{v,ve}} & & & & N_{\theta,\mathrm{ve}} \\
 & & N_{\mathrm{w,we}} & & N_{\psi,\mathrm{we}} & \\
 & & & N_{\varphi,\varphi\mathrm{e}} & & \\
 & & N_{\mathrm{w},\psi\mathrm{e}} & & N_{\psi,\psi\mathrm{e}} & \\
 & N_{\mathrm{v},\theta\mathrm{e}} & & & & N_{\theta,\theta\mathrm{e}}
\end{bmatrix}^{\mathrm{T}}$$

in which

$$N_{\mathrm{u,u0}} = 1 - \xi,\ N_{\mathrm{u,ue}} = \xi,\ N_{\varphi,\varphi0} = 1 - \xi,\ N_{\varphi,\varphi e} = \xi$$

$$N_{\mathrm{v,v0}} = \beta_{\mathrm{y}}\left[2\xi^3 - 3\xi^2 - \alpha_{\mathrm{y}}\xi + (\alpha_{\mathrm{y}} + 1)\right],\ N_{\mathrm{v,ve}} = \beta_{\mathrm{y}}\left(-2\xi^3 + 3\xi^2 + \alpha_{\mathrm{y}}\xi\right)$$

$$N_{\mathrm{v},\theta0} = \beta_{\mathrm{y}}L\left[\xi^3 - \left(\frac{\alpha_{\mathrm{y}}}{2} + 2\right)\xi^2 + \left(\frac{\alpha_{\mathrm{y}}}{2} + 1\right)\xi\right],\ N_{\mathrm{v},\theta\mathrm{e}} = \beta_{\mathrm{y}}L\left(\xi^3 + \frac{\alpha_{\mathrm{y}} - 2}{2}\xi^2 - \frac{\alpha_{\mathrm{y}}}{2}\xi\right)$$

$$N_{\theta,\mathrm{v0}} = 6\beta_{\mathrm{y}}/L\left(\xi^2 - \xi\right),\ N_{\theta,\mathrm{ve}} = -6\beta_{\mathrm{y}}/L\left(\xi^2 - \xi\right)$$

$$N_{\psi,\mathrm{w0}} = -6\beta_{\mathrm{z}}/L\left(\xi^2 - \xi\right),\ N_{\psi,\mathrm{we}} = 6\beta_{\mathrm{z}}/L\left(\xi^2 - \xi\right)$$

$$N_{\theta,\theta0} = \beta_{\mathrm{y}}\left[3\xi^2 - (\alpha_{\mathrm{y}} + 4)\xi + (\alpha_{\mathrm{y}} + 1)\right],\ N_{\theta,\theta\mathrm{e}} = \beta_{\mathrm{y}}\left[3\xi^2 + (\alpha_{\mathrm{y}} - 2)\xi\right]$$

$$N_{\mathrm{w,w0}} = \beta_{\mathrm{z}}\left[2\xi^3 - 3\xi^2 - \alpha_{\mathrm{z}}\xi + (\alpha_{\mathrm{z}} + 1)\right],\ N_{\mathrm{w,we}} = \beta_{\mathrm{z}}\left(-2\xi^3 + 3\xi^2 + \alpha_{\mathrm{z}}\xi\right)$$

$$N_{\mathrm{w},\psi0} = \beta_{\mathrm{z}}L\left[-\xi^3 + \left(\frac{\alpha_{\mathrm{z}}}{2} + 2\right)\xi^2 - \left(\frac{\alpha_{\mathrm{z}}}{2} + 1\right)\xi\right],\ N_{\mathrm{w},\psi\mathrm{e}} = -\beta_{\mathrm{z}}L\left(\xi^3 + \frac{\alpha_{\mathrm{z}} - 2}{2}\xi^2 - \frac{\alpha_{\mathrm{z}}}{2}\xi\right)$$

$$N_{\psi,\psi0} = \beta_{\mathrm{z}}\left[3\xi^2 - (\alpha_{\mathrm{z}} + 4)\xi + (\alpha_{\mathrm{z}} + 1)\right],\ N_{\psi,\psi\mathrm{e}} = \beta_{\mathrm{z}}\left[3\xi^2 + (\alpha_{\mathrm{z}} - 2)\xi\right]$$

with $\xi = \frac{s}{L}$, $\alpha_{\mathrm{y}} = \frac{48EI_{\mathrm{y}}}{AGL^2}$, $\alpha_{\mathrm{z}} = \frac{48EI_{\mathrm{z}}}{AGL^2}$, $\beta_{\mathrm{y}} = \frac{1}{\alpha_{\mathrm{y}}+1}$ and $\beta_{\mathrm{z}} = \frac{1}{\alpha_{\mathrm{z}}+1}$. Here it should be noticed that the shape function is only related to the arc-coordinate and the shape and material of the beam are irrelevant with the generalized coordinate $\boldsymbol{q}_{\mathrm{e}}$.

By using the co-rotational coordinate and the deformation at that point, the kinematics properties (translational position, velocity, and acceleration) of any point ${}_{\mathrm{I}}\boldsymbol{r}$, ${}_{\mathrm{I}}\dot{\boldsymbol{r}}$, ${}_{\mathrm{I}}\ddot{\boldsymbol{r}}$ on the beam can be calculated [7]. The equivalent kinematic properties and the equivalent stiffness of the beam section including the translational part and the rotational part can be written based on the energy conservation law as

$$\frac{1}{2}\rho\iint_{A}{}_{\mathrm{I}}\dot{\boldsymbol{r}}^{\mathrm{T}}{}_{\mathrm{I}}\dot{\boldsymbol{r}}\mathrm{dA}=\frac{1}{2}\rho A_{\mathrm{I}}\dot{\boldsymbol{r}}^{\mathrm{cT}}{}_{\mathrm{I}}\dot{\boldsymbol{r}}^{\mathrm{c}}+\frac{1}{2}\rho_{\mathrm{c}}\boldsymbol{\omega}^{\mathrm{cT}}\boldsymbol{J}_{\mathrm{c}}\boldsymbol{\omega}^{\mathrm{c}} \tag{2}$$

$$\iint_{A(\mathrm{c})}\sum_{\mathrm{i}}\sum_{\mathrm{j}}{}_{\mathrm{B}}\varepsilon_{\mathrm{ij}\mathrm{B}}\sigma_{\mathrm{ij}}\mathrm{dA}={}_{\mathrm{B}}\boldsymbol{\varepsilon}^{\mathrm{cT}}\boldsymbol{K}_{\mathrm{cB}}\boldsymbol{\varepsilon}^{\mathrm{c}} \tag{3}$$

with

$${}_{\mathrm{I}}\dot{\boldsymbol{r}}^{\mathrm{c}}=\boldsymbol{T}_{\mathrm{r,c}}\mathrm{d}\boldsymbol{q}_{\mathrm{e}},\ \boldsymbol{T}_{\mathrm{r,c}}=[\boldsymbol{I}\ 0]\boldsymbol{T}_{\mathrm{B}}-\boldsymbol{R}_{\mathrm{B}}\left({}_{\mathrm{B}}\tilde{\boldsymbol{r}}^{\mathrm{r,c}}+{}_{\mathrm{B}}\tilde{\boldsymbol{u}}^{\mathrm{c}}\right)[0\ \boldsymbol{I}]\boldsymbol{T}_{\mathrm{B}}+\boldsymbol{R}_{\mathrm{B}}\boldsymbol{N}_{\mathrm{c,r}}\boldsymbol{T}_{\mathrm{d,end}}$$

$${}_{\mathrm{c}}\boldsymbol{\omega}^{\mathrm{c}}=\boldsymbol{T}_{\varphi,\mathrm{c}}\mathrm{d}\boldsymbol{q}_{\mathrm{e}},\ \boldsymbol{T}_{\varphi,\mathrm{c}}=\boldsymbol{R}_{\mathrm{d,c}}^{\mathrm{T}}[0\ \boldsymbol{I}]\boldsymbol{T}_{\mathrm{B}}+\boldsymbol{N}_{\mathrm{c},\varphi}\boldsymbol{T}_{\mathrm{d,end}},\ \boldsymbol{J}=\mathrm{diag}\left(\left[I_{\mathrm{y}}+I_{\mathrm{z}}\ I_{\mathrm{z}}\ I_{\mathrm{y}}\right]\right)$$

$${}_{\mathrm{B}}\boldsymbol{\varepsilon}^{\mathrm{c}}=\left(\boldsymbol{N}_{\mathrm{c}}^{'}+\boldsymbol{T}^{\mathrm{d,c}}\boldsymbol{N}_{\mathrm{c}}\right)\boldsymbol{q}_{\mathrm{d,end}},\ \boldsymbol{K}_{\mathrm{c}}=\mathrm{diag}\left(\left[EA\ \tfrac{GA}{4}\ \tfrac{GA}{4}\ \tfrac{G(I_{\mathrm{y}}+I_{\mathrm{z}})}{4}\ EI_{\mathrm{z}}\ EI_{\mathrm{y}}\right]\right)$$

where $\boldsymbol{T}^{\mathrm{d,c}}$ is almost a zero matrix except $\boldsymbol{T}^{\mathrm{d,c}}(2,6)=-1$ and $\boldsymbol{T}^{\mathrm{d,c}}(3,5)=1$. Then the virtual inertial power, the virtual internal power and the virtual external power caused by gravity can be written as

$$\delta p_{\mathrm{ine}}=-\rho\delta\int_{0}^{L}\left(A_{\mathrm{I}}\ddot{\boldsymbol{r}}^{\mathrm{cT}}{}_{\mathrm{I}}\dot{\boldsymbol{r}}^{\mathrm{c}}+{}_{\mathrm{c}}\dot{\boldsymbol{\omega}}^{\mathrm{cT}}\boldsymbol{J}_{\mathrm{c}}\boldsymbol{\omega}^{\mathrm{c}}\right)\mathrm{ds}=-\delta\mathrm{d}\boldsymbol{q}_{\mathrm{e}}^{\mathrm{T}}\left(\boldsymbol{M}_{\mathrm{ine}}\mathrm{d}\dot{\boldsymbol{q}}_{\mathrm{e}}+\boldsymbol{F}_{\mathrm{ine}}\right) \tag{4}$$

$$\delta p_{\mathrm{int}}=-\delta\int_{0}^{L}{}_{\mathrm{B}}\dot{\boldsymbol{\varepsilon}}^{\mathrm{cT}}\boldsymbol{K}_{\mathrm{cB}}\boldsymbol{\varepsilon}^{\mathrm{c}}\mathrm{ds}=-\delta\mathrm{d}\boldsymbol{q}_{\mathrm{e}}^{\mathrm{T}}\boldsymbol{F}_{\mathrm{int}} \tag{5}$$

$$\delta p_{\mathrm{ext}}=\rho A\int_{0}^{L}{}_{\mathrm{I}}\ddot{\boldsymbol{r}}^{\mathrm{cT}}{}_{\mathrm{I}}\boldsymbol{g}\mathrm{ds}=-\delta\mathrm{d}\boldsymbol{q}_{\mathrm{e}}^{\mathrm{T}}\boldsymbol{F}_{\mathrm{ext,g}} \tag{6}$$

with the shape function, the generalized internal force is only related to the values at the end section of the beam [9]. The generalized force can be written as

$$\boldsymbol{F}_{\mathrm{int}}=\boldsymbol{T}_{\mathrm{B,end}}^{\mathrm{T}}\boldsymbol{N}_{\mathrm{c}}^{\mathrm{T}}\left(\boldsymbol{H}_{1}\boldsymbol{N}_{\mathrm{c}}^{'}+\boldsymbol{H}_{2}\boldsymbol{N}_{\mathrm{c}}\right)\Big|_{0}^{L}\boldsymbol{q}_{\mathrm{d,end}}$$

$$\boldsymbol{M}_{\mathrm{ine}}=\rho\int_{0}^{L}\left(A\boldsymbol{T}_{\mathrm{r,c}}^{\mathrm{T}}\boldsymbol{T}_{\mathrm{r,c}}+\boldsymbol{T}_{\varphi,\mathrm{c}}^{\mathrm{T}}\boldsymbol{J}\boldsymbol{T}_{\varphi,\mathrm{c}}\right)\mathrm{ds},\ \boldsymbol{F}_{\mathrm{ine}}=\boldsymbol{D}_{\mathrm{ine}}\mathrm{d}\boldsymbol{q}_{\mathrm{e}}$$

$$\boldsymbol{D}_{\mathrm{ine}}=\rho\int_{0}^{L}\left(A\boldsymbol{T}_{\mathrm{r,c}}^{\mathrm{T}}\dot{\boldsymbol{T}}_{\mathrm{r,c}}+\boldsymbol{T}_{\varphi,\mathrm{c}}^{\mathrm{T}}\boldsymbol{J}\dot{\boldsymbol{T}}_{\varphi,\mathrm{c}}\right)\mathrm{ds},\ \boldsymbol{F}_{\mathrm{ext,g}}=-\rho A\int_{0}^{L}\boldsymbol{T}_{\mathrm{r,c}}^{\mathrm{T}}\mathrm{ds}_{\mathrm{I}}\boldsymbol{g}$$

in which

$$\dot{\boldsymbol{T}}_{\mathrm{r,c}}=\boldsymbol{R}_{\mathrm{BB}}\tilde{\boldsymbol{\omega}}^{\mathrm{B}}\boldsymbol{N}_{\mathrm{c,r}}\boldsymbol{T}_{\mathrm{d,end}}+\boldsymbol{R}_{\mathrm{B}}\boldsymbol{N}_{\mathrm{c,r}}\dot{\boldsymbol{T}}_{\mathrm{d,end}}-\boldsymbol{R}_{\mathrm{BB}}\tilde{\boldsymbol{\omega}}^{\mathrm{B}}\left({}_{\mathrm{B}}\tilde{\boldsymbol{r}}^{\mathrm{r,c}}+{}_{\mathrm{B}}\tilde{\boldsymbol{u}}^{\mathrm{c}}\right)[0\ \boldsymbol{I}]\boldsymbol{T}_{\mathrm{B}}$$

$$\dot{\boldsymbol{T}}_{\varphi,\mathrm{c}} = \boldsymbol{N}_{\mathrm{c},\varphi}\dot{\boldsymbol{T}}_{\mathrm{d,end}} + \boldsymbol{R}_{\mathrm{d,cB}}^{\mathrm{T}}\overset{\sim\mathrm{B}}{\boldsymbol{\omega}}\boldsymbol{R}_{\mathrm{d,c}}\boldsymbol{N}_{\mathrm{c},\varphi}\boldsymbol{T}_{\mathrm{d,end}}$$

The total virtual power of Timoshenko beam can be written as

$$\delta p_{\mathrm{e}} = \delta p_{\mathrm{ine}} + \delta p_{\mathrm{int}} + \delta p_{\mathrm{ext,g}} = -\delta\dot{\boldsymbol{q}}_{\mathrm{e}}^{\mathrm{T}}\left(\boldsymbol{M}_{\mathrm{e}}\mathrm{d}\dot{\boldsymbol{q}}_{\mathrm{e}} + \boldsymbol{F}_{\mathrm{e}}\right) \tag{7}$$

with $\boldsymbol{M}_{\mathrm{e}} = \boldsymbol{M}_{\mathrm{ine}}$ and $\boldsymbol{F}_{\mathrm{e}} = \boldsymbol{F}_{\mathrm{ine}} + \boldsymbol{F}_{\mathrm{int}} + \boldsymbol{F}_{\mathrm{ext,g}}$.

Ideal Cubic Spline Rope Model. It is assumed that the axis of the rope after the deformation remains geometrically continuous. Hence the cubic spline rope model is selected. The assumptions for the ideal cubic spline rope are as follows:

1. the axis after the deformation is smooth and continuous in the absolute coordinate system, and satisfies the positions and postures of the two end points;
2. only the axial strain along the rope is considered;
3. the density and total mass of the rope remain unchanged.

According to the third assumption, in addition to the position and the orientation angle of the end point, the generalized coordinates also include the norm of the first derivative of the position with respect to the arc length coordinate s. This norm can represent the axial strain of the end point. Therefore, the generalized coordinates and the generalized velocity of the cubic spline beam are defined as follows:

$$\boldsymbol{q}_{\mathrm{e}}^{\mathrm{T}} = \left[\boldsymbol{q}_1^{\mathrm{T}}\ \boldsymbol{q}_2^{\mathrm{T}}\right], \boldsymbol{q}_{\mathrm{i}}^{\mathrm{T}} = \left[{}_{\mathrm{I}}\boldsymbol{r}^{\mathrm{i}^{\mathrm{T}}}\ \boldsymbol{\varphi}^{\mathrm{i}^{\mathrm{T}}}\ \|{}_{\mathrm{I}}\boldsymbol{r}'^{\mathrm{i}}\|\right], \mathrm{i} = 1, 2$$

$$\mathrm{d}\boldsymbol{q}_{\mathrm{e}}^{\mathrm{T}} = \left[\mathrm{d}\boldsymbol{q}_1^{\mathrm{T}}\ \mathrm{d}\boldsymbol{q}_2^{\mathrm{T}}\right], \mathrm{d}\boldsymbol{q}_{\mathrm{i}}^{\mathrm{T}} = \left[{}_{\mathrm{I}}\dot{\boldsymbol{r}}^{\mathrm{i}^{\mathrm{T}}}\ {}_{\mathrm{i}}\boldsymbol{\omega}^{\mathrm{i}^{\mathrm{T}}}\ \|{}_{\mathrm{I}}\dot{\boldsymbol{r}}'^{\mathrm{i}}\|\right], \mathrm{i} = 1, 2 \tag{8}$$

The relative rotation between two end sections, ${}_1\boldsymbol{\varphi}^{12}$, can be formulated through cardan angle with the sequence of $1 \to z\left({}_1\theta^{12}\right) \to y\left({}_1\psi^{12}\right) \to x\left({}_1\varphi^{12}\right) \to 2$. The detailed expression is written in [7]. The velocity and the acceleration of the cardan angle can be written as

$${}_1\dot{\boldsymbol{\varphi}}^{12} = \boldsymbol{T}_{\varphi 12}\mathrm{d}\boldsymbol{q}_{\mathrm{e}}, {}_1\ddot{\boldsymbol{\varphi}}^{12} = \boldsymbol{T}_{\varphi 12}\mathrm{d}\dot{\boldsymbol{q}}_{\mathrm{e}} + \dot{\boldsymbol{T}}_{\varphi 12}\mathrm{d}\boldsymbol{q}_{\mathrm{e}} \tag{9}$$

with $\boldsymbol{R}_1 = \boldsymbol{R}(\boldsymbol{\varphi}^1), \boldsymbol{R}_2 = \boldsymbol{R}(\boldsymbol{\varphi}^2)$ and

$$\boldsymbol{T}_{12}^{\mathrm{c}} = \boldsymbol{T}^{\mathrm{c}}\left({}_1\boldsymbol{\varphi}^{12}\right), \dot{\boldsymbol{T}}_{12}^{\mathrm{c}} = \dot{\boldsymbol{T}}^{\mathrm{c}}\left({}_1\boldsymbol{\varphi}^{12}, {}_1\dot{\boldsymbol{\varphi}}^{12}\right)$$

$$\boldsymbol{T}_{\varphi 12} = \boldsymbol{T}_{12}^{\mathrm{c}^{-1}}\left(\boldsymbol{T}_{\omega 2} - \boldsymbol{R}_2^{\mathrm{T}}\boldsymbol{R}_1\boldsymbol{T}_{\omega 1}\right), \dot{\boldsymbol{T}}_{\varphi 12} = \boldsymbol{T}_{12}^{\mathrm{c}^{-1}}\left({}_2\overset{\sim 2}{\boldsymbol{\omega}}\boldsymbol{R}_2^{\mathrm{T}}\boldsymbol{R}_1\boldsymbol{T}_{\omega 1} - \dot{\boldsymbol{T}}_{12}^{\mathrm{c}}\boldsymbol{T}_{\varphi 12}\right)$$

$$\boldsymbol{T}_{\omega 1} = \left[\boldsymbol{0}\,\boldsymbol{I}\,\boldsymbol{0}\,\boldsymbol{0}\,\boldsymbol{0}\,\boldsymbol{0}\right], \boldsymbol{T}_{\omega 2} = \left[\boldsymbol{0}\,\boldsymbol{0}\,\boldsymbol{0}\,\boldsymbol{0}\,\boldsymbol{I}\,\boldsymbol{0}\right]$$

where $\boldsymbol{T}^{\mathrm{c}}$ and $\dot{\boldsymbol{T}}^{\mathrm{c}}$ should be calculated with the definition of cardan angle. In order to ensure the geometric continuity of the cubic spline rope in the absolute coordinate system, Hermite interpolation is used to determine the position of the center of the section. Therefore, the position vector of section c and its derivative along the arc length coordinate can be written as

$$_{\mathrm{I}}\boldsymbol{r}^{(\mathrm{s})\mathrm{c}} = N_0^{1\,(\mathrm{s})}\,{}_{\mathrm{I}}\boldsymbol{r}^1 + N_0^{2\,(\mathrm{s})}\,{}_{\mathrm{I}}\boldsymbol{r}^2 + N_1^{1\,(\mathrm{s})}\,{}_{\mathrm{I}}\boldsymbol{r}'^1 + N_1^{2\,(\mathrm{s})}\,{}_{\mathrm{I}}\boldsymbol{r}'^2 \tag{10}$$

with $(*)^{(\mathrm{s})} = \mathrm{d}^{(\mathrm{s})}(*)/\mathrm{d}s^{(\mathrm{s})}$, $\mathrm{s} \in \mathbb{N}$, $\xi = x/L$ and

$$N_0^1 = 1 - 3\xi^2 + 2\xi^3,\ N_0^2 = 3\xi^2 - 2\xi^3,\ N_1^1 = L\left(\xi - 2\xi^2 + \xi^3\right),\ N_1^2 = L\left(\xi^3 - \xi^2\right)$$

$$_{\mathrm{I}}\boldsymbol{r}'^1 = \|_{\mathrm{I}}\boldsymbol{r}'^1\|_{\mathrm{I}}\boldsymbol{n}_{\mathrm{x}}^1,\ {}_{\mathrm{I}}\boldsymbol{r}'^2 = \|_{\mathrm{I}}\boldsymbol{r}'^2\|_{\mathrm{I}}\boldsymbol{n}_{\mathrm{x}}^2,\ {}_{\mathrm{I}}\boldsymbol{n}_{\mathrm{x}}^1 = \boldsymbol{R}_1\boldsymbol{g}_{\mathrm{x}},\ {}_{\mathrm{I}}\boldsymbol{n}_{\mathrm{x}}^2 = \boldsymbol{R}_2\boldsymbol{g}_{\mathrm{x}}$$

in which, $\boldsymbol{g}_{\mathrm{x}} = \begin{bmatrix}1\ 0\ 0\end{bmatrix}^{\mathrm{T}}$.The velocity and the acceleration of the section c can be calculated through

$$_{\mathrm{I}}\dot{\boldsymbol{r}}^{(\mathrm{s})\mathrm{c}} = \boldsymbol{N}^{(\mathrm{s})}\boldsymbol{D}\mathrm{d}\boldsymbol{q}_{\mathrm{e}},\ {}_{\mathrm{I}}\ddot{\boldsymbol{r}}^{(\mathrm{s})\mathrm{c}} = \boldsymbol{N}^{(\mathrm{s})}\dot{\boldsymbol{D}}\mathrm{d}\boldsymbol{q}_{\mathrm{e}} + \boldsymbol{N}^{(\mathrm{s})}\boldsymbol{D}\mathrm{d}\dot{\boldsymbol{q}}_{\mathrm{e}} \tag{11}$$

where $\boldsymbol{N}^{(\mathrm{s})} = \left[N_0^{1\,(\mathrm{s})}\boldsymbol{I}\ N_1^{1\,(\mathrm{s})}\boldsymbol{I}\ N_0^{2\,(\mathrm{s})}\boldsymbol{I}\ N_1^{2\,(\mathrm{s})}\boldsymbol{I}\right]$ and

$$\boldsymbol{D} = \begin{bmatrix} \boldsymbol{I} & & & \\ & -{}_{\mathrm{I}}\tilde{\boldsymbol{r}}'^1\boldsymbol{R}_1\ {}_{\mathrm{I}}\boldsymbol{n}_{\mathrm{x}}^1 & & \\ & & \boldsymbol{I} & \\ & & & -{}_{\mathrm{I}}\tilde{\boldsymbol{r}}'^2\boldsymbol{R}_2\ {}_{\mathrm{I}}\boldsymbol{n}_{\mathrm{x}}^2 \end{bmatrix}$$

$$\dot{\boldsymbol{D}} = \begin{bmatrix} \boldsymbol{0} & & & \\ & -\|_{\mathrm{I}}\tilde{\boldsymbol{r}}'^1\|\boldsymbol{R}_{1\,1}\tilde{\boldsymbol{\omega}}^1\tilde{\boldsymbol{g}}_{\mathrm{x}}\ -2_{\mathrm{I}}\boldsymbol{n}_{\mathrm{x}}^1{}_{\mathrm{I}}\boldsymbol{\omega}^1 & & \\ & & \boldsymbol{0} & \\ & & & -\|_{\mathrm{I}}\tilde{\boldsymbol{r}}'^2\|\boldsymbol{R}_{2\,2}\tilde{\boldsymbol{\omega}}^2\tilde{\boldsymbol{g}}_{\mathrm{x}}\ -2_{\mathrm{I}}\boldsymbol{n}_{\mathrm{x}}^2{}_{\mathrm{I}}\boldsymbol{\omega}^2 \end{bmatrix}$$

The relative rotation cardan angle from Sect. 1 to any section in the rope.

$_1\boldsymbol{\varphi} = \left[{}_1\varphi\ {}_1\psi\ {}_1\theta\right]^{\mathrm{T}}$ can be obtained through the normal vector of the section ${}_{\mathrm{I}}\boldsymbol{n}_{\mathrm{x}}$ and linear interpolation for torsion. The detailed expression is shown in [7]. The velocity and the acceleration of the relative cardan angle and its derivative with respect to the arc length can be written as

$$_1\dot{\boldsymbol{\varphi}} = \boldsymbol{T}_{\mathrm{Ang}}\mathrm{d}\boldsymbol{q}_{\mathrm{e}},\ {}_1\dot{\boldsymbol{\varphi}}' = \boldsymbol{T}_{\mathrm{dAng}}\mathrm{d}\boldsymbol{q}_{\mathrm{e}},\ {}_1\ddot{\boldsymbol{\varphi}} = \boldsymbol{T}_{\mathrm{Ang}}\mathrm{d}\dot{\boldsymbol{q}}_{\mathrm{e}} + \dot{\boldsymbol{T}}_{\mathrm{Ang}}\mathrm{d}\boldsymbol{q}_{\mathrm{e}} \tag{12}$$

with

$$\boldsymbol{T}_{\mathrm{Ang}}^{\mathrm{T}} = \left[\boldsymbol{T}_{\varphi}\ \boldsymbol{T}_{\psi}\ \boldsymbol{T}_{\theta}\right],\ \dot{\boldsymbol{T}}_{\mathrm{Ang}}^{\mathrm{T}} = \left[\dot{\boldsymbol{T}}_{\varphi}\ \dot{\boldsymbol{T}}_{\psi}\ \dot{\boldsymbol{T}}_{\theta}\right],\ \boldsymbol{T}_{\mathrm{dAng}}^{\mathrm{T}} = \left[\boldsymbol{T}_{\mathrm{d}\varphi}\ \boldsymbol{T}_{\mathrm{d}\psi}\ \boldsymbol{T}_{\mathrm{d}\theta}\right]$$

where

$$\boldsymbol{T}_{\varphi}^{\mathrm{T}} = \xi\boldsymbol{g}_{\mathrm{x}}^{\mathrm{T}}\boldsymbol{T}_{\varphi 12},\ \dot{\boldsymbol{T}}_{\varphi}^{\mathrm{T}} = \xi\boldsymbol{g}_{\mathrm{x}}^{\mathrm{T}}\dot{\boldsymbol{T}}_{\varphi 12},\ \boldsymbol{T}_{\mathrm{d}\varphi}^{\mathrm{T}} = \frac{1}{L}\boldsymbol{g}_{\mathrm{x}}^{\mathrm{T}}\boldsymbol{T}_{\varphi 12}$$

$$\boldsymbol{T}_{\psi}^{\mathrm{T}} = -\boldsymbol{g}_{\mathrm{z}}^{\mathrm{T}}\boldsymbol{R}_{1}^{\mathrm{T}}\left(\boldsymbol{T}_{\mathrm{nx}} + {}_{\mathrm{I}}\tilde{\boldsymbol{n}}_{\mathrm{x}}\boldsymbol{R}_{1}\boldsymbol{T}_{\omega 1}\right)/\sqrt{1-\sin^{2}{}_{1}\psi}$$

$$\dot{\boldsymbol{T}}_{\psi}^{\mathrm{T}} = -\frac{\boldsymbol{g}_{\mathrm{z}}^{\mathrm{T}}\boldsymbol{R}_{1}^{\mathrm{T}}}{\sqrt{1-\sin^{2}{}_{1}\psi}}\left[\dot{\boldsymbol{T}}_{\mathrm{nx}} + \left(2{}_{\mathrm{I}}\dot{\tilde{\boldsymbol{n}}}_{\mathrm{x}} - {}_{\mathrm{I}}\tilde{\boldsymbol{\omega}}^{1}\,{}_{\mathrm{I}}\tilde{\boldsymbol{n}}_{\mathrm{x}}\right)\boldsymbol{R}_{1}\boldsymbol{T}_{\omega 1} + \left({}_{\mathrm{I}}\dot{\boldsymbol{n}}_{\mathrm{x}} + {}_{\mathrm{I}}\tilde{\boldsymbol{n}}_{\mathrm{x}\mathrm{I}}\boldsymbol{\omega}^{1}\right)\boldsymbol{T}_{\psi}^{\mathrm{T}}\tan_{1}\psi\right]$$

$$\boldsymbol{T}_{\mathrm{d}\psi}^{\mathrm{T}} = -\boldsymbol{g}_{\mathrm{z}}^{\mathrm{T}}\boldsymbol{R}_{1}^{\mathrm{T}}\left(\boldsymbol{T}_{\mathrm{dnx}} + {}_{\mathrm{I}}\tilde{\boldsymbol{n}}_{\mathrm{x}}^{'}\boldsymbol{R}_{1}\boldsymbol{T}_{\omega 1} + {}_{\mathrm{I}}\boldsymbol{n}_{\mathrm{x}}^{'}\boldsymbol{T}_{\psi}^{\mathrm{T}}\tan_{1}\psi\right)/\sqrt{1-\sin^{2}{}_{1}\psi}$$

$$\boldsymbol{T}_{\theta}^{\mathrm{T}} = \boldsymbol{g}_{\mathrm{y}}^{\mathrm{T}}\boldsymbol{R}_{1}^{\mathrm{T}}\left(\boldsymbol{T}_{\mathrm{nx}} + {}_{\mathrm{I}}\tilde{\boldsymbol{n}}_{\mathrm{x}}\boldsymbol{R}_{1}\boldsymbol{T}_{\omega 1} + {}_{\mathrm{I}}\boldsymbol{n}_{\mathrm{x}}\boldsymbol{T}_{\psi}^{\mathrm{T}}\tan_{1}\psi\right)/\left(\cos_{1}\psi\sqrt{1-\sin^{2}{}_{1}\theta}\right)$$

$$\begin{aligned}\dot{\boldsymbol{T}}_{\theta}^{\mathrm{T}} = {} & 1/\left(\cos_{1}\psi\sqrt{1-\sin^{2}{}_{1}\theta}\right)\boldsymbol{g}_{\mathrm{y}}^{\mathrm{T}}\boldsymbol{R}_{1}^{\mathrm{T}}(\dot{\boldsymbol{T}}_{\mathrm{nx}} + {}_{\mathrm{I}}\boldsymbol{n}_{\mathrm{x}}\dot{\boldsymbol{T}}_{\psi}^{\mathrm{T}}\tan_{1}\psi - {}_{\mathrm{I}}\tilde{\boldsymbol{\omega}}^{1}\,{}_{\mathrm{I}}\tilde{\boldsymbol{n}}_{\mathrm{x}}\boldsymbol{R}_{1}\boldsymbol{T}_{\omega 1} - 2{}_{\mathrm{I}}\tilde{\boldsymbol{\omega}}^{1}\boldsymbol{T}_{\mathrm{nx}} \\ & + \left(2\tan_{1}\psi_{\mathrm{I}}\dot{\boldsymbol{n}}_{\mathrm{x}} - 2\tan_{1}\psi_{\mathrm{I}}\tilde{\boldsymbol{\omega}}^{1}\,{}_{\mathrm{I}}\boldsymbol{n}_{\mathrm{x}} + {}_{1}\dot{\psi}\sec^{2}{}_{1}\psi_{\mathrm{I}}\boldsymbol{n}_{\mathrm{x}} + {}_{1}\dot{\psi}\tan^{2}{}_{1}\psi_{\mathrm{I}}\boldsymbol{n}_{\mathrm{x}}\right)\boldsymbol{T}_{\psi}^{\mathrm{T}} \\ & + \left({}_{\mathrm{I}}\dot{\boldsymbol{n}}_{\mathrm{x}} + {}_{1}\dot{\psi}\tan_{1}\psi_{\mathrm{I}}\boldsymbol{n}_{\mathrm{x}} - {}_{\mathrm{I}}\tilde{\boldsymbol{\omega}}^{1}\,{}_{\mathrm{I}}\boldsymbol{n}_{\mathrm{x}}\right)\tan_{1}\theta\boldsymbol{T}_{\theta}^{\mathrm{T}})\end{aligned}$$

$$\begin{aligned}\boldsymbol{T}_{\mathrm{d}\theta}^{\mathrm{T}} = {} & 1/\left(\cos_{1}\psi\sqrt{1-\sin^{2}{}_{1}\theta}\right) \\ & \boldsymbol{g}_{\mathrm{y}}^{\mathrm{T}}\boldsymbol{R}_{1}^{\mathrm{T}}\left[\boldsymbol{T}_{\mathrm{dnx}} + {}_{\mathrm{I}}\tilde{\boldsymbol{n}}_{\mathrm{x}}^{'}\boldsymbol{R}_{1}\boldsymbol{T}_{\omega 1} + {}_{\mathrm{I}}\boldsymbol{n}_{\mathrm{x}}^{'}\boldsymbol{T}_{\psi}^{\mathrm{T}}\tan_{1}\psi + \sec^{2}{}_{1}\psi_{1}\psi^{'}\,{}_{\mathrm{I}}\boldsymbol{n}_{\mathrm{x}}\boldsymbol{T}_{\psi}^{\mathrm{T}} + {}_{\mathrm{I}}\boldsymbol{n}_{\mathrm{x}}\boldsymbol{T}_{\mathrm{d}\psi}^{\mathrm{T}}\tan_{1}\psi\right] \\ & + \left(\tan_{1}\psi_{1}\psi^{'} + \tan_{1}\theta_{1}\theta^{'}\right)\boldsymbol{T}_{\theta}^{\mathrm{T}}\end{aligned}$$

in which

$$\boldsymbol{T}_{\mathrm{nx}} = 1/\|{}_{\mathrm{I}}\boldsymbol{r}^{'\mathrm{c}}\|\left(\boldsymbol{I} - {}_{\mathrm{I}}\boldsymbol{n}_{\mathrm{x}\mathrm{I}}\boldsymbol{n}_{\mathrm{x}}^{\mathrm{T}}\right)\boldsymbol{N}^{'}\boldsymbol{D},\ {}_{\mathrm{I}}\dot{\boldsymbol{n}}_{\mathrm{x}} = \boldsymbol{T}_{\mathrm{nx}}\mathrm{d}\boldsymbol{q}_{\mathrm{e}},\ {}_{\mathrm{I}}\boldsymbol{n}_{\mathrm{x}}^{'} = 1/\|{}_{\mathrm{I}}\boldsymbol{r}^{'\mathrm{c}}\|\left(\boldsymbol{I} - {}_{\mathrm{I}}\boldsymbol{n}_{\mathrm{x}\mathrm{I}}\boldsymbol{n}_{\mathrm{x}}^{\mathrm{T}}\right){}_{\mathrm{I}}\boldsymbol{r}^{'\mathrm{c}}$$

$$\dot{\boldsymbol{T}}_{\mathrm{nx}} = -1/\|{}_{\mathrm{I}}\boldsymbol{r}^{'\mathrm{c}}\|\left[\left(2{}_{\mathrm{I}}\dot{\boldsymbol{n}}_{\mathrm{x}\mathrm{I}}\boldsymbol{n}_{\mathrm{x}}^{\mathrm{T}} + {}_{\mathrm{I}}\boldsymbol{n}_{\mathrm{x}\mathrm{I}}\dot{\boldsymbol{n}}_{\mathrm{x}}^{\mathrm{T}}\right)\boldsymbol{N}^{'}\boldsymbol{D} - \left(\boldsymbol{I} - {}_{\mathrm{I}}\boldsymbol{n}_{\mathrm{x}\mathrm{I}}\boldsymbol{n}_{\mathrm{x}}^{\mathrm{T}}\right)\boldsymbol{N}^{'}\dot{\boldsymbol{D}}\right]$$

$$\boldsymbol{T}_{\mathrm{dnx}} = -1/\|{}_{\mathrm{I}}\boldsymbol{r}^{'\mathrm{c}}\|\left[{}_{\mathrm{I}}\boldsymbol{n}_{\mathrm{x}\mathrm{I}}^{'}\boldsymbol{n}_{\mathrm{x}}^{\mathrm{T}}\boldsymbol{N}^{'}\boldsymbol{D} + \left({}_{\mathrm{I}}\boldsymbol{n}_{\mathrm{x}}^{\mathrm{T}}{}_{I\mathrm{I}}\boldsymbol{r}^{''\mathrm{c}}\boldsymbol{I} + {}_{\mathrm{I}}\boldsymbol{n}_{\mathrm{x}I\mathrm{I}}\boldsymbol{r}^{''\mathrm{cT}}\right)\boldsymbol{T}_{\mathrm{nx}} - \left(\boldsymbol{I} - {}_{\mathrm{I}}\boldsymbol{n}_{\mathrm{x}\mathrm{I}}\boldsymbol{n}_{\mathrm{x}}^{\mathrm{T}}\right)\boldsymbol{N}^{''}\boldsymbol{D}\right]$$

with the derivative of the cardan angle, the angular velocity and the angular acceleration of section c can be easily calculated [7]. According to the characteristics of the shape function and the ideal rope assumption, all other strain except normal strain will be neglected. The normal strain can be written as

$$ {}_{\mathrm{I}}\varepsilon = \|{}_{\mathrm{I}}\boldsymbol{r}^{'\mathrm{c}}\| - 1,\ {}_{\mathrm{I}}\dot{\varepsilon} = \boldsymbol{T}_{\varepsilon}^{\mathrm{T}}\mathrm{d}\boldsymbol{q}_{\mathrm{e}} \tag{13}$$

with $\boldsymbol{T}_{\varepsilon}^{\mathrm{T}} = 1/\|{}_{\mathrm{I}}\boldsymbol{r}^{'\mathrm{c}}\|{}_{\mathrm{I}}\boldsymbol{r}^{'\mathrm{c}\mathrm{T}}\boldsymbol{N}^{'}\boldsymbol{D}$. The virtual power then can be formulated as

$$\delta p_{\mathrm{ine,tra}} = -\rho A\int_{0}^{L}\delta_{\mathrm{I}}\dot{\boldsymbol{r}}^{\mathrm{cT}}{}_{\mathrm{I}}\ddot{\boldsymbol{r}}^{\mathrm{c}}\mathrm{d}s = -\delta\mathrm{d}\boldsymbol{q}_{\mathrm{e}}^{\mathrm{T}}\left(\boldsymbol{M}_{\mathrm{ine,tra}}\mathrm{d}\dot{\boldsymbol{q}}_{\mathrm{e}} + \boldsymbol{F}_{\mathrm{ine,tra}}\right) \tag{14}$$

$$\delta p_{\text{ine,rot}} = -\rho \int_0^L \delta_{\text{c}} \boldsymbol{\omega}^{\text{cT}} \left(\boldsymbol{J}_{\text{c}} \dot{\boldsymbol{\omega}}^{\text{c}} + {}_{\text{c}}\tilde{\boldsymbol{\omega}}^{\text{c}} \boldsymbol{J}_{\text{c}} \boldsymbol{\omega}^{\text{c}} \right) \text{ds} = -\delta \text{d} \boldsymbol{q}_{\text{e}}^{\text{T}} \left(\boldsymbol{M}_{\text{ine,rot}} \text{d} \dot{\boldsymbol{q}}_{\text{e}} + \boldsymbol{F}_{\text{ine,rot}} \right) \quad (15)$$

$$\delta p_{\text{int}} = -E \int_0^L \delta_{\text{I}} \dot{\varepsilon}_{\text{I}} \varepsilon \text{ds} = -\delta \text{d} \boldsymbol{q}_{\text{e}}^{\text{T}} \boldsymbol{F}_{\text{int}} \quad (16)$$

$$\delta p_{\text{ext,g}} = \rho A \int_0^L \delta_{\text{I}} \dot{\boldsymbol{r}}^{\text{cT}} \text{ds}_{\text{I}} \boldsymbol{g} = -\delta \text{d} \boldsymbol{q}_{\text{e}}^{\text{T}} \boldsymbol{V}_{\text{e}} {}_{\text{I}}\boldsymbol{g} \quad (17)$$

where

$$\boldsymbol{M}_{\text{ine,tra}} = \rho A \boldsymbol{D}^{\text{T}} \int_0^L \boldsymbol{N}^T \boldsymbol{N} \text{ds} \boldsymbol{D}, \boldsymbol{M}_{\text{ine,rot}} = \rho \int_0^L \boldsymbol{T}_{\omega}^{\text{T}} \boldsymbol{J} \boldsymbol{T}_{\omega} \text{ds}$$

$$\boldsymbol{F}_{\text{ine,tra}} = \rho A \boldsymbol{D}^{\text{T}} \int_0^L \boldsymbol{N}^T \boldsymbol{N} \text{ds} \dot{\boldsymbol{D}} \text{d} \boldsymbol{q}_{\text{e}}, \boldsymbol{F}_{\text{ine,rot}} = \rho \int_0^L \boldsymbol{T}_{\omega}^{\text{T}} \left(\boldsymbol{J} \dot{\boldsymbol{T}}_{\omega} + {}_{\text{c}}\tilde{\boldsymbol{\omega}}^{\text{c}} \boldsymbol{J} \boldsymbol{T}_{\omega} \right) \text{ds} \, \text{d} \boldsymbol{q}_{\text{e}}$$

$$\boldsymbol{F}_{\text{int}} = EA \int_0^L \boldsymbol{T}_{\varepsilon \text{I}} \varepsilon \text{ds}, \boldsymbol{V}_{\text{e}} = \rho A \boldsymbol{D}^{\text{T}} \int_0^L \boldsymbol{N}^T \text{ds}$$

in which $\boldsymbol{T}_{\omega}$ and $\dot{\boldsymbol{T}}_{\omega}$ are shown in [7]. The virtual power of rope model is

$$\delta p_{\text{e}} = \delta p_{\text{ine,tra}} + \delta p_{\text{ine,rot}} + \delta p_{\text{int}} + \delta p_{\text{ext,g}} = -\delta \dot{\boldsymbol{q}}_{\text{e}}^{\text{T}} \left(\boldsymbol{M}_{\text{e}} \ddot{\boldsymbol{q}}_{\text{e}} + \boldsymbol{F}_{\text{e}} \right) \quad (18)$$

with $\boldsymbol{M}_{\text{e}} = \boldsymbol{M}_{\text{ine,tra}} + \boldsymbol{M}_{\text{ine,rot}}$ and $\boldsymbol{F}_{\text{e}} = \boldsymbol{F}_{\text{ine,tra}} + \boldsymbol{F}_{\text{ine,rot}} + \boldsymbol{F}_{\text{int}} + \boldsymbol{F}_{\text{ext,g}}$.

Strut Tie (Pendant) Model. The strut tie (pendant) model described here satisfies the following assumptions:

1. only the normal stress from axial tension and compression is considered, and other internal forces that may exist in the real rod are ignored;
2. the normal stress is evenly distributed with the length of the rod;
3. the cross section of the rod is symmetrical about the y- and z-axis of the section local coordinate;
4. the density and total mass of the rod remain unchanged.

Based on these assumptions, the generalized coordinates of the strut tie model can be expressed by the position of two end points.

$$\boldsymbol{q}_{\text{e}}^{\text{T}} = \left[{}_{\text{I}}\boldsymbol{r}^{1\text{T}} \; {}_{\text{I}}\boldsymbol{r}^{2\text{T}} \right] \quad (19)$$

The requirement of the posture of the strut tie model is, that the x-axis should be parallel to the line connecting two end points. Although there are infinite posture possibilities (rotate along x-axis), it will not affect the following computation. Here shows one definition method

$$\boldsymbol{\varphi}^{\text{B}} = \varphi^{\text{B}} \boldsymbol{n}_{\varphi}^{\text{B}}, \boldsymbol{R}_{\text{B}} = \boldsymbol{R}\left(\boldsymbol{\varphi}^{\text{B}}\right) \quad (20)$$

with $\boldsymbol{g}_{\mathrm{x}} = \begin{bmatrix} 1 & 0 & 0 \end{bmatrix}^{\mathrm{T}}$ and

$$\sin\varphi^{\mathrm{B}} = \|\widetilde{\boldsymbol{g}}_{\mathrm{x}}\boldsymbol{n}_{\mathrm{x}}\|/l_{\varphi,\mathrm{B}}, \cos\varphi^{\mathrm{B}} = \boldsymbol{g}_{\mathrm{x}}^{\mathrm{T}}\boldsymbol{n}_{\mathrm{x}}/l_{\varphi,\mathrm{B}}, \boldsymbol{n}_{\varphi}^{\mathrm{B}} = \widetilde{\boldsymbol{g}}_{\mathrm{x}}\boldsymbol{n}_{\mathrm{x}}/\|\widetilde{\boldsymbol{g}}_{\mathrm{x}}\boldsymbol{n}_{\mathrm{x}}\|$$

$$l_{\varphi,\mathrm{B}} = \sqrt{\|\widetilde{\boldsymbol{g}}_{\mathrm{x}}\boldsymbol{n}_{\mathrm{x}}\|^2 + \left(\boldsymbol{g}_{\mathrm{x}}^{\mathrm{T}}\boldsymbol{n}_{\mathrm{x}}\right)^2}, \boldsymbol{n}_{\mathrm{x}} = {}_{\mathrm{I}}\boldsymbol{r}^{12}/\|{}_{\mathrm{I}}\boldsymbol{r}^{12}\|, {}_{\mathrm{I}}\boldsymbol{r}^{12} = {}_{\mathrm{I}}\boldsymbol{r}^{2} - {}_{\mathrm{I}}\boldsymbol{r}^{1}$$

The angular velocity and velocity of the axial deformation can be obtained through the relationship between the translational velocities of two end points.

$$_{\mathrm{B}}\dot{\boldsymbol{r}}^{12} = \boldsymbol{T}_{\mathrm{r}}\dot{\boldsymbol{q}}_{\mathrm{e}}, {}_{\mathrm{B}}\boldsymbol{\omega}^{\mathrm{B}} = \boldsymbol{T}_{\varphi}\dot{\boldsymbol{q}}_{\mathrm{e}} \tag{21}$$

in which

$$\boldsymbol{T}_{\mathrm{r}} = \boldsymbol{g}_{\mathrm{x}}\boldsymbol{n}_{\mathrm{x}}^{\mathrm{T}}\begin{bmatrix} -\boldsymbol{I} & \boldsymbol{I} \end{bmatrix}, \boldsymbol{R}_{\mathrm{x},90^{\circ}} = \boldsymbol{R}\left(\begin{bmatrix} \pi/2 & 0 & 0 \end{bmatrix}^{\mathrm{T}}\right)$$

$$\boldsymbol{T}_{\varphi} = 1/\|{}_{\mathrm{I}}\boldsymbol{r}^{12}\|\boldsymbol{R}_{\mathrm{x},90^{\circ}}\left(\boldsymbol{R}_{\mathrm{B}}^{\mathrm{T}} - \boldsymbol{g}_{\mathrm{x}}\boldsymbol{n}_{\mathrm{x}}^{\mathrm{T}}\right)\begin{bmatrix} -\boldsymbol{I} & \boldsymbol{I} \end{bmatrix}$$

The angular acceleration can be written as

$$_{\mathrm{B}}\dot{\boldsymbol{\omega}}^{\mathrm{B}} = \left[\boldsymbol{R}_{\mathrm{x},90^{\circ}}\left(\frac{{}_{\mathrm{I}}\boldsymbol{r}^{12\mathrm{T}}\,{}_{\mathrm{I}}\dot{\boldsymbol{r}}^{12}}{\|{}_{\mathrm{I}}\boldsymbol{r}^{12}\|^2}\boldsymbol{I} - {}_{\mathrm{B}}\widetilde{\boldsymbol{\omega}}^{\mathrm{B}}\right)\widetilde{\boldsymbol{g}}_{\mathrm{x}}\boldsymbol{T}_{\varphi}\right]\dot{\boldsymbol{q}}_{\mathrm{e}} + \boldsymbol{T}_{\varphi}\ddot{\boldsymbol{q}}_{\mathrm{e}} = \dot{\boldsymbol{T}}_{\varphi}\dot{\boldsymbol{q}}_{\mathrm{e}} + \boldsymbol{T}_{\varphi}\ddot{\boldsymbol{q}}_{\mathrm{e}} \tag{22}$$

where ${}_{\mathrm{I}}\dot{\boldsymbol{r}}^{12} = {}_{\mathrm{I}}\dot{\boldsymbol{r}}^{2} - {}_{\mathrm{I}}\dot{\boldsymbol{r}}^{1}$.

Since only the axial deformation is considered, there is no relative rotation with the local coordinate on any cross section. Hence, the position, the velocity, and the acceleration of any point on any section of the rod can be expressed as

$${}_{\mathrm{I}}\boldsymbol{r} = {}_{\mathrm{I}}\boldsymbol{r}^{\mathrm{c}} + \boldsymbol{R}_{\mathrm{B}}{}_{\mathrm{c}}\boldsymbol{t}, {}_{\mathrm{I}}\dot{\boldsymbol{r}} = {}_{\mathrm{I}}\dot{\boldsymbol{r}}^{\mathrm{c}} + \boldsymbol{R}_{\mathrm{B}}{}_{\mathrm{B}}\widetilde{\boldsymbol{\omega}}^{\mathrm{B}}{}_{\mathrm{c}}\boldsymbol{t}, {}_{\mathrm{I}}\ddot{\boldsymbol{r}} = {}_{\mathrm{I}}\ddot{\boldsymbol{r}}^{\mathrm{c}} + \boldsymbol{R}_{\mathrm{B}}\left({}_{\mathrm{B}}\dot{\widetilde{\boldsymbol{\omega}}}^{\mathrm{B}} + {}_{\mathrm{B}}\widetilde{\boldsymbol{\omega}}^{\mathrm{B}}{}_{\mathrm{B}}\widetilde{\boldsymbol{\omega}}^{\mathrm{B}}\right){}_{\mathrm{c}}\boldsymbol{t} \tag{23}$$

where ${}_{\mathrm{I}}\boldsymbol{r}^{\mathrm{c}}$, ${}_{\mathrm{I}}\dot{\boldsymbol{r}}^{\mathrm{c}}$ and ${}_{\mathrm{I}}\ddot{\boldsymbol{r}}^{\mathrm{c}}$ represent the position, the velocity, and the acceleration of the center of the cross section. ${}_{\mathrm{c}}\boldsymbol{t}$ is the relative position from the center of the section in the local coordinate. To satisfy the second assumption, the first derivative of the position vector of the center of any section on the rod should be constant. Therefore, its position vector can be expressed by the linear interpolation of the position vectors of two end section.

$${}_{\mathrm{I}}\boldsymbol{r}^{\mathrm{c}} = \boldsymbol{N}\boldsymbol{q}_{\mathrm{e}}, {}_{\mathrm{I}}\dot{\boldsymbol{r}}^{\mathrm{c}} = \boldsymbol{N}\dot{\boldsymbol{q}}_{\mathrm{e}}, {}_{\mathrm{I}}\ddot{\boldsymbol{r}}^{\mathrm{c}} = \boldsymbol{N}\ddot{\boldsymbol{q}}_{\mathrm{e}} \tag{24}$$

where $\boldsymbol{N} = \begin{bmatrix} (1-\xi)\boldsymbol{I} & \xi\boldsymbol{I} \end{bmatrix}$, $\xi = s/L$ is the shape function.

According to the definition of strain in absolute coordinates, the axial strain of the rod can be written as

$${}_{\mathrm{I}}\varepsilon = \|{}_{\mathrm{I}}\boldsymbol{r}'\| - 1, {}_{\mathrm{I}}\dot{\varepsilon} = \boldsymbol{T}_{\varepsilon}^{\mathrm{T}}\dot{\boldsymbol{q}}_{\mathrm{e}}, {}_{\mathrm{I}}\ddot{\varepsilon} = \boldsymbol{T}_{\varepsilon}^{\mathrm{T}}\ddot{\boldsymbol{q}}_{\mathrm{e}} + \dot{\boldsymbol{T}}_{\varepsilon}^{\mathrm{T}}\dot{\boldsymbol{q}}_{\mathrm{e}} \tag{25}$$

with $(*)' = \mathrm{d}(*)/\mathrm{d}s$, ${}_{\mathrm{I}}\boldsymbol{r}' = {}_{\mathrm{I}}\boldsymbol{r}^{12}/L$ and

$$\boldsymbol{T}_{\varepsilon}^{\mathrm{T}} = \left(1/\|{}_{\mathrm{I}}\boldsymbol{r}'\|\right)\boldsymbol{q}_{\mathrm{e}}^{\mathrm{T}}\boldsymbol{N}'^{\mathrm{T}}\boldsymbol{N}'\dot{\boldsymbol{q}}_{\mathrm{e}}, \dot{\boldsymbol{T}}_{\varepsilon}^{\mathrm{T}} = \left(1/\|{}_{\mathrm{I}}\boldsymbol{r}'\|\right)\dot{\boldsymbol{q}}_{\mathrm{e}}^{\mathrm{T}}\boldsymbol{N}'^{\mathrm{T}}\left[\boldsymbol{I} - \left(1/\|{}_{\mathrm{I}}\boldsymbol{r}'\|^{2}\right){}_{\mathrm{I}}\boldsymbol{r}^{\mathrm{c}'}{}_{\mathrm{I}}\boldsymbol{r}^{\mathrm{c}'T}\right]\boldsymbol{N}'$$

The stress can be obtained from the constitutive relationship with the strain. In order to avoid the high-frequency oscillation of the rod caused by the strain, the smooth factor h is introduced to replace the instantaneous stress by the average stress. The average stress in inertial coordinate can be expressed as

$${}_{\mathrm{I}}\tilde{\sigma} = E_{\mathrm{I}}\bar{\varepsilon} = E\left({}_{\mathrm{I}}\varepsilon + h_{\mathrm{I}}\dot{\varepsilon}/2 + h^{2}{}_{\mathrm{I}}\ddot{\varepsilon}/6\right) \tag{26}$$

The virtual inertial power, the virtual internal power and the virtual external power caused by the gravity can be written as

$$\delta p_{\mathrm{ine}} = -\iiint_{V} \delta_{\mathrm{I}}\dot{\boldsymbol{r}}^{\mathrm{T}}{}_{\mathrm{I}}\ddot{\boldsymbol{r}}\rho \mathrm{d}V = -\delta\dot{\boldsymbol{q}}_{\mathrm{e}}^{\mathrm{T}}\left(\boldsymbol{M}_{\mathrm{ine}}\ddot{\boldsymbol{q}}_{\mathrm{e}} + \boldsymbol{F}_{\mathrm{ine}}\right) \tag{27}$$

$$\delta p_{\mathrm{int}} = -\int_{0}^{L} \delta_{\mathrm{I}}\dot{\varepsilon}_{\mathrm{I}}\bar{\sigma}A\mathrm{d}s = -\delta\dot{\boldsymbol{q}}_{\mathrm{e}}^{\mathrm{T}}\left(\boldsymbol{M}_{\mathrm{int}}\ddot{\boldsymbol{q}}_{\mathrm{e}} + \boldsymbol{F}_{\mathrm{int}}\right) \tag{28}$$

$$\delta p_{\mathrm{ext,g}} = \iiint_{V} \delta_{\mathrm{I}}\dot{\boldsymbol{r}}^{\mathrm{T}}{}_{\mathrm{I}}\boldsymbol{g}\rho \mathrm{d}V = -\delta\dot{\boldsymbol{q}}_{\mathrm{e}}^{\mathrm{T}}\boldsymbol{F}_{\mathrm{ext,g}} \tag{29}$$

where

$$\boldsymbol{M}_{\mathrm{ine}} = \rho L\left(\frac{A}{6}\begin{bmatrix} 2\boldsymbol{I} & \boldsymbol{I} \\ \boldsymbol{I} & 2\boldsymbol{I} \end{bmatrix} + \boldsymbol{T}_{\varphi}^{\mathrm{T}}\boldsymbol{J}\boldsymbol{T}_{\varphi}\right), \boldsymbol{M}_{\mathrm{int}} = \frac{h^{2}}{6}EAL\boldsymbol{T}_{\varepsilon}\boldsymbol{T}_{\varepsilon}^{\mathrm{T}}, \boldsymbol{F}_{\mathrm{ext,g}} = \frac{1}{2}\rho AL\left[\boldsymbol{I}\,\boldsymbol{I}\right]^{T}{}_{\mathrm{I}}\boldsymbol{g}$$

$$\boldsymbol{D}_{\mathrm{e,ine}} = -\rho L\boldsymbol{T}_{\varphi}^{\mathrm{T}}\left(\boldsymbol{J}\dot{\boldsymbol{T}}_{\varphi} + {}_{\mathrm{B}}\overset{\sim\mathrm{B}}{\boldsymbol{\omega}}\boldsymbol{J}\boldsymbol{T}_{\varphi}\right), \boldsymbol{D}_{\mathrm{e,int}} = EAL\left(h\boldsymbol{T}_{\varepsilon}\boldsymbol{T}_{\varepsilon}^{\mathrm{T}}/2 + h^{2}\boldsymbol{T}_{\varepsilon}\dot{\boldsymbol{T}}_{\varepsilon}^{\mathrm{T}}/6\right)$$

$$\boldsymbol{F}_{\mathrm{ine}} = \boldsymbol{D}_{\mathrm{e,ine}}\dot{\boldsymbol{q}}_{\mathrm{e}}, \boldsymbol{F}_{\mathrm{int}} = \boldsymbol{D}_{\mathrm{e,int}}\dot{\boldsymbol{q}}_{\mathrm{e}} + \boldsymbol{K}_{\mathrm{e}}\boldsymbol{q}_{\mathrm{e}}, \boldsymbol{K}_{\mathrm{e}} = \left(1/\|{}_{\mathrm{I}}\boldsymbol{r}'\|\right)\boldsymbol{N}'^{\mathrm{T}}\boldsymbol{N}'EA_{\mathrm{I}}\varepsilon L$$

The virtual power of strut tie model is

$$\delta p_{\mathrm{e}} = \delta p_{\mathrm{ine}} + \delta p_{\mathrm{int}} + \delta p_{\mathrm{ext,g}} = -\delta\dot{\boldsymbol{q}}_{\mathrm{e}}^{\mathrm{T}}\left(\boldsymbol{M}_{\mathrm{e}}\ddot{\boldsymbol{q}}_{\mathrm{e}} + \boldsymbol{F}_{\mathrm{e}}\right) \tag{30}$$

with $\boldsymbol{M}_{\mathrm{e}} = \boldsymbol{M}_{\mathrm{e,ine}} + \boldsymbol{M}_{\mathrm{e,int}}$ and $\boldsymbol{F}_{\mathrm{e}} = \boldsymbol{F}_{\mathrm{ine}} + \boldsymbol{F}_{\mathrm{int}} + \boldsymbol{F}_{\mathrm{ext,g}}$.

2.2 Super Truss Element

Considering the structural asymmetry and discontinuity of the lattice boom, above mentioned continuous model cannot describe some of the characteristics, such as shorten while twisting. However, modeling all the elements in the lattice structure will significantly increase the number of DoFs and reduce the calculation speed. Here, a super truss element is proposed to describe the lattice boom with lower numbers of DoFs and acceptable characteristics. The super truss element is defined by nodes (cross section nodes, internal nodes), planes (cross sections, sub-beam planes) and beam elements (cross section beams, main beams, sub-beams), which are shown in Fig. 1.

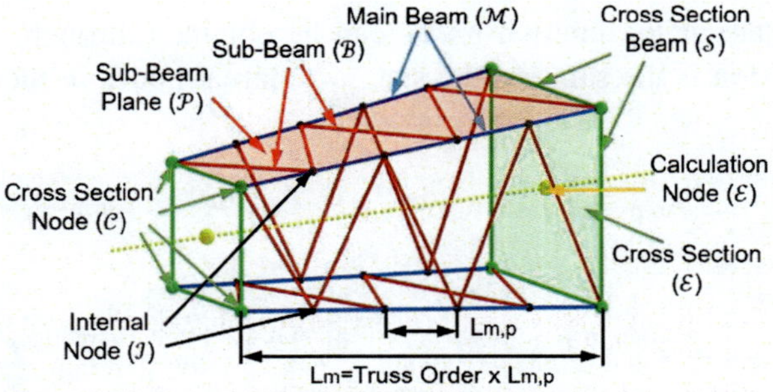

Fig. 1. Definition of truss elements and truss order [7].

The super truss element is established based on 3 assumptions.

Assumption 1: The two end sections of the super truss element are rigid.

Assumption 2: The main beams connecting two end sections are geometrically continuous after the deformation.

Assumption 3: All the connections are rigid, which means the relative pose between two elements remains unchanged after deformation.

In addition, all truss elements here are regular, which means the shape satisfies the following conditions:

- The end section of the truss element is a plane, and the two sections are parallel to each other and perpendicular to the virtual main axis;
- All beam members are straight before the deformation.

The calculation of super truss elements can be divided into two phases: parameterization phase for the initialization and dynamic calculation phase. In the parameterization phase, all the reference and the constant relative position and the posture of each beam element are defined. In the dynamic calculation phase, the state of the cross sections will be firstly calculated based on the state of the cross section and assumption 1 (the constant relative pose $\boldsymbol{T}_{\mathrm{cn,k}}^{\mathrm{c,i}}$, $\dot{\boldsymbol{T}}_{\mathrm{cn,k}}^{\mathrm{c,i}}$, $\mathrm{i} = 1, 2$). To avoid the unnecessary calculation error between the rotation vector and the rotation matrix, here the rotation matrix is used as the rotational part of the generalized coordinate of each node.

$$\boldsymbol{q}_{\mathrm{n}}^{\mathrm{T}} = \left[{}_{\mathrm{I}}\boldsymbol{r}^{\mathrm{nT}}\ \boldsymbol{R}_{\mathrm{n}}^{\mathrm{T}}\right], \mathrm{d}\boldsymbol{q}_{\mathrm{n}}^{\mathrm{T}} = \left[{}_{\mathrm{I}}\dot{\boldsymbol{r}}^{\mathrm{nT}}\ {}_{\mathrm{n}}\boldsymbol{\omega}^{\mathrm{nT}}\right], \mathrm{n} \in \text{All nodes} \tag{31}$$

$$\mathrm{d}\boldsymbol{q}_{\mathrm{cn,k}} = \boldsymbol{T}_{\mathrm{cn,k}}\mathrm{d}\boldsymbol{q}_{\mathrm{e}}, \mathrm{d}\dot{\boldsymbol{q}}_{\mathrm{cn,k}} = \boldsymbol{T}_{\mathrm{cn,k}}\mathrm{d}\dot{\boldsymbol{q}}_{\mathrm{e}} + \dot{\boldsymbol{T}}_{\mathrm{cn,k}}\mathrm{d}\boldsymbol{q}_{\mathrm{e}} \tag{32}$$

in which

$$\boldsymbol{T}_{\mathrm{cn,k}} = \boldsymbol{T}_{\mathrm{cn,k}}^{\mathrm{c,i}}\boldsymbol{T}_{\mathrm{c,i}}, \dot{\boldsymbol{T}}_{\mathrm{cn,k}} = \dot{\boldsymbol{T}}_{\mathrm{cn,k}}^{\mathrm{c,i}}\boldsymbol{T}_{\mathrm{c,i}}$$

where $\boldsymbol{T}_{\mathrm{c,i}}$ reflects the relationship between the cross section i with the generalized state and $\mathbf{T}_{\mathrm{c,1}} = \left[\boldsymbol{I}\ \mathbf{0}\right]$, $\mathbf{T}_{\mathrm{c,2}} = \left[\mathbf{0}\ \boldsymbol{I}\right]$.

Next, the state of the internal nodes can be obtained through assumption 2. The detailed expression is the same as the state of internal nodes in the ideal cubic spline rope model.

$$\mathrm{d}\boldsymbol{q}_{\mathrm{in,p}} = \boldsymbol{T}_{\mathrm{in,p}}\mathrm{d}\boldsymbol{q}_{\mathrm{e}},\ \mathrm{d}\dot{\boldsymbol{q}}_{\mathrm{in,p}} = \boldsymbol{T}_{\mathrm{in,p}}\mathrm{d}\dot{\boldsymbol{q}}_{\mathrm{e}} + \dot{\boldsymbol{T}}_{\mathrm{in,p}}\mathrm{d}\boldsymbol{q}_{\mathrm{e}} \tag{33}$$

in which

$$\boldsymbol{T}_{\mathrm{in,p}} = \boldsymbol{T}_{\mathrm{in,p}}^{\mathrm{mb,m}}\boldsymbol{T}_{\mathrm{mb,m}},\ \dot{\boldsymbol{T}}_{\mathrm{in,p}} = \dot{\boldsymbol{T}}_{\mathrm{in,p}}^{\mathrm{mb,m}}\boldsymbol{T}_{\mathrm{mb,m}} + \boldsymbol{T}_{\mathrm{in,p}}^{\mathrm{mb,m}}\dot{\boldsymbol{T}}_{\mathrm{mb,m}}$$

where $\boldsymbol{T}_{\mathrm{mb,m}}$, $\dot{\boldsymbol{T}}_{\mathrm{mb,m}}$ reflect the relationship between the end nodes of the main beam with the generalized state, which can be calculated as

$$\boldsymbol{T}_{\mathrm{mb,m}} = \left[\boldsymbol{T}_{\mathrm{c,1}}^{\mathrm{T}}\boldsymbol{T}_{\mathrm{cn,k}}^{\mathrm{c,1T}}\ \boldsymbol{T}_{\mathrm{c,2}}^{\mathrm{T}}\boldsymbol{T}_{\mathrm{cn,l}}^{\mathrm{c,2T}}\right]^{\mathrm{T}},\ \dot{\boldsymbol{T}}_{\mathrm{mb,m}} = \left[\boldsymbol{T}_{\mathrm{c,1}}^{\mathrm{T}}\dot{\boldsymbol{T}}_{\mathrm{cn,k}}^{\mathrm{c,1T}}\ \boldsymbol{T}_{\mathrm{c,2}}^{\mathrm{T}}\dot{\boldsymbol{T}}_{\mathrm{cn,l}}^{\mathrm{c,2T}}\right]^{\mathrm{T}}$$

Based on the state of the cross-section nodes and the internal nodes, the generalized coordinates of each beam element can be determined. According to the definition of the beam elements, the cross-section beam s is determined by two cross section nodes k and l belonging to one cross section; the main beam g is determined by two internal nodes p and q belonging to one main beam; the sub beam h is determined by two internal nodes p* and q* belonging to different main beams and connecting the sub beam.

$$\boldsymbol{q}_{\mathrm{cb,s}} = \left[\boldsymbol{q}_{\mathrm{cn,k}}^{\mathrm{T}}\ \boldsymbol{q}_{\mathrm{cn,l}}^{\mathrm{T}}\right]^{\mathrm{T}},\ \mathrm{d}\boldsymbol{q}_{\mathrm{cb,s}} = \boldsymbol{T}_{\mathrm{cb,s}}\mathrm{d}\boldsymbol{q}_{\mathrm{e}},\ \mathrm{d}\dot{\boldsymbol{q}}_{\mathrm{cb,s}} = \boldsymbol{T}_{\mathrm{cb,s}}\mathrm{d}\dot{\boldsymbol{q}}_{\mathrm{e}} + \dot{\boldsymbol{T}}_{\mathrm{cb,s}}\mathrm{d}\boldsymbol{q}_{\mathrm{e}} \tag{34}$$

$$\boldsymbol{q}_{\mathrm{mb,g}} = \left[\boldsymbol{q}_{\mathrm{in,p}}^{\mathrm{T}}\ \boldsymbol{q}_{\mathrm{in,q}}^{\mathrm{T}}\right]^{\mathrm{T}},\ \mathrm{d}\boldsymbol{q}_{\mathrm{mb,g}} = \boldsymbol{T}_{\mathrm{mb,g}}\mathrm{d}\boldsymbol{q}_{\mathrm{e}},\ \mathrm{d}\dot{\boldsymbol{q}}_{\mathrm{mb,g}} = \boldsymbol{T}_{\mathrm{mb,g}}\mathrm{d}\dot{\boldsymbol{q}}_{\mathrm{e}} + \dot{\boldsymbol{T}}_{\mathrm{mb,g}}\mathrm{d}\boldsymbol{q}_{\mathrm{e}} \tag{35}$$

$$\boldsymbol{q}_{\mathrm{sb,h}} = \left[\boldsymbol{q}_{\mathrm{in,p^*}}^{\mathrm{T}}\ \boldsymbol{q}_{\mathrm{in,q^*}}^{\mathrm{T}}\right]^{\mathrm{T}},\ \mathrm{d}\boldsymbol{q}_{\mathrm{sb,h}} = \boldsymbol{T}_{\mathrm{sb,h}}\mathrm{d}\boldsymbol{q}_{\mathrm{e}},\ \mathrm{d}\dot{\boldsymbol{q}}_{\mathrm{sb,h}} = \boldsymbol{T}_{\mathrm{sb,h}}\mathrm{d}\dot{\boldsymbol{q}}_{\mathrm{e}} + \dot{\boldsymbol{T}}_{\mathrm{sb,h}}\mathrm{d}\boldsymbol{q}_{\mathrm{e}} \tag{36}$$

where

$$\boldsymbol{T}_{\mathrm{cb,s}} = \left[\boldsymbol{T}_{\mathrm{cn,k}}^{\mathrm{T}}\ \boldsymbol{T}_{\mathrm{cn,l}}^{\mathrm{T}}\right]^{\mathrm{T}},\ \dot{\boldsymbol{T}}_{\mathrm{cb,s}} = \left[\dot{\boldsymbol{T}}_{\mathrm{cn,k}}^{\mathrm{T}}\ \dot{\boldsymbol{T}}_{\mathrm{cn,l}}^{\mathrm{T}}\right]^{\mathrm{T}}$$

$$\boldsymbol{T}_{\mathrm{mb,g}} = \left[\boldsymbol{T}_{\mathrm{in,p}}^{\mathrm{T}}\ \boldsymbol{T}_{\mathrm{in,q}}^{\mathrm{T}}\right]^{\mathrm{T}},\ \dot{\boldsymbol{T}}_{\mathrm{mb,g}} = \left[\dot{\boldsymbol{T}}_{\mathrm{in,p}}^{\mathrm{T}}\ \dot{\boldsymbol{T}}_{\mathrm{in,q}}^{\mathrm{T}}\right]^{\mathrm{T}}$$

$$\boldsymbol{T}_{\mathrm{sb,h}} = \left[\boldsymbol{T}_{\mathrm{in,p^*}}^{\mathrm{T}}\ \boldsymbol{T}_{\mathrm{in,q^*}}^{\mathrm{T}}\right]^{\mathrm{T}},\ \dot{\boldsymbol{T}}_{\mathrm{sb,h}} = \left[\dot{\boldsymbol{T}}_{\mathrm{in,p^*}}^{\mathrm{T}}\ \dot{\boldsymbol{T}}_{\mathrm{in,q^*}}^{\mathrm{T}}\right]^{\mathrm{T}}$$

By applying the beam model, the generalized mass $\boldsymbol{M}_{\mathrm{e}}^{\mathrm{b}}$ and the generalized force $\boldsymbol{F}_{\mathrm{e}}^{\mathrm{b}}$ in the beam coordinate can be obtained, and then transferred to the generalized mass $\boldsymbol{M}_{\mathrm{e}}$ and the generalize force $\boldsymbol{F}_{\mathrm{e}}$ based on the generalized coordinates of super truss elements. The dynamic model for modeling elements in the super truss element theoretically can be any beam model. Here the spatial Timoshenko beam is selected because of its fast calculation speed and relative short length of the elements.

$$\boldsymbol{M}_{\mathrm{e}} = \sum\nolimits_{\mathrm{b}\in\mathrm{Allbeams}} \boldsymbol{T}_{\mathrm{e}}^{\mathrm{bT}}\boldsymbol{M}_{\mathrm{e}}^{\mathrm{b}}\boldsymbol{T}_{\mathrm{e}}^{\mathrm{b}},\ \boldsymbol{F}_{\mathrm{e}} = \sum\nolimits_{\mathrm{b}\in\mathrm{Allbeams}} \left(\boldsymbol{T}_{\mathrm{e}}^{\mathrm{bT}}\boldsymbol{M}_{\mathrm{e}}^{\mathrm{b}}\dot{\boldsymbol{T}}_{\mathrm{e}}^{\mathrm{b}} + \boldsymbol{T}_{\mathrm{e}}^{\mathrm{bT}}\boldsymbol{F}_{\mathrm{e}}^{\mathrm{b}}\right) \tag{37}$$

The whole dynamic calculation phase can be summarized with the following flow chart Fig. 2. After calculation of the states of internal nodes, the computing of the general-ized mass and the generalized force will not affect each other. Hence, parallel com-puting method can be introduced to further increase the calculation speed of the super truss elements.

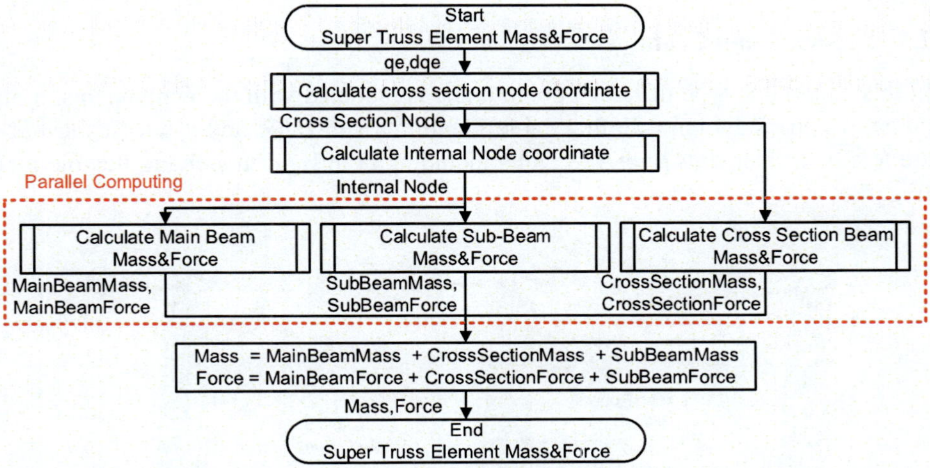

Fig. 2. Flow chart of dynamic calculation of super truss element [7].

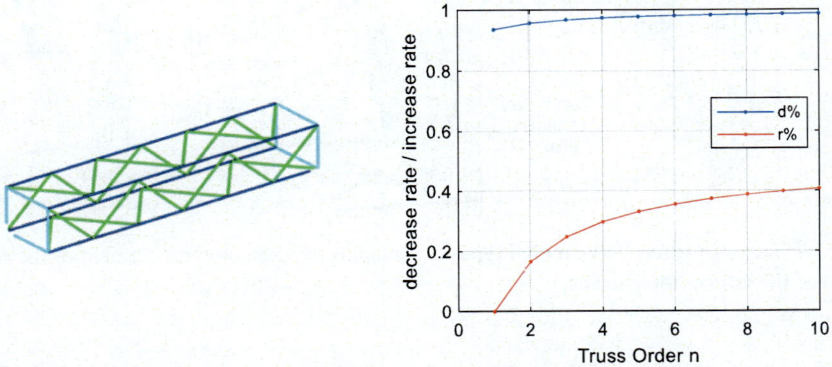

Fig. 3. One example of truss structure (left), decrease rate for number of degrees of freedom and increase rate for amount of calculation in one dynamic calculation over truss order (right).

Comparing with the traditional dynamic calculation methods, super truss elements use far less generalized coordinates to describe the truss. However, due to the state estimation of the internal nodes, the amount of calculation in one dynamic calculation increases. The decrease rate $d\%$ and increase rate $r\%$ can be different for different truss structure. For the following type of truss with truss order $n \in \mathbb{Z}^+$ in Fig. 3, they can be expressed as

$$d\% = \frac{\text{Nr.DoF}_{\text{tra}} - \text{Nr.DoF}_{\text{sup}}}{\text{Nr.DoF}_{\text{tra}}} = 1 - \frac{1}{8}\frac{1}{n+1} \tag{38}$$

$$r\% = \frac{\mathrm{Nr.Cal_{sup}} - \mathrm{Nr.Cal_{tra}}}{\mathrm{Nr.Cal_{tra}}} = \frac{1}{2} - \frac{1}{n+1} \tag{39}$$

From the Fig. 3 (right), the amount of calculation in one dynamic calculation increases maximal 50% and the reduction rate of number of DoFs reaches over 90%. Overall, introducing the super truss element can save calculation time.

2.3 Lattice Boom Crane

In this paper, one type of lattice boom crane is modeled with the components: a main boom, a derrick boom, a luffing cable module, a sub-cable and a hoist cable. There are totally 16 elements with 168 DoFs including both rigid model and flexible model (Fig. 4).

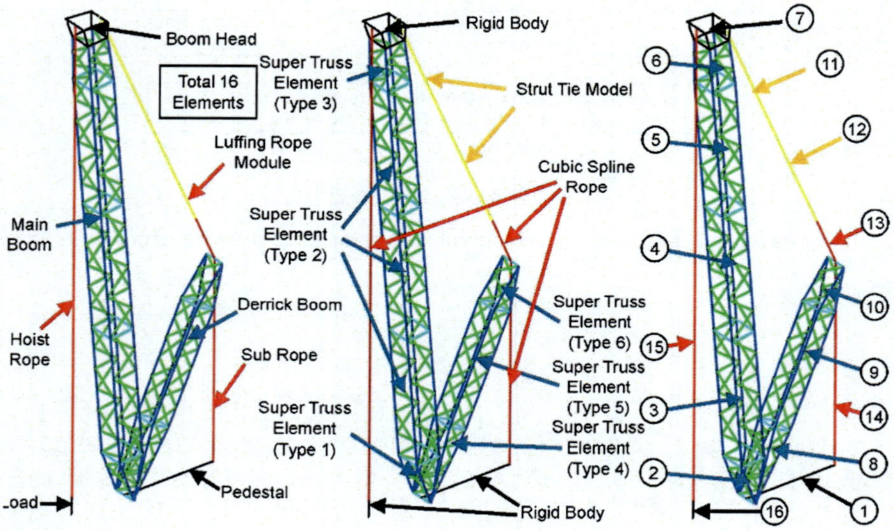

Fig. 4. The composition (left), model types of elements (middle) and the number of the elements (right) of the lattice boom crane.

Main Boom. For the main boom, 5 super truss elements in 3 different types are used to substitute 227 beam elements, which decreases the number of DoFs from 1362 to 60. After eliminating the redundant DoF at the connection nodes between two elements, there are totally 36 DoF (reduced by 97.4%). However, the required amount of computation increases from 227 to 339 (increased by 49.3%) due to the additional estimation of the state of the internal nodes.

Derrick Boom. The derrick boom is composed of 3 super truss elements in 3 different types with totally 126 beam elements. For traditional dynamic modeling method, 756 DoFs are needed. By using super truss elements, the number of DoFs of the derrick boom becomes 24 (reduced by 96.8%) and the amount of calculation in single-step is increased from 126 to 190 (increased by 50.8%) (Table 1).

Table 1. Properties of different types of super truss elements.

Main Boom	Type 1	Type 2	Type 3
Structure	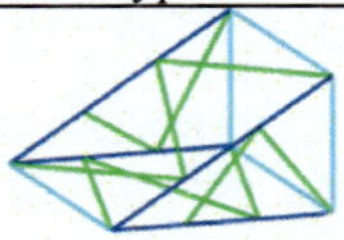	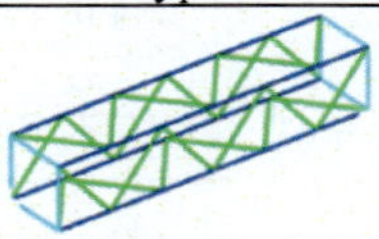	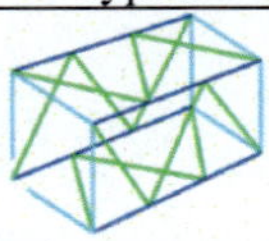
Type	Triangular Prism	Rectangle	Shrink Rectangle
Derrick Boom	**Type 4**	**Type 5**	**Type 6**
Structure	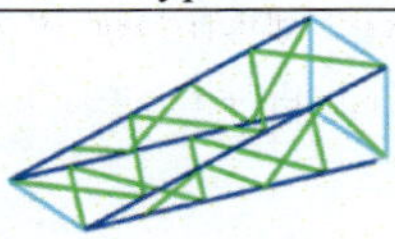	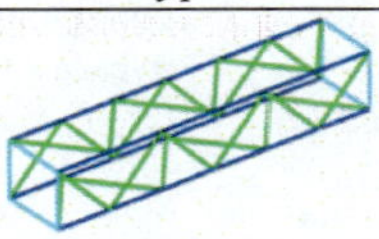	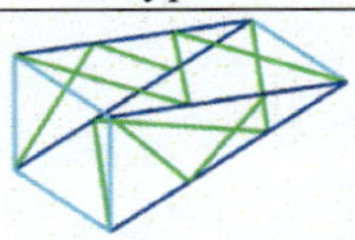
Type	Triangular Prism	Rectangle	Anti Triangular Prism

Luffing Module, Sub-cable and Hoist Cable. The lattice boom crane contains 3 rope systems: luffing module, hoist cable and sub-cable. According to the actual model, the luffing rope module consists of two strut tie models and one rope element. The hoist cable and the sub-cable each consist of one ideal cubic spline rope model.

Turntable and Load. The boom system is mounted on a turntable (pedestal) and will rotate with the turntable relative to the inertial coordinate system. The overall stiffness of the turntable is always relatively large. Therefore, here a rigid body model is used to simulate the turntable. The sway of the load is also one of the research directions of lattice boom cranes. In this model a small rigid body is added at the end of the hoist cable to simulate the load. The position and the posture of the load during the moving of the lattice boom crane will be studied.

Constraints and Drives. The lattice boom crane has original totally 18 constraints. by using public computing nodes, the 6 fixed constraints connecting the super truss elements and the spherical constraints in the rope model can be eliminated. The lattice boom crane model finally has 120 degrees of freedom, 240 state variables and 8 constraints (Fig. 5).

In reality, the lattice boom crane is controlled through the winch driven by the hydraulic motor to change the length of hoisting and luffing ropes. The overall rotation of the crane is also realized by the hydraulic motor. However, due to the complexity of the pulley rope model and the hydraulic system, these systems are not considered in this model. Here the angle of the pedestal and the length of the hoist cable and the luffing rope are directly set to control variables.

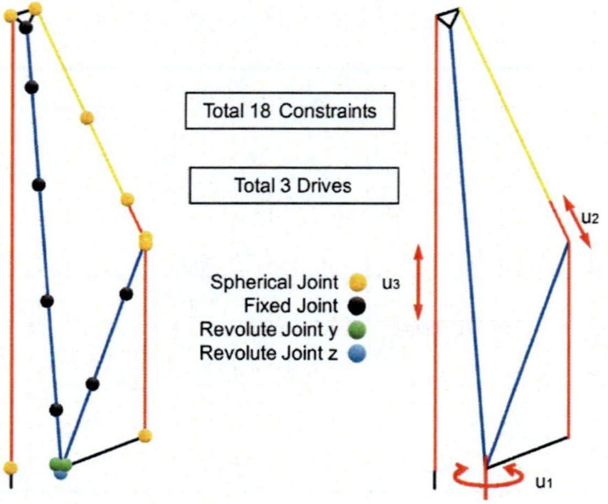

Fig. 5. Locations and types of constraints and drive configuration displayed on the calculation model.

3 Quasi-static Optimal Control

3.1 Quasi-static Optimal Trajectory Tracking Strategy

The lattice boom crane model is a complex rigid-flexible coupled multibody model with large degrees of freedom, which makes it difficult to implement commonly used optimal control methods due to the large computation requirement. On the other hand, because the change of the rope element length affects the Gaussian integral and the calculation of stress and strain, the influence of the control variable on the system is very complicated, and it is almost impossible to separate the control variable from the state variable. Here, the quasi-static optimal control method is carried out to generate offline optimal control strategy for the lattice boom crane. This method combines the optimal trajectory of the end point with the control variables through the quasi-static mapping relationship. The generated control trajectory can track the optimal trajectory of the end point, thereby achieving optimal control of complex systems. The specific implementation process is shown as follows (Fig. 6).

The quasi-static optimal control method is composed of three steps: the establishment of a static mapping relationship, the generation of the optimal trajectory at the end point and the generation of the optimal trajectory of the control inputs.

3.2 Detailed Implementation Steps

Static State Mapping. The establishment of the static mapping relationship requires to calculate the relationship between the control variable and the control object (position of the end point) under all possible static states. However, calculating relationship of all possible positions for continuous dynamic system is not realizable. Here some discrete states are selected, and the mapping relationship is obtained by fitting. The detailed process is shown as follows (Fig. 7).

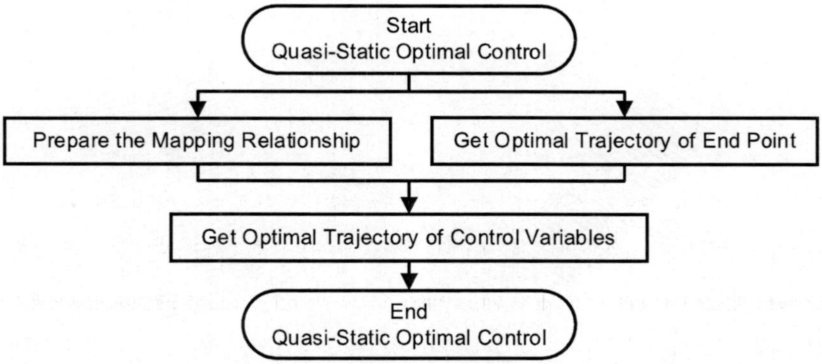

Fig. 6. Quasi-static optimal trajectory tracking strategy.

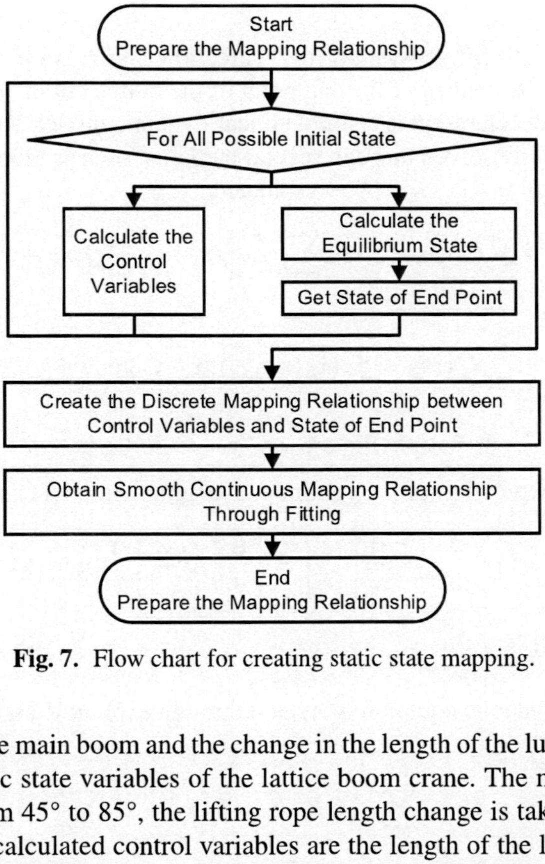

Fig. 7. Flow chart for creating static state mapping.

The angle of the main boom and the change in the length of the luffing rope are taken as the characteristic state variables of the lattice boom crane. The main boom angle is taken every 1° from 45° to 85°, the lifting rope length change is taken every 1 m from 0 m to 20 m. The calculated control variables are the length of the luffing rope and the lifting rope in the initial state before the deformation. The coordinates of the end point in the equilibrium state are calculated as the control object. The final experimental results can generate a discrete mapping table, which is shown as following.

In order to obtain a continuous mapping relationship, the surface fitting needs to be performed on discrete data points. Considering the fitting accuracy and the error sensitivity, a polynomial surface of order 2–2 is sufficient and recommended.

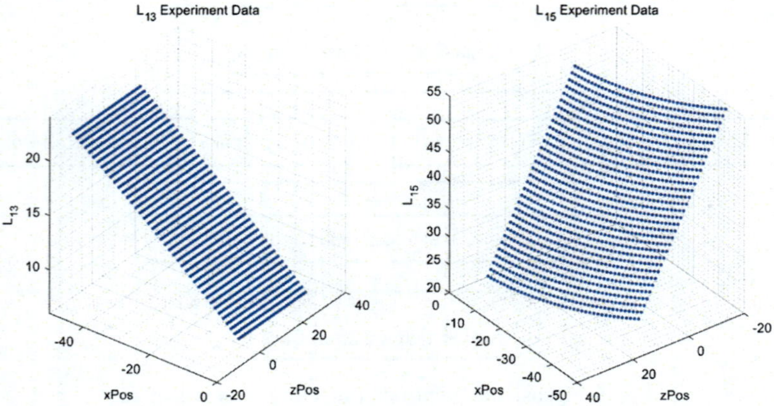

Fig. 8. Discrete experimental data from statics calculation for lattice boom crane with 400t load.

Optimal Control of Simple System for Trajectory Generation. Here, a kinematic model is used for the control of the end point of the lattice boom. The kinematic model controlled by the acceleration is a simple linear dynamic model, whose optimal control problem can be easily solved through several methods, such as model predictive control (MPC). The control problem can be formulated as

$$\min J = (\boldsymbol{x}_N - \boldsymbol{x}_\mathrm{e})^\mathrm{T}\boldsymbol{P}(\boldsymbol{x}_N - \boldsymbol{x}_\mathrm{e}) + \sum\nolimits_{k=0}^{N-1}\left[(\boldsymbol{x}_k - \boldsymbol{x}_\mathrm{e})^\mathrm{T}\boldsymbol{Q}(\boldsymbol{x}_k - \boldsymbol{x}_\mathrm{e}) + \boldsymbol{u}_k^\mathrm{T}\boldsymbol{R}\boldsymbol{u}_k\right] \quad (40)$$

with the constraints

$$\boldsymbol{x}_{k+1} = \boldsymbol{f}_k(\boldsymbol{x}_k, \boldsymbol{u}_k, t_k), \boldsymbol{x}_0, \boldsymbol{x}_\mathrm{e} \text{ known}$$

$$\boldsymbol{x}_\mathrm{min} \le \boldsymbol{x}_\mathrm{k} \le \boldsymbol{x}_\mathrm{max}, \boldsymbol{u}_\mathrm{min} \le \boldsymbol{u}_\mathrm{k} \le \boldsymbol{u}_\mathrm{max}$$

The detailed expression of the continuous kinematic model can be formulated as

$$(\boldsymbol{x}, \boldsymbol{u}) = \frac{\mathrm{d}}{\mathrm{d}t}\boldsymbol{x} = \frac{\mathrm{d}}{\mathrm{d}t}\begin{bmatrix} {}_\mathrm{I}\boldsymbol{r} \\ {}_\mathrm{I}\dot{\boldsymbol{r}} \end{bmatrix} = \begin{bmatrix} \boldsymbol{0} & \boldsymbol{I} \\ \boldsymbol{0} & \boldsymbol{0} \end{bmatrix}\boldsymbol{x} + \begin{bmatrix} \boldsymbol{0} \\ \boldsymbol{I} \end{bmatrix}\boldsymbol{u} = \boldsymbol{A}\boldsymbol{x} + \boldsymbol{b}\boldsymbol{u} \quad (41)$$

with $\boldsymbol{A} = \begin{bmatrix} \boldsymbol{0} & \boldsymbol{I} \\ \boldsymbol{0} & \boldsymbol{0} \end{bmatrix}, \boldsymbol{b} = \begin{bmatrix} \boldsymbol{0} \\ \boldsymbol{I} \end{bmatrix}, \boldsymbol{u} = {}_\mathrm{I}\ddot{\boldsymbol{r}}$.

The discrete dynamic equation is created through explicit Runge Kutta 4, which can be written as

$$\boldsymbol{x}_{k+1} = \boldsymbol{x}_k + \Delta t\sum\nolimits_{i=1}^{n} b_i\boldsymbol{k}_i, \boldsymbol{k}_i = \boldsymbol{f}\left(\boldsymbol{x}_k + \Delta t\sum\nolimits_{j=1}^{n} a_{ij}\boldsymbol{k}_j, \boldsymbol{u}_k\right) \quad (42)$$

where for 4th order Runge Kutta $n = 4$ and

$$\boldsymbol{b}^\mathrm{T} = \{b_i\} = \begin{bmatrix} 1 & 1 & 1 & 1 \end{bmatrix}/6, \boldsymbol{A} = \{a_{ij}\} = \begin{bmatrix} 0 & & & \\ 1/2 & 0 & & \\ & 1/2 & 0 & \\ & & 1 & 0 \end{bmatrix}$$

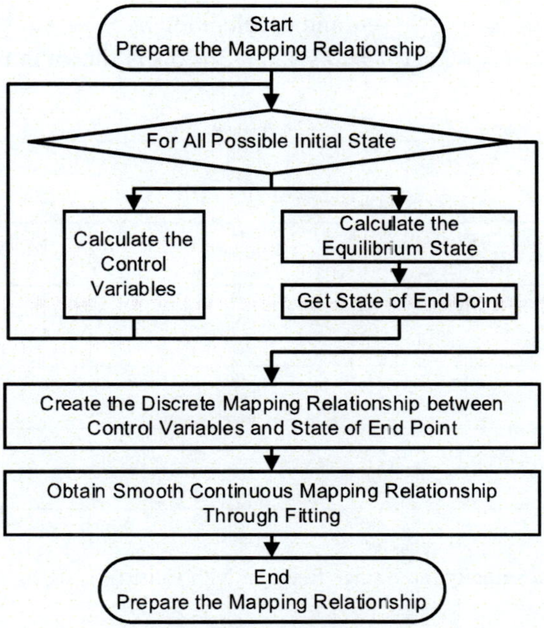

Fig. 9. Generation of optimal trajectory.

Quasi-Static Tracking Strategy. Through the optimal trajectory of end point and the mapping relationship, the optimal trajectories of the control inputs can be obtained. To create more smooth control trajectory and to meet the requirements of the control variables at the initial and the end stage, the discrete control trajectory will be resampled and fitted. The specific process is shown in Fig. 9.

4 Simulation and Analysis

4.1 Simulation for Lattice Boom Crane

The motions of the mobile cranes in the operation can be specified as three kinds, lifting, slewing, and luffing corresponding to 3 control variables. The slewing means the boom system and the turntable (super-structure) rotates along the vertical slewing axis. The luffing means to change the distance between the payload and the slewing axis by changing the elevation angle of the boom through luffing module. In order to make the movement of the crane relatively stable, the crane will start from the equilibrium position. The driven function is a second-order smooth continuous function, which can be written as

$$s(t) = \begin{cases} 0, & t < 0 \\ 1/2a_1t^2, & 0 \le t < t_1 \\ a_1t_1t + \left(3/2a_1t_1^2 - a_1t_1t_0 + a_1t_1t_3\right), & t_1 \le t < t_1 + t_2 \\ 1/2a_3t^2 - a_3t_0t + \left[a_3(t_0 - t_3)^2 - a_1t_1^2\right]/2, & t_1 \le t < t_0 \\ s_0, t_0 \le t & \end{cases} \quad (43)$$

with acceleration $a_1 = \frac{2s_0}{t_1(2t_0 - t_1 - t_3)}$ and deceleration $a_3 = -\frac{2s_0}{t_3(2t_0 - t_1 - t_3)}$. The example with s_0 = 10 m, t_0 = 40 s, t_1 = 5 s and t_3 = 5 s is shown in Fig. 10.

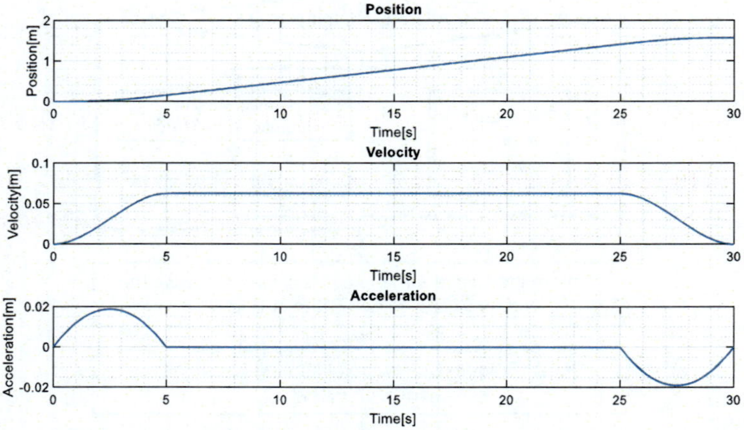

Fig. 10. Trapezoidal velocity three stage function with value $s_0 = 10\,\text{m}$, $t_0 = 40\,\text{s}$, $t_1 = 5\,\text{s}$ and $t_3 = 5\,\text{s}$.

Lifting Stage. The lifting stage requires the lifting rope to be shortened by 20 m within 40 s, including 5 s acceleration time and 5 s deceleration time. The Fig. 11 shows the translational movement of the load. It can be seen from the figure that the main movement trend is the lifting in the z-axis direction, and the vibration of the crane is smaller due to the reduction in the mass and elastic deformation of the lifting rope.

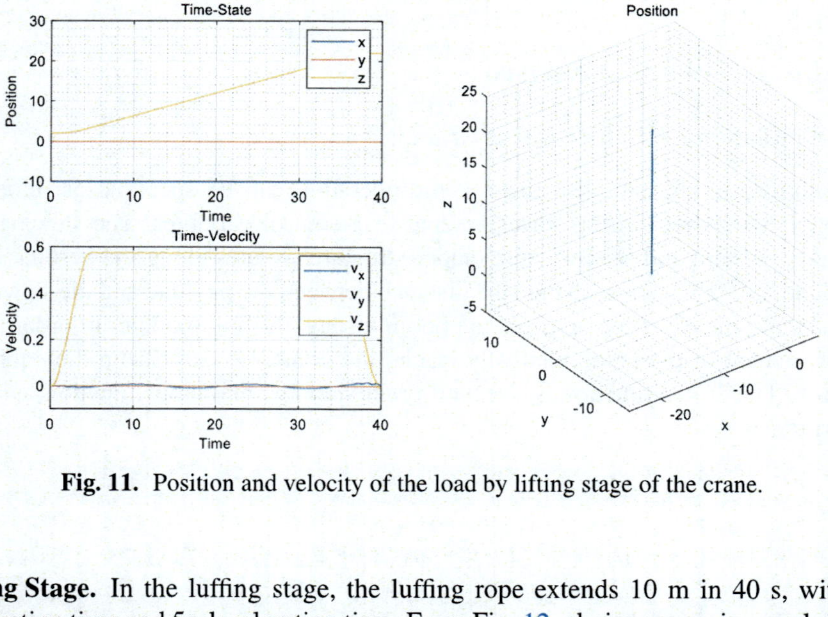

Fig. 11. Position and velocity of the load by lifting stage of the crane.

Luffing Stage. In the luffing stage, the luffing rope extends 10 m in 40 s, with 5 s acceleration time and 5s deceleration time. From Fig. 12, obvious sway in x-z plane can be detected, which has huger effect rather than elastic vibration. In order to reduce the sway of the load, the change of the length of the luffing rope should be limited.

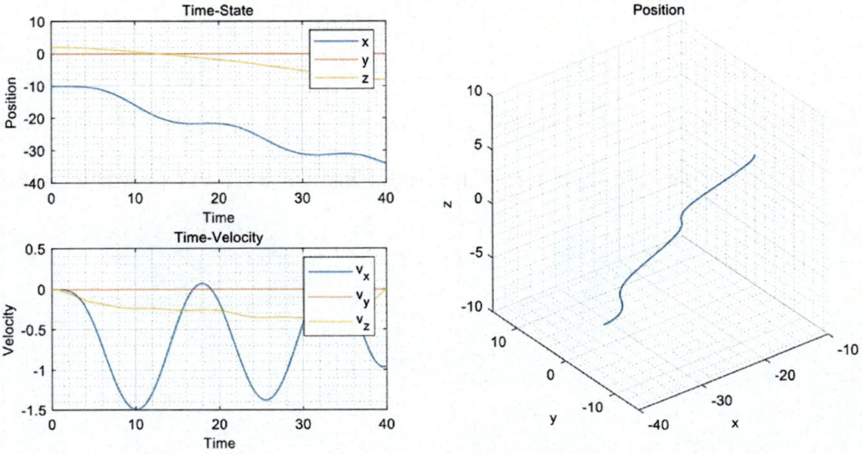

Fig. 12. Position and velocity of the load by luffing stage of the crane.

Slewing Stage. During the slewing stage, the crane will rotate 90° along z-axis within 40 s, with 5 s acceleration time and 5 s deceleration time. Because of the centrifugal effect caused by the slewing, the load will sway. Due to the long lifting rope, the sway has little influence on the movement along z-axis. Hence, the sway caused by the slewing can be considered on the x–y plane (Fig. 13).

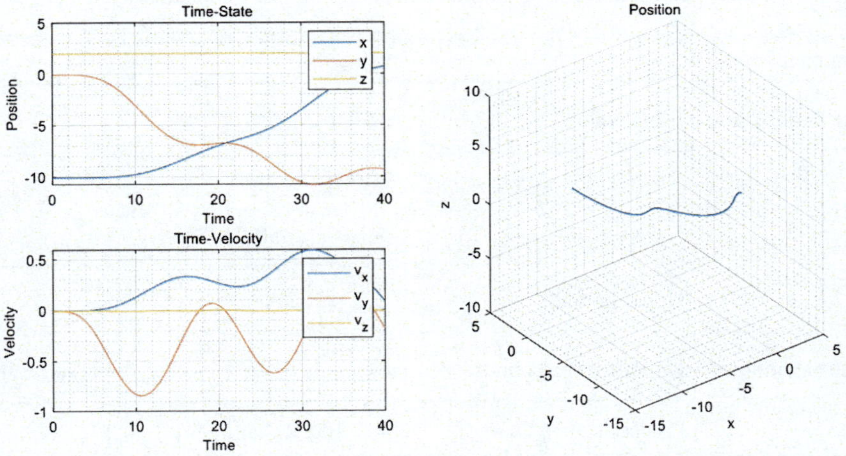

Fig. 13. Position and velocity of the load by slewing stage of the crane.

4.2 Quasi-static Optimal Control of Lattice Boom Crane

After applying 2-2 polynomial fitting on the point in Fig. 8, the continuous mapping relationship can be obtained and shown in Fig. 14.

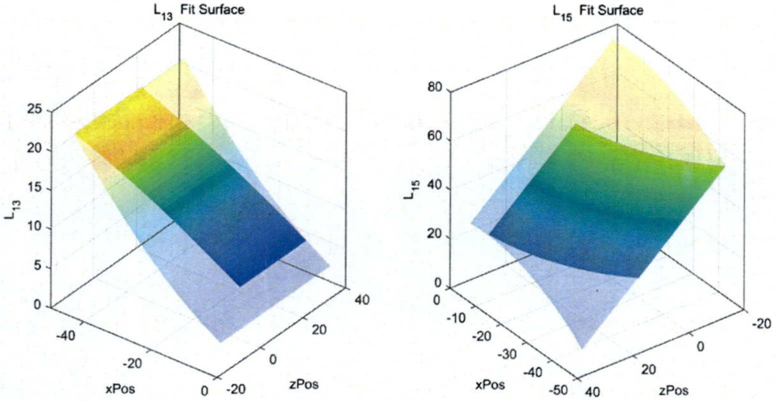

Fig. 14. Experimental data and fitted surface as actual mapping relationship.

The solid surface is where the experimental data locates, and the transparent surface the expanded area through flat quadratic surface fitting. The fitted surface is used as the actual mapping relationship to reduce the influence of experimental errors. However, considering the fitting error, it is suggested to use the solid area. The parameters of QR optimal control for simple kinematics model are shown in Table 2.

Table 2. Parameters of optimal control for simple kinematics system.

Parameter		Value
State variables	Initial value	$\boldsymbol{x}_0 = \begin{bmatrix} -10\ 0\ 0\ 0 \end{bmatrix}^{\mathrm{T}}$
	Nr. of state	$\mathrm{n_x} = 4$
	Lower bound	$\boldsymbol{x}_{\min} = \begin{bmatrix} -50\ 0\ -1\ -1 \end{bmatrix}^{\mathrm{T}}$
	Upper bound	$\boldsymbol{x}_{\max} = \begin{bmatrix} -5\ 30\ 1\ 1 \end{bmatrix}^{\mathrm{T}}$
Control variables	Upper/Lower bound	$\boldsymbol{u}_{\max} = \begin{bmatrix} 0.2\ 0.2 \end{bmatrix}^{\mathrm{T}}, \boldsymbol{u}_{\min} = \begin{bmatrix} -0.2\ -0.2 \end{bmatrix}^{\mathrm{T}}$
	Process state gain	$\boldsymbol{Q} = 100 \times \mathrm{diag}\left(\begin{bmatrix} 1\ 1\ 1\ 1 \end{bmatrix}\right)$
	Control gain	$\boldsymbol{R} = \mathrm{diag}\left(\begin{bmatrix} 1\ 1 \end{bmatrix}\right)$

(continued)

Table 2. *(continued)*

Parameter		Value
	End state gain	$\boldsymbol{P} = 100 \times \mathrm{diag}\left(\begin{bmatrix} 1 & 1 & 1 & 1 \end{bmatrix}\right)$
	Terminal State	$\boldsymbol{x}_{\mathrm{e}} = \begin{bmatrix} -35 & 10 & 0 & 0 \end{bmatrix}^{\mathrm{T}}$
Optimal Control	Time horizon	N = 400
	Sampling/Simulation time	$t_{\mathrm{s}} = 0.1$, $t_{\mathrm{f}} = 40$

The obtained optimal trajectory will be projected on the mapping surface to get the trajectory of control variables. The discrete control trajectory is shown in Fig. 15(b) with blue lines. To get the continuous smooth control trajectory fulfilled the requirement in initial and end stage, the discrete control trajectory is resampled (20 interpolation points), and 5 additional point is added at each initial and end stage. The projection process and control trajectory are shown in.

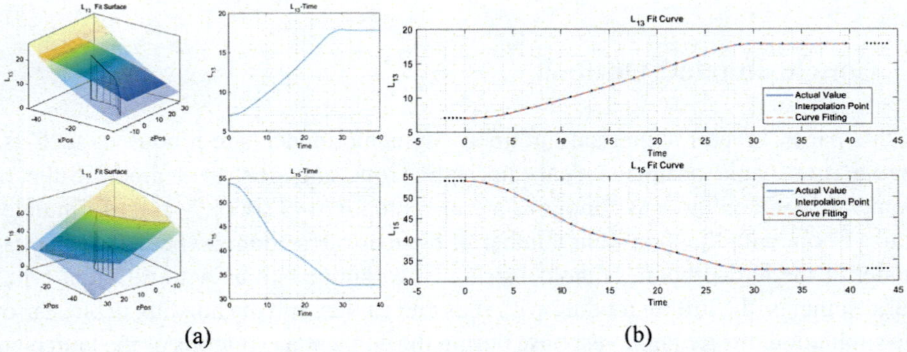

Fig. 15. (a) Projection of trajectory on mapping surface and optimal trajectories of control inputs (b) Expansion and fitting of optimal trajectories of control inputs.

In order to verify the quasi-static optimal control method, the generated continuous control trajectory is applied to the crane model to obtain the dynamic response. According to the experimental results in Fig. 16, the actual trajectory is basically close to the theoretical trajectory with slight vibration. This proves that the quasi-static control strategy can be used to track the optimal path. The vibration is mainly caused by the inertia of the crane's boom and the payload.

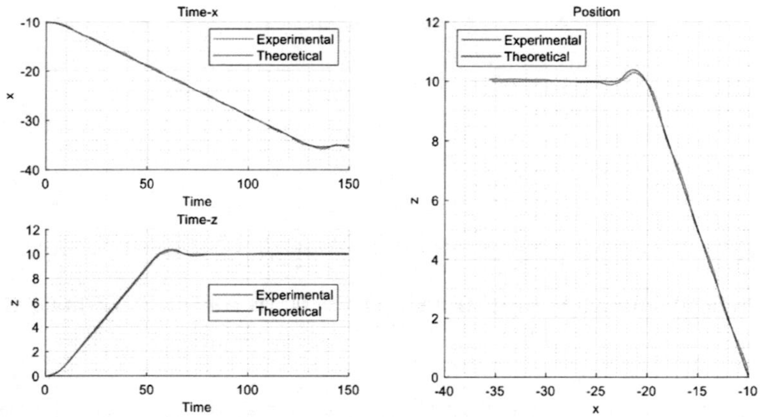

Fig. 16. Theoretical optimal trajectory and experimental trajectory using quasi-static optimal control strategy.

To suppress vibration, the additional feedback control of the lattice boom crane is required, such as PID, real-time MPC etc. The feedback controller for suppressing flexible vibration is the future research direction.

5 Conclusion and Outlook

In this paper, several numerical multibody dynamic models are proposed, such as the spatial Timoshenko beam, the ideal cubic spline rope, and the strut tie model. Super truss elements based on three assumptions are also proposed to simplify and to simulate the lattice boom with the minimum number of degrees of freedoms. The detailed dynamic model of the lattice boom crane consists of two lattice booms with different types of truss elements, the luffing module with rods and cables, and the hoisting cable. Through the simulation, the dynamic response during the elementary motions of the lattice boom crane is obtained.

According to the special control variables and the control requirements of the lattice boom crane, a quasi-static control method is designed. This method combines the static mapping relationship with the target optimal trajectory to generate the optimal trajectories of the control variables. The effect of the quasi-static optimal control strategy on the lattice boom crane is also verified through the simulation.

The following topics can be considered as further research:

1) The dynamic model is still large-scale with high nonlinearity. The further simplification requirement is raised, which may use the data-driven methods to get the dynamic through the input and output of models from the experiment or simulation.
2) The current control method is difficult to control the sway of the payload during the operations. Therefore, it is necessary to add the feedback loop between the payload and the control inputs into the control strategy and to carry out real-time adjustment by using real-time MPC or PID controller.

Acknowledgement. The work of this paper is supported by Deutsche Forschungsgemeinschaft (DFG) (FO 1180 1-1). We are grateful to Mr. Zhenxiao Yin for his work in MPC development.

References

1. Kleeberger, M.: Nichtlineare dynamische Berechnung von Gittermast-Fahrzeugkranen (en: Nonlinear dynamic calculation of lattice boom mobile cranes). Ph.D. Dissertation, Technical University of Munich (1996)
2. Günthner, W.A., Kleeberger, M.: Zum Stand der Berechnung von Gitter-mast-Fahrzeugkranen. dhf (03), 56–61 (1997)
3. Kleeberger, M., Hübner, K.T.: Using superelements in the calculation of lattice-boom cranes. Logist. J. Referierte Veröffentlichungen **2006**. Dezember (2006)
4. Koutsovasilis, P., Beitelschmidt, M.: Model reduction of large elastic systems: A comparison study on the elastic piston rod. In: Proceedings of XII World Congress in Mechanism and Machine Science, IFToMM, Besancon (2007)
5. Arnold, E., Sawodny, O., Neupert, J., Schneider, K.: Anti-sway system for boom cranes based on a model predictive control approach. In: 2005 IEEE International Conference Mechatronics and Automation. IEEE (2005)
6. Neupert, J., Arnold, E., Schneider, K., Sawodny, O.: Tracking and anti-sway control for boom cranes. Control Eng. Pract. **18**(1), 31–44 (2010)
7. Gao, L., Dai, X., Kleeberger, M., Fottner, J.: Dynamics modelling and simulation of super truss element based on non-linear beam element. In: Proceedings of the 11th International Conference on Simulation and Modeling Methodologies, Technologies and Applications. SCITEPRESS – Science and Technology Publications, Lda (2021)
8. Bazoune, A., Khulief, Y.A., Stephen, N.G.: Shape functions of three-dimensional Timoshenko beam element. J. Sound Vibr. **259**(2), 473–480 (2003)
9. Luo, Y.: An efficient 3D Timoshenko beam element with consistent shape functions. Adv. Theor. Appl. Mech **1**(3), 95–106 (2008)

Simulation Conditionally to a Subvariety and Application to Bayesian Optimization: A Dichotomous Approach

Frédéric Dambreville(✉)

ONERA/DTIS, Université Paris Saclay, 91123 Palaiseau Cedex, France
frederic.dambreville@onera.fr

Abstract. This chapter concerns the issue of simulating a random vector conditionally to a subvariety. In practice, the subvariety is approximated by a thin shell surrounding it. These simulations are unusual cases of rare event simulation, which are made even more difficult with the dimension, the thinness of the shell and the nonlinearity of the subvariety. We propose a generic approach dedicated to the simulation of this type of rare event, which is dichotomous and inspired by interval analysis, with a particular effort to reduce the dimension curse. Non-incremental and incremental versions of the algorithm are designed in order to handle possible incremental constraint sets. Details on a multithread implementation are given especially in terms of data structure. These simulation methods are then used to propose a non-linear approach to Bayesian optimization. Examples of simulations and of Bayesian optimization are given in order to illustrate and compare the performances of the methods.

Keywords: Rare event simulation · Subvariety · Bayesian optimization · Interval analysis

Notations

- $\mathbb{R}$ is the set of reals and x, y, z are real variables,
- $[x] \triangleq [x^-, x^+]$, $[y]$, $[z]$ are notations for intervals,
- Bold notations $[\boldsymbol{x}] \triangleq \prod_{k=1}^n [x_k] = \prod_{k=1}^n [x_k^-, x_k^+]$ and $[\boldsymbol{x}[\triangleq \prod_{k=1}^n [x_k^-, x_k^+[$ are used for boxes and half-open boxes respectively,
- $[\mathbb{R}] \triangleq \{[x^-, x^+] \,:\, [x^-, x^+] \subset \mathbb{R}\}$ is the set of interval subsets of $\mathbb{R}$.
  $[\mathbb{R}^n] \triangleq \{\prod_{k=1}^n [x_k] \,:\, \forall k,\, [x_k] \in [\mathbb{R}]\}$ is the set of box subsets of $\mathbb{R}^n$,
- $\rho([x]) = x^+ - x^-$ and $\rho([\boldsymbol{x}]) = \max\limits_{1 \leq k \leq n} \rho([x_k])$ are the length of interval & box,
- $g : \mathbb{R}^k \to \mathbb{R}^j$ is a (multivariate) real function,
- $g([\boldsymbol{x}]) \triangleq \{g(\boldsymbol{y}) : \boldsymbol{y} \in [\boldsymbol{x}]\}$. *Set* $g([\boldsymbol{x}])$ *is not necessarily a box,*
- $[g] : [\mathbb{R}^k] \to [\mathbb{R}^j]$ is an interval function. *Set* $[g]([\boldsymbol{x}])$ *is a box.*

G. Wagner et al. (Eds.): SIMULTECH 2021, LNNS 601, pp. 178–203, 2023.
https://doi.org/10.1007/978-3-031-23149-0_10

1 Introduction

This chapter is the long version of conference paper [4]. As a main topic, it addresses the issue of sampling a random vector defined on an n-dimension box conditionally to a subvariety of this box. As a secondary topic and a main application of the simulation, the chapter proposes a non-linear algorithmic approach for Bayesian optimisation [10].

Thus, given a random vector $\boldsymbol{X}$ of density $f_{\boldsymbol{X}}$ defined on $\mathbb{R}^n$, given box $[\boldsymbol{b}] \in [\mathbb{R}^n]$, *small* box $[\boldsymbol{\epsilon}] \in [\mathbb{R}^m]$ and map $g : [\boldsymbol{b}] \to \mathbb{R}^m$, our main objective is to sample conditional vector $[\boldsymbol{X} \,|g(\boldsymbol{X}) \in [\boldsymbol{\epsilon}]]$[1].

When space dimension n and constraint dimension m increase and length $\rho([\boldsymbol{\epsilon}])$ of $[\boldsymbol{\epsilon}]$ is small, then event $[g(\boldsymbol{X}) \in [\boldsymbol{\epsilon}]]$ has very low probability. We are dealing here with a particular case of rare event simulation. Moreover, as $[g(\boldsymbol{X}) \in [\boldsymbol{\epsilon}]]$ approximates a subvariety, it is foreseeable that conditional random vector $[\boldsymbol{X} \,|g(\boldsymbol{X}) \in [\boldsymbol{\epsilon}]]$ is essentially and "continuously" multimodal. Sketch Fig. 1 illustrates this well: to sample conditionally to the blue curve by a mixture of Gaussians, a large number of Gaussians are needed.

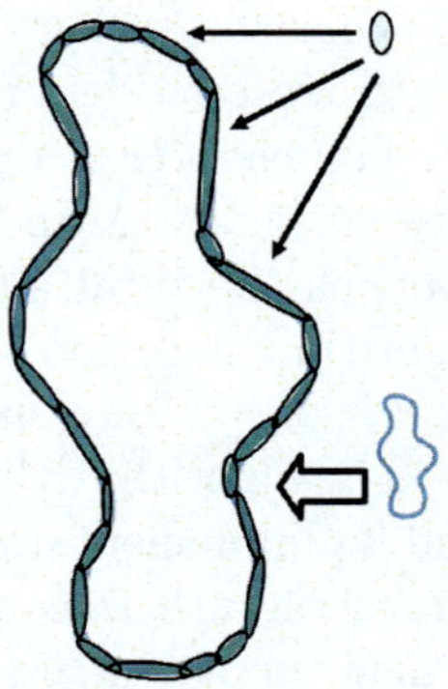

Fig. 1. Gaussian mixture

In survey [11], Morio, Balesdent et al. have evaluated the advantages and drawbacks of various rare event sampling methods. At this point, a comment may be done. In the literature, rare events are generally modelled by a function of risk being above an acceptable threshold. This formalism is quite general, but it suggests that the simulation of a rare event is tightly related to the maximization of a function. In practice, the maximizing set of a function is unimodal or somewhat multimodal. It is uncommon that this maximizing set is a subvariety. Not surprisingly, many methods evaluated in [11] are working well when the rare event is unimodal or moderately multimodal. In the case of conditional vector $[\boldsymbol{X} \,|g(\boldsymbol{X}) \in [\boldsymbol{\epsilon}]]$, these approaches do not work properly or need additional studies to take beneficially into account the subvariety structure.

This chapter promotes an alternative approach based on a dichotomous exploration in order to sample conditionally to a subvariety characterized by a known function. By design, this approach produces independent samples and avoids the phenomenon of sample impoverishement. These properties make it particularly well suited for an accurate approximation of a law conditionally to the subvariety. However, the method must fight the curse of dimensionality and we will be faced with two orthogonal dimensional problems: the exponential increase in sampling exploration and the degeneracy of particle weights. This paper presents an approach as well as generic and weakly parameterized algorithms, allowing good sampling performance for moderate dimensions (up to 11 for now) with balanced management of dimensional issues.

[1] $[\boldsymbol{X} \,|g(\boldsymbol{X}) \in [\boldsymbol{\epsilon}]]$ is used as an abbreviation for $[\boldsymbol{X} \,|\boldsymbol{X} \in [\boldsymbol{b}] \;\&\; g(\boldsymbol{X}) \in [\boldsymbol{\epsilon}]]$.

One main applicative purpose of this work is to contribute to the domain of Bayesian optimization from a nonlinear point of view. Bayesian optimization is among the key techniques for black-box optimization. In this kind of problems, one have to optimize fonctions which do not have usable formalization and for which the request to an evaluator is generally particularly expensive. Accordingly, an inherent goal of such optimization is to save as much as possible on the number of candidate solutions to be evaluated. Approaches based on Gaussian surrogate models [8] have demonstrated the great interest of these Bayesian methods, but suffer from a drop in efficiency with the increase in dimension. Our contribution aims to address this dimensional difficulty by considering a known non-linear criterion depending on a model noise (a random vector) rather than a surrogate model of the criterion.

The chapter is divided in four main parts. Section 2 introduces concepts of interval analysis and derives some first ideas for a generic sampler. Section 3 deepens the intuitions introduced in Sect. 2 and presents the working generic algorithms. Section 4 applies the sampling method for the design of a nonlinear algorithm for Bayesian optimization. Section 5 presents tests and analyses.

The algorithm already presented in conference paper [4] has slightly evolved and its implementation has been deeply revisited in terms of multithreading and data structure. In comparison to the conference paper, this chapter also describes a new incremental sampling algorithm, includes a new chapter dedicated to Bayesian optimization and presents rebuilt and supplemented tests and results.

2 Toward a Generic Sampling Approach

This section presents some useful concepts and ideas for defining actually working sampling algorithms. Interval analysis is one of these tools. Interval analysis is a performing and accurate tool for dealing with constraints, and it provides a precise control on the approximation errors. Another interesting point is that intervals and boxes combine well with probability distributions.

2.1 Condiderations About Interval Analysis

It is not our purpose to perform a good introduction on interval analysis [1,7]. However, we refer to some key concepts, which are inspiring ideas for this work.

Functions and Operators. As a main ingredient throughout this chapter, it is reasonable to assume that for most common real functions g, it is possible to derive an interval function $[g]$ that satisfies properties:

- $g([\boldsymbol{x}]) \subset [g]([\boldsymbol{x}])$ for any $[\boldsymbol{x}] \in [\mathbb{R}^n]$,[2]
- $[g]([\boldsymbol{x}]) \subset [g]([\boldsymbol{y}])$ when $[\boldsymbol{x}] \subset [\boldsymbol{y}]$,

[2] In particular, it is better if $[g]$ is minimal, that is $[g]([\boldsymbol{x}])$ is the minimal box containing $g([\boldsymbol{x}])$ as subset for all $[\boldsymbol{x}] \in [\mathbb{R}^n]$.

– If $\rho([\boldsymbol{x}])$ vanishes, then $\rho([g]([\boldsymbol{x}]))$ vanishes:

$$\rho\left([g]([\boldsymbol{x}])\right) \xrightarrow[\rho([\boldsymbol{x}])\to 0]{} 0 \tag{1}$$

These properties express that $[g]$ implies a bound on the error propagated by g, and this bound has good convergence behavior in regards to the error. Incidentally, we assume from now on that notation $[g]$ refers to an interval function, which verifies these good properties with respect to the function g.

Let us see how to build the interval functions on some common examples:

1. Reference functions, $g \in \{\ln, \exp, \sin, \cos, .^n, \dots\}$, are continuous onto $\mathbb{R}$, so that $g([x]) \in [\mathbb{R}]$ for all $[x] \in [\mathbb{R}]$. Then, it is optimal to set $[g]([x]) = g([x])$ for all $[x] \in [\mathbb{R}]$. As a consequence, the interval functions are easily optimaly implementable for most reference functions. Here are some incomplete examples of definitions:

$$[\ln]([x]) \triangleq [\ln(x^-), \ln(x^+)] \text{ and } [x]^n \triangleq \begin{cases} [x_-^n, x_+^n] \text{ for } n > 0, \\ [x_+^n, x_-^n] \text{ for } n < 0 \;\&\; 0 \notin [x], \end{cases} \tag{2}$$

$$[\cos][x] \triangleq \begin{cases} [\cos(x^-), \cos(x^+)] \text{ for } [x] \subset [-\pi, 0] \\ [\cos(x^+), \cos(x^-)] \text{ for } [x] \subset [0, \pi] \\ [\min(\cos(x^-), \cos(x^+)), 1] \quad \text{for} \quad 0 \in [x] \subset [-\pi, \pi] \\ \text{etc.} \end{cases} \tag{3}$$

2. Minimal interval functions for classical operators $+$, $\cdot$, $-$, $/$ are also easily defined. For example:

$$[x] + [y] \triangleq [x^- + y^-, x^+ + y^+] \quad \text{and} \quad [x] - [y] \triangleq [x^- - y^+, x^+ - y^-]. \tag{4}$$

3. Function g defined by $g(\theta) = (\cos(\theta), \sin(\theta))$ is an example, which is such that $g([\theta]) \notin \left[\mathbb{R}^2\right]$. One would rather define $[g]([\theta]) = [\cos]([\theta]) \times [\sin]([\theta])$ ($\times$ is the Carthesian product) which is a strict supset of $g([\theta])$ in general. By the way, this construction is an illustration on how the interval functions of reference are used to define complex interval functions straightforwardly.

There is no uniqueness in the construction of $[g]$, unless it is chosen minimal. Unfortunately, constructing a minimal $[g]$ is not always easy or automatic. Let us consider the case of function $g : \theta \mapsto \cos^2\theta + \sin^2\theta$. Then, there are two obvious definitions for $[g]$:

1. By using the reference functions cos, sin, $.^2$ and $+$ one may derive:

$$[g]([\theta]) = ([\cos]([\theta]))^2 + ([\sin]([\theta]))^2. \tag{5}$$

   For example, we compute $[g]\left(\left[0, \frac{\pi}{2}\right]\right) = [0, 2]$ which is a bad error bound on theoretical value 1. Now, we also compute $[g]\left(\left[-\frac{1}{10}, \frac{1}{10}\right]\right) \simeq [0.99, 1.01]$ which is a tight error bound on 1. This example holds confirmation that $[g]([\theta])$ has a good behavior for small boxes $[\theta]$.
2. By noticing that $\cos^2\theta + \sin^2\theta = 1$, it is optimal to define $[g]([\theta]) = [1, 1]$.

Although approach 2 gives the best solution, in practice approach 1 is prefered since it is generic, it is based on already implemented functions of reference, and it provides a way to construct automatically $[g]$ without any specific knowledge.

Subpaving and Set Inversion. As $[g]$ implies a bound on errors propagated by g with good convergence behavior, it may be used combined with a dichotomous process to produce a subpaving which efficiently approximates a set inversion $g^{-1}([\boldsymbol{y}])$. The example on Fig. 2 is kindly given by professor Jaulin and is also taken from [4]. It shows a resulting subpaving which approximates a set inversion. The decomposition is clearly dichotomous.

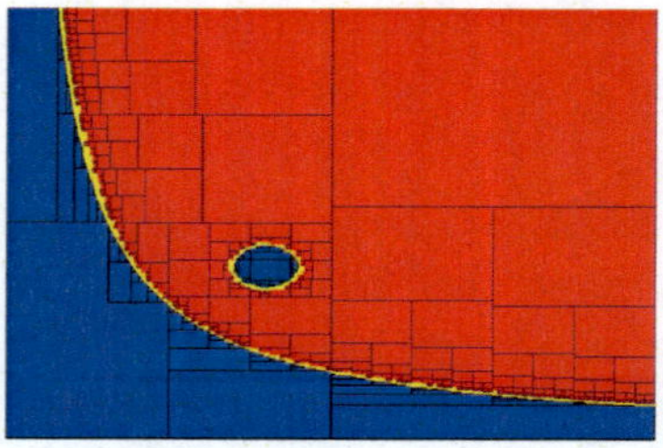

Fig. 2. Subpaving

A bisection process is iterated starting from main box $[\boldsymbol{b}]$; at each iteration, sub-boxes $[\boldsymbol{x}]$ are tested against the constraint $g([\boldsymbol{x}]) \subset [\boldsymbol{y}]$. Three cases arise:

case a: $[g]([\boldsymbol{x}]) \subset [\boldsymbol{y}]$, then $[\boldsymbol{x}]$ is among *red* boxes, which constitute a *subpaving* of $g^{-1}([\boldsymbol{y}])$.

case b: $[g]([\boldsymbol{x}]) \cap [\boldsymbol{y}] = \emptyset$, then $[\boldsymbol{x}]$ is among *blue* boxes, which constitute a *subpaving* of $\mathbb{R}^n \setminus g^{-1}([\boldsymbol{y}])$.

case c: Otherwise, bisection has to be repeated on $[\boldsymbol{x}]$ until sufficient convergence (*yellow* color).

Property (1) plays a key role in the decomposition process, ensuring that case **a** or case **b** are finally achieved when sub-boxes $[\boldsymbol{x}]$ are sufficiently small and sufficiently far from the frontier of set $g^{-1}([\boldsymbol{y}])$.

2.2 Naive Dichotomous Approach for Sampling

Let μ be Borel measure on $\mathbb{R}^n$. It is given from now on:

- a random vector $\boldsymbol{X}$ on $\mathbb{R}^n$ characterized by a *bounded* density $f_{\boldsymbol{X}}$. Cumulative distribution function $F_{\boldsymbol{X}}$ of $\boldsymbol{X}$, defined by $F_{\boldsymbol{X}}(\boldsymbol{x}) = P(\boldsymbol{X} \leq \boldsymbol{x})$ for all $\boldsymbol{x} \in \mathbb{R}^n$, is assumed to be easily computable,
- a box $[\boldsymbol{b}] \in [\mathbb{R}^n]$ and a small box $[\boldsymbol{\epsilon}] \in [\mathbb{R}^m]$,
- a continuous map $g : [\boldsymbol{b}] \to \mathbb{R}^m$ built of functions and operators of reference,
- $[g]$ derived from g and related interval functions and operators of reference.

Sampling Within Boxes. We point out that it is easy to compute $P(\boldsymbol{X} \in [\boldsymbol{y}])$ or to sample $[\boldsymbol{X} \,|\boldsymbol{X} \in [\boldsymbol{y}]]$ when $F_{\boldsymbol{X}}$ is available, especially when the components of $\boldsymbol{X}$ are jointly independent. These results are well known, and details are given in [4]. Thus, these features are taken for granted in this paper.

Sampling by Means of a Subpaving. Assume that set $g^{-1}([\boldsymbol{\epsilon}])$ has been approximated (by excess) by a subpaving. Thus, there is $\mathbb{P} \subset [R^n]$ such that:

- For all $[\boldsymbol{x}], [\boldsymbol{y}] \in \mathbb{P}$, boxes $[\boldsymbol{x}[$ and $[\boldsymbol{y}[$ are disjoint,
- $g^{-1}([\boldsymbol{\epsilon}]) \subseteq \bigcup_{[\boldsymbol{x}]\in\mathbb{P}} [\boldsymbol{x}]$ and $\sqcup_{\mathbb{P}} \simeq g^{-1}([\boldsymbol{\epsilon}])$, where $\sqcup_{\mathbb{P}} \triangleq \bigsqcup_{[\boldsymbol{x}]\in\mathbb{P}} [\boldsymbol{x}[$.

Set $\mathbb{P}$ is typically composed from boxes of case **a** and case **c** (red and yellow colors) after a dichotomous subpaving construction of the set inversion.

The quality of the approximation may be quantified by set measures:

$$\alpha_{\mathbb{P}} = \mu\Big(\sqcup_{\mathbb{P}} \setminus g^{-1}([\epsilon])\Big) = \sum_{[\boldsymbol{x}]\in\mathbb{P}} \mu\left([\boldsymbol{x}[\right) - \mu\left(g^{-1}([\epsilon])\right). \tag{6}$$

Smaller is $\alpha_{\mathbb{P}}$, better is the approximation. Interval based set inversions are able to reach arbitrary precision for small dimensions (2 or 3 typically).

Now, for all $\boldsymbol{y} \in \mathbb{R}^n$, it happens that:

$$f_{\boldsymbol{X}\,|\,\boldsymbol{X}\in\sqcup_{\mathbb{P}}}(\boldsymbol{y}) = \frac{f_{\boldsymbol{X}}(\boldsymbol{y})\delta_{\boldsymbol{y}\in\sqcup_{\mathbb{P}}}}{P\left(\boldsymbol{X}\in\sqcup_{\mathbb{P}}\right)} \tag{7}$$

$$= \frac{f_{\boldsymbol{X}}(\boldsymbol{y})\sum_{[\boldsymbol{x}]\in\mathbb{P}}\delta_{\boldsymbol{y}\in[\boldsymbol{x}[}}{\sum_{[\boldsymbol{x}]\in\mathbb{P}} P\left(\boldsymbol{X}\in[\boldsymbol{x}[\right)} = \frac{\sum_{[\boldsymbol{x}]\in\mathbb{P}} P\left(\boldsymbol{X}\in[\boldsymbol{x}[\right) f_{\boldsymbol{X}\,|\,\boldsymbol{X}\in[\boldsymbol{x}[}(\boldsymbol{y})}{\sum_{[\boldsymbol{x}]\in\mathbb{P}} P\left(\boldsymbol{X}\in[\boldsymbol{x}[\right)}$$

$$= \sum_{[\boldsymbol{x}]\in\mathbb{P}} \frac{P\left(X\in[\boldsymbol{x}]\right)}{\sum_{[\boldsymbol{x}]\in\mathbb{P}} P\left(X\in[\boldsymbol{x}]\right)} f_{\boldsymbol{X}\,|\,\boldsymbol{X}\in[\boldsymbol{x}[}(\boldsymbol{y}), \tag{8}$$

where $\delta_{\mathbf{true}} = 1$ and $\delta_{\mathbf{false}} = 0$ else. Since $f_{\boldsymbol{X}}$ is bounded, Eq. (7) implies:

$$\left[\boldsymbol{X}\,|\,\boldsymbol{X}\in\sqcup_{\mathbb{P}}\right] \xrightarrow[\alpha_{\mathbb{P}}\to 0]{\mathcal{L}} \left[\boldsymbol{X}\,\middle|\,\boldsymbol{X}\in g^{-1}([\epsilon]\right] \tag{9}$$

Now, Eq. (8) shows clearly that $f_{\boldsymbol{X}\,|\,\boldsymbol{X}\in\sqcup_{\mathbb{P}}}$ may be sampled by applying two steps: first sample a box $[\boldsymbol{x}] \in \mathbb{P}$ according to the discrete probability $\frac{P(\boldsymbol{X}\in[\boldsymbol{x}])}{\sum_{[\boldsymbol{x}]\in\mathbb{P}} P(\boldsymbol{X}\in[\boldsymbol{x}])}$, then sample $\boldsymbol{y}$ by the conditional law $f_{\boldsymbol{X}\,|\,\boldsymbol{X}\in[\boldsymbol{x}[}$. At last, we have here an efficient method for sampling $[\boldsymbol{X}\,|g(\boldsymbol{X})\in[\epsilon]]$ (Algorithm 1).

The approach is efficient on conditional events like $[\boldsymbol{X}\,|g(\boldsymbol{X})\in[\epsilon]]$. But this application of the interval-based inversion is only applicable to rather small dimensions. Taking inspiration of this preliminary approach, we address now the sampling problem in higher dimensions.

Function *Sampling* $[\boldsymbol{X}\,|g(\boldsymbol{X})\in[\epsilon]]$
    input : α,g,N **output:** $\boldsymbol{y}_{1:N}$
    Build subpaving $\mathbb{P}$ such that $\alpha_{\mathbb{P}} < \alpha$
    for $k \leftarrow 1$ **to** N **do**
        Select $[\boldsymbol{x}] \in \mathbb{P}$ with proba. $\frac{P(\boldsymbol{X}\in[\boldsymbol{x}])}{\sum_{[\boldsymbol{x}]\in\mathbb{P}} P(\boldsymbol{X}\in[\boldsymbol{x}])}$
        Build $\boldsymbol{y}_k$ by sampling $[\boldsymbol{X}\,|\,\boldsymbol{X}\in[\boldsymbol{x}]]$
    end
end

Algorithm 1: Based on a subpaving.

Toward an Improved Approach. A key point of Algorithm 1 is to be able to sample a box $[\boldsymbol{x}]$ of a subpaving of $g^{-1}([\epsilon])$. It is noticeable that an entire build of the subpaving is not needed here. Indeed, if we were able to construct a box $[\boldsymbol{x}]$ of the subpaving *on demand*, together with its *relative weight* within the subpaving, then we would be able to build sample $\boldsymbol{y}$.

Therefore, it may be opportune to merge the sampling process with the dichotomous construction of the subpaving itself. Now, a main ingredient of a dichotomous approach is also how the algorithm divides and conquers. In general, bisections are often used in dichotomous processes, as there is a garanty of exponential volume decrease of the search area. Our approach is less constrained, so as to better tune the exploration strategy.

Function *Sampling* $[\boldsymbol{X}\,|g(\boldsymbol{X})\in[\boldsymbol{\epsilon}]]$
  input : r, g, ω, N **output:** $(\boldsymbol{y}_k, w_k)_{1:N}$
  for $k \leftarrow 1$ **to** N **do**
    $([\boldsymbol{x}_0], \pi_k, j) \leftarrow ([\boldsymbol{b}], 1, 0)$
    while $\rho([\boldsymbol{x}_j]) > r$ **and** $[g]([\boldsymbol{x}_j]) \not\subset [\boldsymbol{\epsilon}]$ **do**
      $([\boldsymbol{l}_{j+1}], [\boldsymbol{r}_{j+1}]) \leftarrow \texttt{Cut}([\boldsymbol{x}_j])$
      $j \leftarrow j+1$
      $[\boldsymbol{x}_j] \leftarrow \texttt{Bern}(([\boldsymbol{l}_j], \omega_{[\boldsymbol{l}_j]}), ([\boldsymbol{r}_j], \omega_{[\boldsymbol{r}_j]}))$
      $\pi_k \leftarrow \frac{\omega_{[\boldsymbol{x}_j]}}{\omega_{[\boldsymbol{l}_j]}+\omega_{[\boldsymbol{r}_j]}}\pi_k$
    end
    $w_k \leftarrow P(\boldsymbol{X}\in[\boldsymbol{x}_j]) \,/\, \pi_k$
    Build $\boldsymbol{y}_k$ by sampling $[\boldsymbol{X}\,|\boldsymbol{X}\in[\boldsymbol{x}_j]]$
  end
end

Algorithm 2: Based on a weighting function.

Here, we speak in terms of cuts, which are more general, *being implied that an appropriate management of the box length is made in order to ensure the convergence.* In this paper, a cut is defined as follows:

– A cut of box $[\boldsymbol{x}] \in [\mathbb{R}^n]$ is a pair $([\boldsymbol{l}], [\boldsymbol{r}]) \in [\mathbb{R}^n]^2$ such that:

$$[\boldsymbol{l}[\cap[\boldsymbol{r}[=\emptyset \quad\text{and}\quad [\boldsymbol{l}[\sqcup[\boldsymbol{r}[=[\boldsymbol{x}[\,, \tag{10}$$

– A bisection is a cut $([\boldsymbol{l}], [\boldsymbol{r}])$ such that $[\boldsymbol{l}]$ and $[\boldsymbol{r}]$ are same-sized.

In order to drive the dichotomous sampling process, we assume that a predictive weighting function is available:

$$\begin{cases} \omega_{[x]} = 0 & \text{if} \quad [g]([\boldsymbol{x}]) \cap [\boldsymbol{\epsilon}] = \emptyset \\ \omega_{[x]} \simeq P(\boldsymbol{X}\in[\boldsymbol{x}]\,\&\,g(\boldsymbol{X})\in[\boldsymbol{\epsilon}]) & \text{otherwise.} \end{cases} \tag{11}$$

Algorithm 2 implicitly builds a partial subpaving, and produces at the same time a weighted particle cloud as a result of the sampling of $[\boldsymbol{X}\,|g(\boldsymbol{X})\in[\boldsymbol{\epsilon}]]$. The algorithm iterates (**for** loop) the same sampling process, that is the following successive steps, until $[\boldsymbol{x}_j]$ is sufficiently small (*i.e.* $\rho([\boldsymbol{x}_j]) \leq r$) or is inside an implied suppaving (*i.e.* $[g]([\boldsymbol{x}_j]) \subset [\boldsymbol{\epsilon}]$)[3]:

– Build a cut of $[\boldsymbol{x}_j]$ by means of function $\texttt{Cut}([\boldsymbol{x}_j])$. This function is designed so as to ensure that $\rho([\boldsymbol{x}_j])$ vanishes,
– Select randomly one box of cut $([\boldsymbol{l}_j], [\boldsymbol{r}_j])$ in proportion to their weight, $\texttt{Bern}(([\boldsymbol{l}_j], \omega_{[\boldsymbol{l}_j]}), ([\boldsymbol{r}_j], \omega_{[\boldsymbol{r}_j]}))$,
– Update π_k which computes the processed probability of $[\boldsymbol{x}_j]$ in regards to the Bernoulli sequence.

[3] Recall that $[g]([\boldsymbol{x}_j])$ is easily computable while $g([\boldsymbol{x}_j])$ is not.

Assume that J is the last value reached by parameter j after the **while** loop. Then, the corrected weight $w_k = \frac{1}{\pi_k} P\left(\boldsymbol{X} \in [\boldsymbol{x}_J]\right)$ is computed for $[\boldsymbol{x}_J]$ and for $\boldsymbol{y}_k$, and $\boldsymbol{y}_k$ is sampled from $[\boldsymbol{x}_J]$.

Notice that $\omega_{[\boldsymbol{x}]} = 0$ when $[g]([\boldsymbol{x}]) \cap [\boldsymbol{\epsilon}] = \emptyset$, so that boxes $[\boldsymbol{x}_J]$ are necessary within a subpaving of $g^{-1}([\boldsymbol{\epsilon}])$ or its border, thanks to the Bernoulli process. Combined with loop constraint $\rho([\boldsymbol{x}_j]) > r$ **and** $[g]([\boldsymbol{x}_j]) \not\subset [\boldsymbol{\epsilon}]$, it follows that a subpaving of $g^{-1}([\boldsymbol{\epsilon}])$ (or its border) is implicitely and partially built during the sampling process.

When $[g]([\boldsymbol{x}_J]) \subset [\boldsymbol{\epsilon}]$, we have $w_k = \frac{P(\boldsymbol{X} \in [\boldsymbol{x}_J])}{\pi_k}$ where π_k evaluates the processed probability for $[\boldsymbol{x}_J]$. As a result, the weighted particles $(\boldsymbol{y}_k, w_k)$ provide an unbiased estimation of $f_{\boldsymbol{X}\,|g(\boldsymbol{X}) \in [\boldsymbol{\epsilon}]}$ in a subpaving of $g^{-1}([\boldsymbol{\epsilon}])$. It is not the same at the border of $g^{-1}([\boldsymbol{\epsilon}])$, but this case is neglected. However, the sampler is not at all efficient when considering its variance.

Assume $\omega_{[\boldsymbol{x}]} = P\left(\boldsymbol{X} \in [\boldsymbol{x}] \,\&\, g(\boldsymbol{X}) \in [\boldsymbol{\epsilon}]\right)$, a case which works perfectly. In this ideal case, the weight along a **while** loop is computed by:

$$\frac{\omega_{[\boldsymbol{x}_j]}}{\omega_{[\boldsymbol{l}_j]} + \omega_{[\boldsymbol{r}_j]}} = \frac{P\left(\boldsymbol{X} \in [\boldsymbol{x}_j] \,\&\, g(\boldsymbol{X}) \in [\boldsymbol{\epsilon}]\right)}{P\left(\boldsymbol{X} \in [\boldsymbol{l}_j] \cup [\boldsymbol{r}_j] \,\&\, g(\boldsymbol{X}) \in [\boldsymbol{\epsilon}]\right)} = \frac{P\left(\boldsymbol{X} \in [\boldsymbol{x}_j] \,\&\, g(\boldsymbol{X}) \in [\boldsymbol{\epsilon}]\right)}{P\left(\boldsymbol{X} \in [\boldsymbol{x}_{j-1}] \,\&\, g(\boldsymbol{X}) \in [\boldsymbol{\epsilon}]\right)}, \tag{12}$$

and then:

$$\pi_k = \prod_{j=1}^{J} \frac{\omega_{[\boldsymbol{x}_j]}}{\omega_{[\boldsymbol{l}_j]} + \omega_{[\boldsymbol{r}_j]}} = \frac{P\left(\boldsymbol{X} \in [\boldsymbol{x}_J] \,\&\, g(\boldsymbol{X}) \in [\boldsymbol{\epsilon}]\right)}{P\left(\boldsymbol{X} \in [\boldsymbol{b}] \,\&\, g(\boldsymbol{X}) \in [\boldsymbol{\epsilon}]\right)}. \tag{13}$$

Three cases potentially arise:

– $[g]([\boldsymbol{x}_J]) \subset [\boldsymbol{\epsilon}]$, *i.e.* $[\boldsymbol{x}_J]$ is in implied subpaving. Since $g([\boldsymbol{x}_J]) \subset [g]([\boldsymbol{x}_J])$, it comes $P\left(\boldsymbol{X} \in [\boldsymbol{x}_J] \,\&\, g(\boldsymbol{X}) \in [\boldsymbol{\epsilon}]\right) = P\left(\boldsymbol{X} \in [\boldsymbol{x}_J]\right)$. Then:

$$\begin{aligned} w_k &= \frac{P\left(\boldsymbol{X} \in [\boldsymbol{x}_J]\right)}{\frac{P(\boldsymbol{X} \in [\boldsymbol{x}_J] \,\&\, g(\boldsymbol{X}) \in [\boldsymbol{\epsilon}])}{P(\boldsymbol{X} \in [\boldsymbol{b}] \,\&\, g(\boldsymbol{X}) \in [\boldsymbol{\epsilon}])}} = \frac{P\left(\boldsymbol{X} \in [\boldsymbol{x}_J]\right)}{\frac{P(\boldsymbol{X} \in [\boldsymbol{x}_J])}{P(\boldsymbol{X} \in [\boldsymbol{b}] \,\&\, g(\boldsymbol{X}) \in [\boldsymbol{\epsilon}])}} = P\left(\boldsymbol{X} \in [\boldsymbol{b}] \,\&\, g(\boldsymbol{X}) \in [\boldsymbol{\epsilon}]\right) \\ &= P\left(g(\boldsymbol{X}) \in [\boldsymbol{\epsilon}]\right) \end{aligned} \tag{14}$$

– $[g]([\boldsymbol{x}_J]) \cap [\boldsymbol{\epsilon}] \neq \emptyset$ but $[g]([\boldsymbol{x}_J]) \not\subset [\boldsymbol{\epsilon}]$, *i.e.* $[\boldsymbol{x}_J]$ is within the border of the implied subpaving. These cases are negligible for small precision r.
– $[g]([\boldsymbol{x}_J]) \cap [\boldsymbol{\epsilon}] = \emptyset$, *i.e.* $[\boldsymbol{x}_J]$ is outside the implied subpaving and its border. This case is simply impossible from the Bernoulli process.

Equation (14) shows that the sampling process results in a cloud of same-weight particles over the implied subpaving. Border cases are negligible. Here we have a sampler of $[\boldsymbol{X}\,|g(\boldsymbol{X}) \in [\boldsymbol{\epsilon}]]$ with the best variance performance in regards to the number of particles. But hypothesis $\omega_{[\boldsymbol{x}]} = P\left(\boldsymbol{X} \in [\boldsymbol{x}] \,\&\, g(\boldsymbol{X}) \in [\boldsymbol{\epsilon}]\right)$ is necessary. Of course such exact weighting function is almost never available.

Why Does It Generally Not Work? When $\omega_{[\boldsymbol{x}]} \neq P\left(\boldsymbol{X} \in [\boldsymbol{x}] \,\&\, g(\boldsymbol{X}) \in [\boldsymbol{\epsilon}]\right)$, the accumulated error will explode with the dimension, which will result in dramatically uneven weights on the particles. The resulting weighted particles cloud is then useless for practical applications.

3 Generic Dichotomous Approaches for Sampling

Algorithms 1 and 2 illustrate the two main dimensional issues, that we have to deal with. These approaches are complementary:

- By building a complete subpaving of $g^{-1}([\boldsymbol{\epsilon}])$, Algorithm 1 makes possible a direct sampling of $[\boldsymbol{X}\,|g(\boldsymbol{X})\in[\boldsymbol{\epsilon}]]$, and incidently an accurate computation of $P(\boldsymbol{X}\in[\boldsymbol{x}]\,\&\,g(\boldsymbol{X})\in[\boldsymbol{\epsilon}])$. However, this construction of a complete subpaving is only possible for small dimensions.
- Algorithm 2 avoids the construction of a complete subpaving. Instead, it builds the boxes of an implied subpaving on demand throughout the sampling iteration. However, the algorithm is inefficient unless the predictive weighting function $\omega_{[\boldsymbol{x}]}$ is a good approximation of $P(\boldsymbol{X}\in[\boldsymbol{x}]\,\&\,g(\boldsymbol{X})\in[\boldsymbol{\epsilon}])$. This condition is not accessible in general.

We propose now an intermediate approach which:

- keeps history of the subpaving construction throughout the sampling process,
- use this history to build an improved estimate of $P(\boldsymbol{X}\in[\boldsymbol{x}]\,\&\,g(\boldsymbol{X})\in[\boldsymbol{\epsilon}])$.

By these tricks, it is expected that the sampling precision will increase with the number of samples. In order to avoid useless exploration, we also truncate the dichotomous process on the basis of some predictive assessment of final weight w_k. Thus, the algorithm tends to favor breadth search instead of depth search at the early stages of the sampling process.

3.1 Some Containment of the Curse of Dimension

From now on, it is assumed that:

$$0\leq\omega_{[\boldsymbol{x}]}\leq P(X\in[\boldsymbol{x}])\ , \tag{15}$$

and that:

$$\omega_{[\boldsymbol{x}]}=\begin{cases}0 \quad \text{if} \quad [g]([\boldsymbol{x}])\cap[\boldsymbol{\epsilon}]=\emptyset\ ,\\ P(\boldsymbol{X}\in[\boldsymbol{x}]) \quad \text{if} \quad [g]([\boldsymbol{x}])\subset[\boldsymbol{\epsilon}]\ .\end{cases} \tag{16}$$

Algorithm 3 is an evolution of Algorithm 2. In addition, it builds an history of cuts, stored in map cuts, and computes dynamically from this history an improved weighting function, stored in map omg.

The lines of this algorithm are colored in black, blue or dark blue. Black lines are inherited from Algorithm 2. Blue lines are new additions to the previous algorithm. Dark blue lines (4, 19, 23 and 2 partially) correspond to the **for** loop of Algorithm 2 rewritten as a **while** loop:

$$k\leftarrow 0\ \ \textbf{while}\ \ k<N\ \ \textbf{do}\ \ \cdots\ \ k\leftarrow k+1\ \ \cdots\ \ \textbf{end}$$

1 **Function** *Sampling* $[\boldsymbol{X}\,|g(\boldsymbol{X}) \in [\epsilon]]$

  input : σ, r, g, ω, N **output:** $(\boldsymbol{y}_k, w_k)_{1:N}$

2   $(\mathsf{cuts}, \mathsf{omg}, k) \leftarrow (\emptyset, \emptyset, 0)$

3   $\mathsf{omg}([\boldsymbol{b}]) \leftarrow \omega_{[\boldsymbol{b}]}$

4   **while** $k < N$ **do**

5     $([\boldsymbol{x}_0], \pi_k, j) \leftarrow ([\boldsymbol{b}], 1, 0)$

6   **while** $\rho([\boldsymbol{x}_j]) > r$ **and** $[g]([\boldsymbol{x}_j]) \not\subset [\epsilon]$ **do**

7   **if** $\left|\log_2\left(\frac{\mathsf{omg}([\boldsymbol{b}])}{\mathsf{omg}([\boldsymbol{x}_j])}\pi_k\right)\right| > \sigma$ **goto** 20

8   **ifundef** $\mathsf{cuts}([\boldsymbol{x}_j]) \leftarrow \mathtt{Cut}([\boldsymbol{x}_j])$

9       $([\boldsymbol{l}_{j+1}], [\boldsymbol{r}_{j+1}]) \leftarrow \mathsf{cuts}([\boldsymbol{x}_j])$

10       $j \leftarrow j + 1$

11   **ifundef** $\mathsf{omg}([\boldsymbol{r}_j]) \leftarrow \omega_{[\boldsymbol{r}_j]}$

12   **ifundef** $\mathsf{omg}([\boldsymbol{l}_j]) \leftarrow \omega_{[\boldsymbol{l}_j]}$

13       $(\nu_{[\boldsymbol{l}_j]}, \nu_{[\boldsymbol{r}_j]}) \leftarrow (\mathsf{omg}([\boldsymbol{l}_j]), \mathsf{omg}([\boldsymbol{r}_j]))$

14       $[\boldsymbol{x}_j] \leftarrow \mathtt{Bern}(([\boldsymbol{l}_j], \nu_{[\boldsymbol{l}_j]}), ([\boldsymbol{r}_j], \nu_{[\boldsymbol{r}_j]}))$

15       $\pi_k \leftarrow \frac{\nu_{[\boldsymbol{x}_j]}}{\nu_{[\boldsymbol{l}_j]} + \nu_{[\boldsymbol{r}_j]}}\pi_k$

16   **end**

17     $w_k \leftarrow P(\boldsymbol{X} \in [\boldsymbol{x}_j]) \;/\; \pi_k$

18   Build $\boldsymbol{y}_k$ by sampling $[\boldsymbol{X}\,|\boldsymbol{X} \in [\boldsymbol{x}_j]]$

19     $k \leftarrow k + 1$

20   **for** $i \leftarrow j$ **to** 1 **do**

21       $\mathsf{omg}([\boldsymbol{x}_{i-1}]) \leftarrow \mathsf{omg}([\boldsymbol{l}_i]) + \mathsf{omg}([\boldsymbol{r}_i])$

22   **end**

23   **end**

24 **end**

Algorithm 3: Based on cuts history.

Variable cuts is a dictionary and is used to register the history of computed cuts. At start, cuts is defined empty (line 2). For a given box $[\boldsymbol{x}_j]$, the cut on $[\boldsymbol{x}_j]$ is computed only once, if it is computed, by line 8:

$$\textbf{ifundef } \mathsf{cuts}([\boldsymbol{x}_j]) \leftarrow \mathtt{Cut}([\boldsymbol{x}_j])$$

Keyword **ifundef** tests if $\mathsf{cuts}([\boldsymbol{x}_j])$ is defined; if still undefined, then $\mathsf{cuts}([\boldsymbol{x}_j])$ is set to $\mathtt{Cut}([\boldsymbol{x}_j])$.

Variable omg is a dictionary which records the predictive weighting function and its possible updates, when needed. At start, omg is only defined for $[\boldsymbol{b}]$ and is set to $\omega_{[\boldsymbol{b}]}$ (lines 2 and 3). Variable $\mathsf{omg}([\boldsymbol{r}_j])$ is set to $\omega_{[\boldsymbol{r}_j]}$, if it has not been initialized yet (line 11). The same is done for variable $\mathsf{omg}([\boldsymbol{l}_j])$ at line 12. When the cuts sequence is done (second **while**), then the weighting function is updated by the **for** loop (lines 20, 21, 22). This ensures the computation of $\mathsf{omg}([\boldsymbol{x}])$ as the sum of the weights $\mathsf{omg}([\boldsymbol{z}])$ of the leaves $[\boldsymbol{z}]$ of the cuts tree rooted on $[\boldsymbol{x}]$. Then, property (16) ensures that $\mathsf{omg}([\boldsymbol{x}])$ gets closer to $P(\boldsymbol{X} \in [\boldsymbol{x}] \,\&\, g(\boldsymbol{X}) \in [\epsilon])$ when the cuts tree rooted on $[\boldsymbol{x}]$ gets more refined.

Algorithm 3 is similar to Algorithm 2, except that:

– cut $([\boldsymbol{l}_{j+1}], [\boldsymbol{r}_{j+1}])$ is recovered from the history, when it is possible (line 9),
– box selection is done by means of $(\nu_{[\boldsymbol{l}_j]}, \nu_{[\boldsymbol{r}_j]}) \triangleq (\mathsf{omg}([\boldsymbol{l}_j]), \mathsf{omg}([\boldsymbol{r}_j]))$.

There is an interesting property here. Assume that J is the last value reached by j and that $J' < J$ *is such that* $\mathsf{omg}([\boldsymbol{l}_j])$ *and* $\mathsf{omg}([\boldsymbol{r}_j])$ *are already defined for all* $1 \leq j \leq J'$. Weighting functions are updated in these cases. Then, it comes for all $1 \leq j \leq J'$ that:

$$\mathsf{omg}([\boldsymbol{x}_{j-1}]) = \mathsf{omg}([\boldsymbol{l}_j]) + \mathsf{omg}([\boldsymbol{r}_j]).$$

The computation of π_k is then simplified:

$$\pi_k = \prod_{j=1}^{J'} \frac{\nu_{[\boldsymbol{x}_j]}}{\nu_{[\boldsymbol{l}_j]} + \nu_{[\boldsymbol{r}_j]}} \prod_{j=J'+1}^{J} \frac{\nu_{[\boldsymbol{x}_j]}}{\nu_{[\boldsymbol{l}_j]} + \nu_{[\boldsymbol{r}_j]}} = \prod_{j=1}^{J'} \frac{\nu_{[\boldsymbol{x}_j]}}{\nu_{[\boldsymbol{x}_{j-1}]}} \prod_{j=J'+1}^{J} \frac{\nu_{[\boldsymbol{x}_j]}}{\nu_{[\boldsymbol{l}_j]} + \nu_{[\boldsymbol{r}_j]}}$$

$$= \frac{\nu_{[\boldsymbol{x}_{J'}]}}{\nu_{[\boldsymbol{b}]}} \prod_{j=J'+1}^{J} \frac{\nu_{[\boldsymbol{x}_j]}}{\nu_{[\boldsymbol{l}_j]} + \nu_{[\boldsymbol{r}_j]}} . \tag{17}$$

Thus, the error on π_k grows exponentially only within the newly explored cuts, that is here from $J'+1$ to J. This is a reason for setting a certain restriction on the depth-oriented aspect of this sampling process. Another good reason is to prevent degenerate particle weights, w_k. Algorithm 3 thus implements some code (line 7) for testing the degeneracy of π_k and eventually restarting the sampling loop (second **while**):

$$\textbf{if} \ \left|\log_2\left(\frac{\mathsf{omg}([\boldsymbol{b}])}{\mathsf{omg}([\boldsymbol{x}_j])}\pi_k\right)\right| > \sigma \ \ \textbf{goto} \ \ 20$$

This code tests the logarithmic distance between the weight of $[\boldsymbol{x}_j]$, $\mathsf{omg}([\boldsymbol{x}_j])$, and the weight resulting from the sampling process, $\mathsf{omg}([\boldsymbol{b}])\ \pi_k$. If it is higher than σ, then the loop is stopped by going to line 20. By doing that, the incrementation of k is skipped, so that the sampling loop is restarted for the same indice k. However, the update of variable omg is done, and of course, the history of cuts stays incremented. So, although the sampling loop has been interrupted in this case, the sampling structure has been upgraded. This results in an adaptive process which will balance depth and breadth explorations when running the sampling. Breadth exploration is favored on the first sampling iterations, but the tendency becomes inverted after several samples.

3.2 Incremental Algorithm

In Sect. 4, we present an application of the conditional simulation for Bayesian optimization. In this applicative context, we have to successively simulate a random vector conditionally to incremental constraints.

More precisely, let $U \geq 1$ and let $[\boldsymbol{\epsilon}_u] \in [\mathbb{R}^{m_u}]$ and $g_u : [\boldsymbol{b}] \to \mathbb{R}^{m_u}$ be defined for $1 \leq u \leq U$. For $1 \leq v \leq U$ are defined:

$$[\boldsymbol{\epsilon}_{1:v}] = \prod_{u=1}^{v} [\boldsymbol{\epsilon}_u], \tag{18}$$

```
Function Sampling [X | g_{1:v}(X) ∈ [ε_{1:v}]]
    input : σ, r, g, ω, N    output: (y_k, w_k)_{1:N}
    (cuts, omg, lev) ← (∅, ∅, ∅)
    for v ← 1 to U do
        k ← 0
        ifundef (omg([b]), lev([b])) ← (ω^{1:v}_{[b]}, v)
        else
            (t, lev([b])) ← (lev([b]), v)
            omg([b]) ← omg([b]) ∏_{t<u≤v} ϖ^u_{[b]}
        while k < N do
            ([x_0], π_k, j) ← ([b], 1, 0)
            while ρ([x_j]) > r and [g_{1:v}]([x_j]) ⊄ [ε_{1:v}] do
                if |log_2( omg([b]) / omg([x_j]) π_k )| > σ goto 31
                ifundef cuts([x_j]) ← Cut([x_j], v)
                ([l_{j+1}], [r_{j+1}]) ← cuts([x_j])
                j ← j + 1
                ifundef (omg([r_j]), lev([r_j])) ← (ω^{1:v}_{[r_j]}, v)
                else
                    (t, lev([r_j])) ← (lev([r_j]), v)
                    omg([r_j) ← omg([r_j]) ∏_{t<u≤v} ϖ^u_{[r_j]}
                ifundef (omg([l_j]), lev([r_j])) ← (ω^{1:v}_{[l_j]}, v)
                else
                    (t, lev([l_j])) ← (lev([l_j]), v)
                    omg([l_j) ← omg([l_j]) ∏_{t<u≤v} ϖ^u_{[l_j]}
                (ν_{[l_j]}, ν_{[r_j]}) ← (omg([l_j]), omg([r_j]))
                [x_j] ← Bern(([l_j], ν_{[l_j]}), ([r_j], ν_{[r_j]}))
                π_k ← ν_{[x_j]} / (ν_{[l_j]} + ν_{[r_j]}) π_k
            end
            w_k ← P(X ∈ [x_j]) / π_k
            Build y_k by sampling [X | X ∈ [x_j]]
            k ← k + 1
            for i ← j to 1 do
                [x_{i-1}] ⇐ Shrink([l_i], [r_i])
                omg([x_{i-1}]) ← omg([l_i]) + omg([r_i])
            end
        end
    end
end
```

Algorithm 4: Incremental sampling.

and $g_{1:v} : [\boldsymbol{b}] \to \prod_{u=1}^{v} \mathbb{R}^{m_u}$ by:

$$g_{1:v}(\boldsymbol{x}) = (g_1(\boldsymbol{x}), \cdots, g_v(\boldsymbol{x})) \quad \text{for all} \quad \boldsymbol{x} \in [\boldsymbol{b}]. \tag{19}$$

Our purpose is to sample successively the sequence of conditional vectors:

$$[\boldsymbol{X}\,|g_{1:1}(\boldsymbol{X})\in[\boldsymbol{\epsilon}_{1:1}]]\,,\cdots,[\boldsymbol{X}\,|g_{1:v}(\boldsymbol{X})\in[\boldsymbol{\epsilon}_{1:v}]]\,,\cdots,[\boldsymbol{X}\,|g_{1:U}(\boldsymbol{X})\in[\boldsymbol{\epsilon}_{1:U}]]\,, \tag{20}$$

in an efficient way. When the sampling is done by means of Algorithm 3, the sampling structure produced for each conditional vector, $[\boldsymbol{X}\,|g_{1:v}(\boldsymbol{X})\in[\boldsymbol{\epsilon}_{1:v}]]$ where $v\in[\![1,U]\!]$, is lost from a simulation process to another. This is inefficient, since it is foreseeable that the sampling structure for vector $[\boldsymbol{X}\,|g_{1:v}(\boldsymbol{X})\in[\boldsymbol{\epsilon}_{1:v}]]$ is certainly informative for sampling vector $[\boldsymbol{X}\,|g_{1:v+1}(\boldsymbol{X})\in[\boldsymbol{\epsilon}_{1:v+1}]]$.

Algorithm 4 is an update of Algorithm 3, with the purpose of incrementally building the sampling structures for the sequence $([\boldsymbol{X}\,|g_{1:v}(\boldsymbol{X})\in[\boldsymbol{\epsilon}_{1:v}]])_{1\leq v\leq U}$. Predictive weight is now incremental and takes multiplicative form:

$$\omega^{1:v}_{[\boldsymbol{x}]}=P\left(\boldsymbol{X}\in[\boldsymbol{x}]\right)\prod_{u=1}^{v}\varpi^{u}_{[\boldsymbol{x}]}\,. \tag{21}$$

Prefix $P\left(\boldsymbol{X}\in[\boldsymbol{x}]\right)$ in (21) is not out of place: recall that $\omega^{1:v}_{[\boldsymbol{x}]}$ should approximate $P\left(\boldsymbol{X}\in[\boldsymbol{x}]\,\&\,g_{1:v}(\boldsymbol{X})\in[\boldsymbol{\epsilon}_{1:v}]\right)=P\left(\boldsymbol{X}\in[\boldsymbol{x}]\right)P\left(g_{1:v}(\boldsymbol{X})\in[\boldsymbol{\epsilon}_{1:v}]\,|\boldsymbol{X}\in[\boldsymbol{x}]\right)$. The multiplicative form is then a consequence of the incremental constraint.

In order to deal with the incremental construction of the samplers, Algorithm 4 introduces new dictionary, lev, complementary to cuts and omg. Constraint level $t=\mathsf{lev}([\boldsymbol{x}])$ indicates that weight $\mathsf{omg}([\boldsymbol{x}])$ has been actually computed with constraints sequence $g_{1:t}(\boldsymbol{X})\in[\boldsymbol{\epsilon}_{1:t}]$. Indeed, the sampling structure is not garanteed to be up-to-date for all subtrees of the sampling structure, and it is thus necessary to trace this upgrade level. On the other hand, function call `Cut`$(v,[\boldsymbol{x}])$, which creates a cut of box $[\boldsymbol{x}]$, now depends on constraint level v. So, news cuts are created by considering all constraints information, although this will of course not be the case for cuts related to past levels $u<v$. At last, new function `Shrink` is introduced. Code $[\boldsymbol{x}]\Leftarrow$ `Shrink`$([\boldsymbol{l}],[\boldsymbol{r}])$ replaces $[\boldsymbol{x}]$ by the smallest box containing both $[\boldsymbol{l}]$ and $[\boldsymbol{r}]$; *this replacement is deep* and thus concerns dictionaries cuts, omg and lev as well. Even root variable $[\boldsymbol{b}]$ in Algorithm 4 can have its value replaced. Nevertheless, deep replacements are cost-free here thanks to the way the structures are implemented.

In Algorithm 4, parts taken unchanged from Algorithm 3 are in black while evolutions are in blue. A **for** loop over the number of constraints, v, is added at lines 3 and 36 and repeats the entire sampling process for a given set of constraints. The sampling process itself does not change much, but incremental computations are now done on omg and lev is updated in consequence: lines 5 to 8, lines 16 to 19 and lines 20 to 23. Incremental computation means that undefined value $\mathsf{omg}([\boldsymbol{x}])$ is set equal to $\omega^{1:v}_{[\boldsymbol{x}]}$, while already defined value $\mathsf{omg}([\boldsymbol{x}])$ is multiplicatively completed by $\prod_{u=t+1}^{v}\varpi^{u}_{[\boldsymbol{x}]}$. At line 32, the shape of box $[\boldsymbol{x}_{i-1}]$ is recomputed from its cut. This is better for improving future incremental computation of omg during the sampling process.

3.3 Practical Implementation and Parallelism

Practical Implementation. Algorithm 3 and 4 draw the main principles of our sampling methods. Some implementation details are described now.

Initial Predictive Weight. We implement the following definition of ω:

$$\omega_{[x]} = \frac{\mu\big([g]([\boldsymbol{x}]) \cap [\epsilon]\big)}{\mu\big([g]([\boldsymbol{x}])\big)} P\left(\boldsymbol{X} \in [\boldsymbol{x}]\right), \tag{22}$$

where μ is Borel measure on $\mathbb{R}^m$. This definition checks properties (15) and (16). It also tries a rough approximation for $P\left(\boldsymbol{X} \in [\boldsymbol{x}] \,\&\, g(\boldsymbol{X}) \in [\epsilon]\right)$. For incremental implementation, definition extends naturally as a product:

$$\omega_{[x]}^{1:v} = P\left(\boldsymbol{X} \in [\boldsymbol{x}]\right) \prod_{u=1}^{v} \varpi_{[x]}^{u} \text{ with } \varpi_{[x]}^{u} = \frac{\mu\big([g_u]([\boldsymbol{x}]) \cap [\epsilon_u]\big)}{\mu\big([g_u]([\boldsymbol{x}])\big)}. \tag{23}$$

Cutting Strategy. The definition of `Cut` is an important choice. Our algorithm selects a cut $([\boldsymbol{y}], [\boldsymbol{z}])$ randomly in regards to the following criteria:

- Favor cuts such that $\omega_{[y]} \ll \omega_{[z]}$ or $\omega_{[y]} \gg \omega_{[z]}$,
- Avoid overly elongated $[\boldsymbol{x}]$, *i.e.* such that $\frac{\max_i (x_i^+ - x_i^-)}{\min_i (x_i^+ - x_i^-)} \gg 1$,

In addition, some cuts history simplifications are implemented (next point).

Reducing Sampling Structure. Our implementation tries to optimize the structure of cut, weight, and level history, when it is possible. Different cases are implying such reductions, which are not mentioned here. A typical example is shown in Fig. 3. Assume that $([\boldsymbol{y}], [\boldsymbol{z}])$ is a cut of $[\boldsymbol{x}]$, $([\boldsymbol{t}], [\boldsymbol{u}])$ is a cut of $[\boldsymbol{z}]$ and $[g]([\boldsymbol{u}]) \cap [\epsilon] = \emptyset$, *i.e.* $[\boldsymbol{u}]$ is outside the subpaving and its border. Then, boxes $[\boldsymbol{z}]$ and $[\boldsymbol{u}]$ are useless and should be removed from the structure.

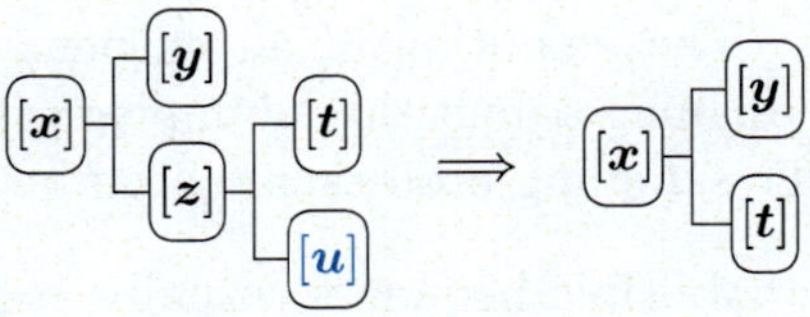

Fig. 3. Structure reduction

Discarding First Samples. M first samples are discarded, so as to initialize the structure of the sampler. After that, N samples are sampled and returned.

Setup. The algorithms are rather simple to set up. Except for the choice of ω, which is structural, r, M and σ are the only parameters to be defined.

Parallelization. This paragraph gives additional information on how incremental Algorithm 4 is implemented with parallelization. Data structures cuts, omg and lev deal rather well with parallel processing, so that our implementation is multithread. Nevertheless, the practical implementation needs some work.

In our implementation, the dictionaries cuts, omg and lev are stored in a same structure. This structure is composed of shared references to vertices which are stored within a Slab allocation [2]. Slab allocation is a structure, which allows easy and efficient deallocation and reallocation, while reducing the risk of framentation of the memory. A vertex is an element of the history of the sampling process, which contains the following fields:

- A box $[\boldsymbol{x}]$ with its weight $\mathsf{omg}([\boldsymbol{x}])$ and its level $v = \mathsf{lev}([\boldsymbol{x}])$, which is an indicator that $\mathsf{omg}([\boldsymbol{x}])$ has been computed from $g_{1:v}$ and $[\boldsymbol{\epsilon}_{1:v}]$,
- The type of vertex: a node or a leaf? Is property $[g_{1:v}]([\boldsymbol{x}]) \subset [\boldsymbol{\epsilon}_{1:v}]$ true?
- When vertex is a node and box $[\boldsymbol{x}]$ has a cut $([\boldsymbol{l}], [\boldsymbol{r}]) = \mathsf{cuts}([\boldsymbol{x}])$, then the vertex is completed with weak references to the vertices related to boxes $[\boldsymbol{l}]$ and $[\boldsymbol{r}]$ respectively.

Shared references allow multiple threads to access a same data; moreover, the existance of a shared reference will prevent the erasure of referenced data. Weak references are derived from shared references and also allow threads to access the data, but they do not prevent the erasure of referenced data. By restricting the shared references to the Slab allocation, we prevent the possibility of memory leak, especially when the sampling structure is reduced by a thread. Since the structure is shared among threads, the access to the vertices are protected by memory locks: the locks allow either many read-only accesses or only one read-write access to the data.

This data structure allowed some fair performance increase in comparison to conference paper [4]. The memory management is now smoother and more efficient. In particular, we no longer have process freezing phenomena following heavy memory management.

The definition of an efficient data structure is a main achievement when designing a multithread implementation. Now, we give some clarification on how the sampling process is shared between the threads:

- Multiple instances (usually as many as the number of processor threads) of the sampling loop are run simultaneously. This includes descending through the sampling structure, possible extension of this structure, possible sampling of box and point if conditions are met, and backward reconstruction of weights and restructuring of the sampling structure,
- If a vertex currently processed by an instanced loop has been deleted by another instance, then the vertex subtree is first deleted by the process, and afterward, the loop is restarted.
- Processes are stop when the desired number of samples is obtained.

4 Application to Bayesian Optimization

Bayesian optimization is a key technique for black-box optimization, and it is actually a great motivation for this work.

Assume that one needs to optimize a function which is not well known, and which may be computed by a highly costly process (a heavy simulation, tests made by human teams on the grounds, etc.). Of course the optimization should be made by sparing at best the number of calls to the costly evaluation.

In [8], Jones, Schonlau and Welch proposed the efficient global optimization method (EGO) for addressing such kind of problem. The idea is to use a surrogate model under the form of a functional random variable. This functional random variable is described by means of a Gaussian process with correlation depending on spatial distance (kriging). Based on such modelling, the construction of an *optimal* parameter sequence to be evaluated is obtained by iterating:

- Compute the posterior law of the functional random variable, according to the past evaluations,
- Compute the expected improvement function, in regards to the posterior law and the already best computed value. This expected improvement is an indicator of the next parameter to be evaluated,
- Find parameter optimizing the expected improvement and evaluate it.

The approach relies on the construction of the Gaussian modelling, and it tends to be less efficient when the dimension of the optimization space increases. Works have been made in order to deal with this dimensional issue and improve the modelling; *e.g.* [6]. But this issue is still a true challenge.

Beside, EGO itself is a form of Bayesian optimization introduced by Mockus [9,10,13]. We proposed in [3] a Bayesian optimization approach in the context of a nonlinear function depending on a model noise; the realization of the noise was the unknown information of the problem. We hoped by a nonlinear approach to better handle some of the difficulty induced by the dimension. The work was not complete, since we were not able to build a good conditional sampling at that time. The purpose of this paper and of [4] was to build such sampler.

It is beyond the scope of this chapter to detail the seminal works of Mockus, or those of Jones, Schonlau and Welch. We will introduce the subject by a short description of EGO algorithm. Then, we present our nonlinear approach.

Formalization: Function $\gamma \mapsto g(\gamma, \boldsymbol{x}^o)$ is to be minimized. Parameter $\boldsymbol{x}^o$ is unknown, but $\boldsymbol{x}^o$ is a realization of random vector $\boldsymbol{X}$, whose law $F_{\boldsymbol{X}}$ is known. In order to optimize γ, we are allowed to request an evaluator for computing $g(\gamma, \boldsymbol{x}^o)$, but each call to this evaluator is costly. The objective is then to solve $\gamma^o \in \arg\min_\gamma g(\gamma, \boldsymbol{x}^o)$ by optimizing at the same time the sequence of evaluated cases γ_u and of their evaluations $e_u = g(\gamma_u, \boldsymbol{x}^o)$.

EGO Method: Main idea of EGO consists in approximating $g(\gamma, \boldsymbol{X})$ by a Gaussian process (GP) $\gamma \mapsto G(\gamma)$. Given past evaluated cases γ_u and evaluations $e_u = g(\gamma_u, \boldsymbol{x}^o)$ for $u \leq v$, posterior variable $[G(\gamma)\,|\forall u \leq v, G(\gamma_u) = e_u\,]$ is Gaussian and mathematically computed. This conditional variable is a main ingredient for computing the expected improvement, $EI(\gamma)$, which is an indicator of value γ for which next evaluation of $g(\gamma, \boldsymbol{x}^o)$ is promizing:

$$EI(\gamma) = E_{G(\gamma)|\forall u \leq v, G(\gamma_u)=e_u} \max\{m_\gamma - G(\gamma), 0\} \text{ where } m_\gamma = \min_{1 \leq u \leq v} e_u. \quad (24)$$

It is computed mathematically. Since $EI(\gamma)$ indicates where evaluations are promising, it is maximized in order to chose next case $\gamma_{v+1} \in \arg\max_\gamma EI(\gamma)$ and request its evaluation $e_{v+1} = g(\gamma_{v+1}, \boldsymbol{x}^o)$.

Nonlinear Algorithm: We took inspiration of EGO, but now, random function $g(\cdot, \boldsymbol{X})$ is used directly instead of GP approximation. Parameters γ_v and evaluations $e_v = g(\gamma_v, \boldsymbol{x}^o)$ are optimized by iterating Algorithm 5. While $EI(\gamma)$ is formed differently in this algorithm, it is essentially the same as in definition (24) modulo an affine transform. It is theoretically defined by:

$$EI(\gamma) = E_{\boldsymbol{X}\,|\forall u \leq v, g(\gamma_u, \boldsymbol{X}) = e_u} \min\{g(\gamma, \boldsymbol{X}), m_\gamma\} \text{ with } m_\gamma = \min_{1 \leq u \leq v} e_u. \tag{25}$$

In practice, $EI(\gamma)$ is obtained by a Monte Carlo derived from an approximated sampling of $[\boldsymbol{X}\,|\forall u \leq v, g(\gamma_u, \boldsymbol{X}) = e_u]$. *As a crucial ingredient:*

- Vector $[\boldsymbol{X}\,|\forall u \leq v, g(\gamma_u, \boldsymbol{X}) = e_u]$ is approximated by $[\boldsymbol{X}\,|g_{1:v}(\boldsymbol{X}) \in [\boldsymbol{\epsilon}_{1:v}]]$, where $g_u = g$ and $\boldsymbol{\epsilon}_u$ is a small box around e_u.
- The sampling is done by the way of this approximation.

This approximated random vector, conditional to a subvariety, incorporates incrementally all constraints related to the past evaluations. The sampling is done by Algorithm 3 or preferably by incremental Algorithm 4.

Algorithm thus consists of:

1. Generating samples of posterior random vector $[\boldsymbol{X}\,|g_{1:v}(\boldsymbol{X}) \in [\boldsymbol{\epsilon}_{1:v}]]$,
2. Building a Monte Carlo approximation of $EI(\gamma)$,
3. Choosing parameter γ_{v+1} by minimizing $EI(\gamma)$ and evaluating it.

1 **Function** *Process next measure*
 input : γ_u and $e_u \triangleq g(\gamma_u, \boldsymbol{x}^o)$ for $1 \leq u \leq v$
 output: γ_{v+1} and $e_{v+1} \triangleq g(\gamma_{v+1}, \boldsymbol{x}^o)$
2 Make samples of $[\boldsymbol{X}\,|g_{1:v}(\boldsymbol{X}) \in [\boldsymbol{\epsilon}_{1:v}]]$
3 Compute $m_\gamma = \min_{1 \leq u \leq v} e_u$ and:
 $EI(\gamma) \simeq E_{\boldsymbol{X}\,|g_{1:v}(\boldsymbol{X}) \in [\boldsymbol{\epsilon}_{1:v}]} \min\{g(\gamma, \boldsymbol{X}), m_\gamma\}$
4 Compute $\gamma_{v+1} \in \arg\min_\gamma EI(\gamma)$
5 Compute $e_{v+1} \triangleq g(\gamma_{v+1}, \boldsymbol{x}^o)$
6 **end**

Algorithm 5: Sampler-based Bayesian optimizer

Implementations of step 4 are not detailed here. Essentially, we used a metaheuristic method, the cross-entropy algorithm for optimization [12], for this task; this method is related to parameterized rare-event simulation.

5 Examples and Tests

Regarding the conference publication [4], all the results presented here have been recomputed or supplemented by recent implementations of the algorithms.

5.1 Simulation: Test Cases

The tests presented here are performed for sampling Algorithm 3 or its incremental version 4. The algorithms have been implemented in `Rust` language (www.rust-lang.org) and were processed on 7 threads. The algorithms are tested on mathematically simple simulation problems, in order to make the statistics of the results clear enough to analyze.

Parameters. All simulations have been achieved with the following parameters:

- $r = 0.001$ is the radius bound for second **while** stop condition,
- $M = 5000$ is the number of samples discarded during sampler initialization,
- $N = 50000$ is the number of sampled particles,
- $\sigma = 10$ on all tests.

Test Cases. Thorough the section, it is assumed that $\boldsymbol{X}$ follows the uniform law on $\boldsymbol{b} = [-2, 2]^n$ with $n \in \{2, 3, \cdots, 11\}$. Three cases are investigated:

Case (a): Are defined $g_a(\boldsymbol{x}) = ||\boldsymbol{x}||_2 = \sqrt{\sum_{j=1}^n x_j^2}$ and $[\boldsymbol{\epsilon}_a] = [0.95, 1.05]$. Then $g_a^{-1}([\boldsymbol{\epsilon}_a])$ is a hyper-spherical shell, which approximates the unit hypersphere of dimension $n - 1$.

Case (b): For $n = 11$ and $0 \leq v \leq 9$, it is considered:

$$g_b(\boldsymbol{x}) = \begin{bmatrix} ||\boldsymbol{x}||_2 & x_3 & \dots & x_{2+v} \end{bmatrix} \text{ and } [\boldsymbol{\epsilon}_b] = [\boldsymbol{\epsilon}_a] \times [\boldsymbol{z}]^v, \tag{26}$$

with $[\boldsymbol{z}] = [-0.05, 0.05]$. Similarly to **(a)**, but with additional constraints, $g_b^{-1}(\boldsymbol{\epsilon}_b)$ approximates a hypersphere of dimension $n - 1 - v$. When $v = 0$, we are back to case **(a)** with $n = 11$. When $v = 9$, then $g_b^{-1}(\boldsymbol{\epsilon}_b)$ approximates the unit circle $\mathcal{C}$ within the first two dimensions, related to coordinates x_1, x_2.

Case (c): For $n = 11$ and $0 \leq v \leq 9$, it is considered:

$$g_c(\boldsymbol{x}) = \begin{bmatrix} ||\boldsymbol{x}||_2 & \min(|x_1|, |x_2|) & x_3 & \dots & x_{2+v} \end{bmatrix}, \tag{27}$$

and $[\boldsymbol{\epsilon}_c] = [\boldsymbol{\epsilon}_a] \times [\boldsymbol{\alpha}] \times [\boldsymbol{z}]^v$ with $[\boldsymbol{\alpha}] = [0, 0.5]$. Thus, this case is obtained by adding constraint $0 \leq \min(|x_1|, |x_2|) \leq 0.5$ to subcases of **(b)**. For subcase $v = 9$ especially, we are approximately sampling on the (disjoint) union of the 4 subsegments $\mathcal{A}_1, \dots, \mathcal{A}_4$ of the unit circle $\mathcal{C}$, defined by:

$$\mathcal{A}_j = \left\{ \begin{bmatrix} x_1, x_2 \end{bmatrix} \in \mathcal{C} \middle/ \arg\left(\begin{bmatrix} x_1, x_2 \end{bmatrix} \right) \in j\frac{\pi}{2} + \left[-\frac{\pi}{6}, \frac{\pi}{6} \right] \mod 2\pi \right\}. \tag{28}$$

These 4 subsegments have the same size so that their probabilities are the same in regard to X.

Purpose of the Test Cases. Subsequently, case **(a)** is used in order to evaluate the performance of the sampling process both in accuracy and in efficiency for different dimensions. Case **(b)** is used in order to evaluate the efficiency of the sampling when the number of constaints increases; Algorithm 3 and incremental Algorithm 4 are also compared. Case **(c)** is used in order to evaluate the accuracy of the sampling in case of complex constaints which introduce disjoint modes.

5.2 Simulation: Case (a)

This case is mathematically easy to predict. In [5], Dezert and Musso proposed a method, which may be used for uniformly sampling on an ellipsoid shell. Whatever, one must keep in mind that our approach is generic and can be applied to an infinite number of configurations.

Histograms. Figure 4 presents some histograms for case (a). For each subcase $n \in \{3, 7, 11\}$, we have computed the radius of all samples $\boldsymbol{x}$ and built the associated histograms – subfigures (1), (3) and (5). For each $n \in \{3, 7, 11\}$ and for all $1 \leq i < j \leq n$, we have computed the angle of all samples (x_i, x_j) and built the associated histograms. From these $\frac{n(n-1)}{2}$ histograms of each subcase, we have computed the minimal, mean and maximal histograms. The results are shown in blue, green and red, respectively, and provide an hint on the error of the estimation – subfigures (2), (4) and (6).

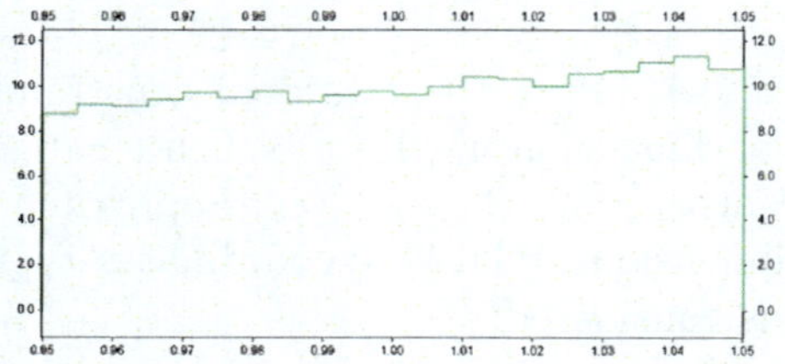

(1) $n = 3$ – radius

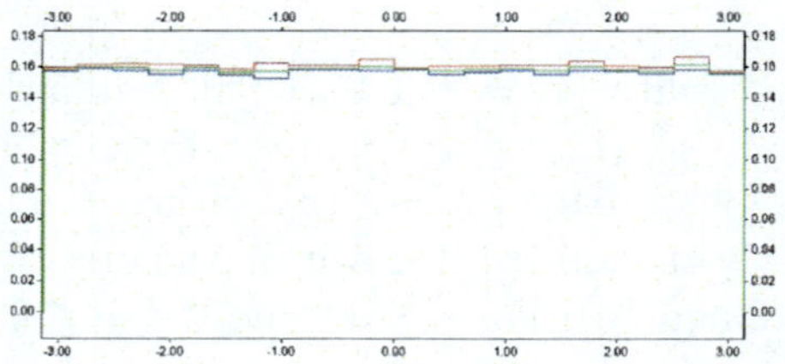

(2) $n = 3$ – angle

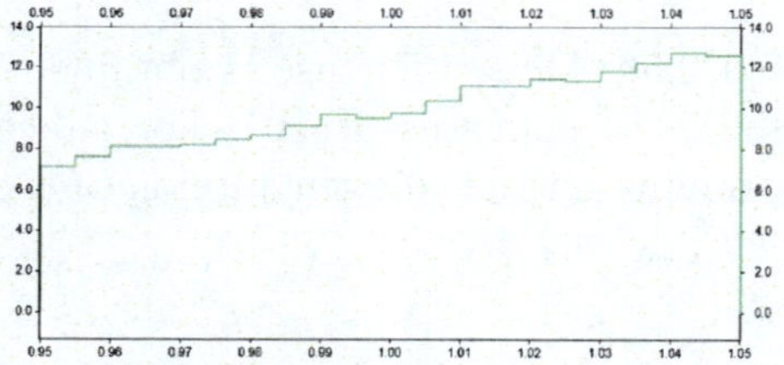

(3) $n = 7$ – radius

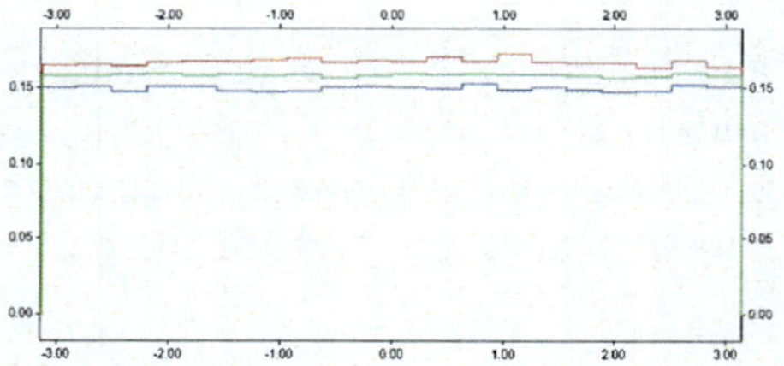

(4) $n = 7$ – angle

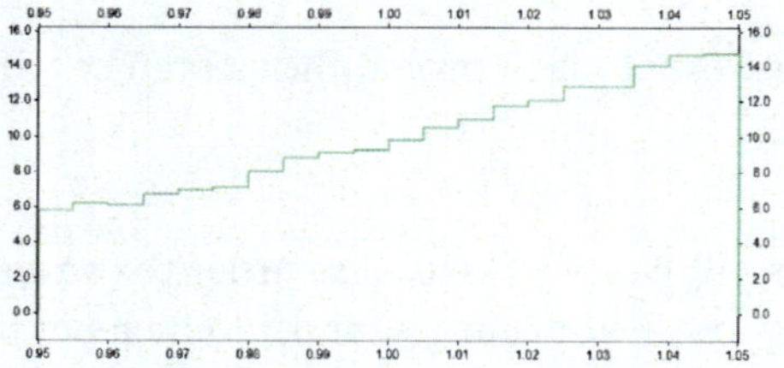

(5) $n = 11$ – radius

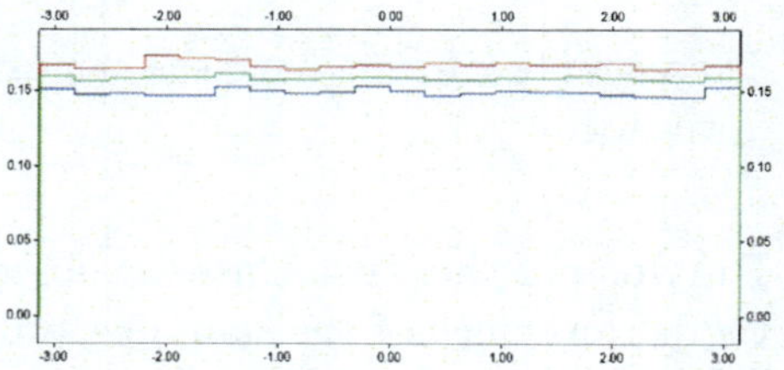

(6) $n = 11$ – angle

Fig. 4. Case **(a)** – histograms – 20 divisions

By symmetry, the *theoretical angle histograms* are uniform. The errors should be of the order of $\sqrt{\frac{20}{50000}} = 0,02$. In comparison, the errors figured in the angular histograms are quite acceptable, even for the highest dimension. The most interesting point is that there is no rupture in the histogram, which shows that the sampler does manage the subvariety structure. We do not have an error estimation for the radius histograms. Actually, the local probability should theoretically increase with the radius, this property being accentuated with the dimension. This is what is obtained on the histograms. It is noteworthy however that the sides of these histograms are subject to additional errors implied by the border of the subvariety.

Process Statistics: Figure 5 presents some statistics graphs for case (a).

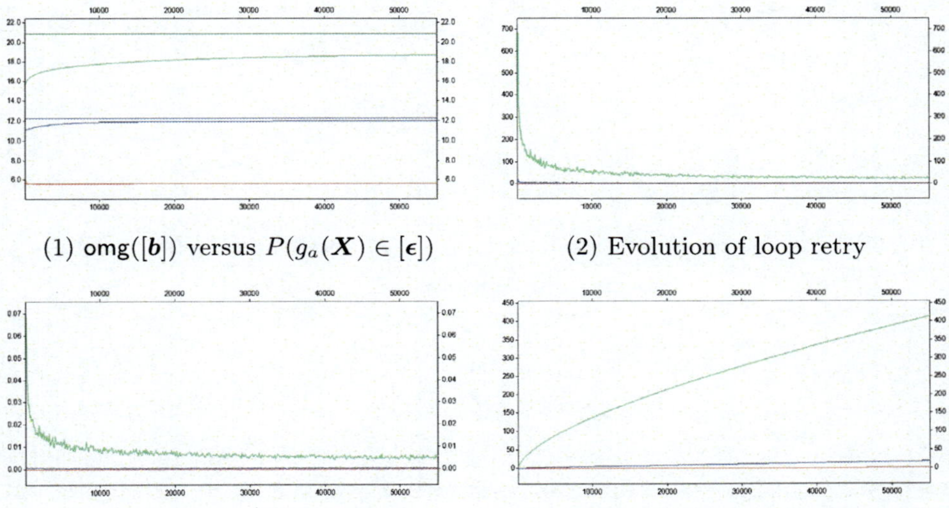

(1) $\mathsf{omg}([\boldsymbol{b}])$ versus $P(g_a(\boldsymbol{X}) \in [\boldsymbol{\epsilon}])$ (2) Evolution of loop retry

(3) Evolution of cpu time (s) (4) Cumulative cpu time (s)

Fig. 5. Case **(a)** – statistics – $n = 3, 7, 11$

Results are plotted against the number of generated samples $k = 1 : 55000$. Cases $n = 3, 7, 11$ are drawn with respective colors, red, blue and green.

Subfigure (1) shows how value $-\log_2(\mathsf{omg}([\boldsymbol{b}]))$ evolves and approximates $-\log_2(P(g_a(\boldsymbol{X}) \in [\boldsymbol{\epsilon}]))$. Each curve increases to theoretical value (same color line), but performance decreases with dimension.

Subfigures (3) and (4) present the cumulative cpu-time per thread and the evolution of the cpu-time per sample and per thread consumed by the process (expressed in second). In comparison to [4], there is no more time discontinuity caused by intermittent memory allocations. Our memory management has been improved as shown in Sect. 3.3. We notice clearly that the sampling efficiency increases with the number of generated samples. However, the cumulative cpu time still increases dramatically with the dimension (the memory use evolves similarly). Although the curse of dimension has been delayed by our approach, it is still there.

The number of loop retries during the sampling is plotted in subfigure (2). It is an interesting indication of the achievement of the sampling structure. It decreases with the number of samples and becomes small, even for the highest dimension (around 28 for $n = 11$). This result should be compared to the probability of the subvariety (around 10^{-6} for $n = 11$).

5.3 Simulation: Case (b)

Figure 6 presents synthetic performance curves, as well as comparative histogram examples for Algorithm 3 and Algorithm 4 on case **(b)**.

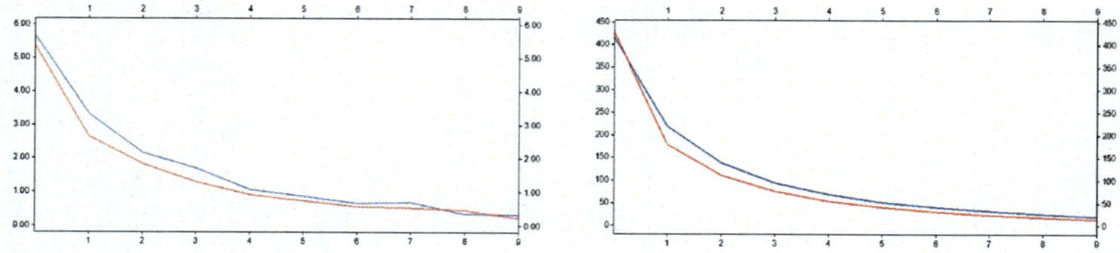

(1) Final cpu time (ms); $v = 0 : 9$ (2) Total cpu time (s); $v = 0 : 9$

algorithm 3 in blue – algorithm 4 in red

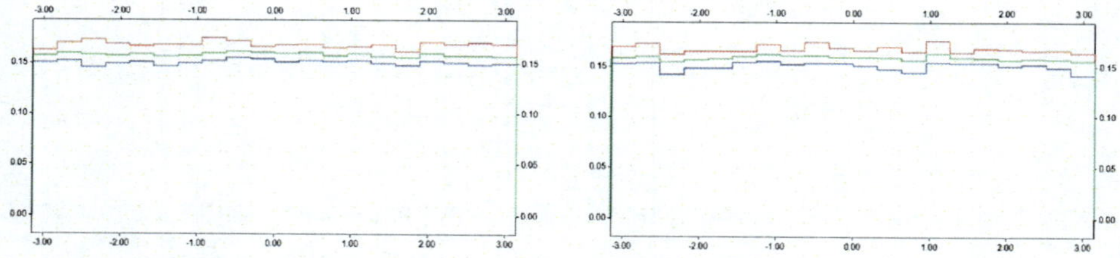

(3) Angle histogram – $v = 5$ (4) Angle histogram – $v = 5$

Fig. 6. Case **(b)** – speed and histograms

Subfigures (1) and (2) are plotted against the number of additional constraints $v = 0 : 9$. Subfigure (1) presents final cpu-time (in ms) per sample and per thread consumed by Algorithm 3 (in blue) and Algorithm 4 (in red). Both curves decrease and approach to zero. Subfigure (2) presents total cpu-time (in s) per thread consumed by these algorithms. Both curves decrease. Red curves are dominated by blue curves, which means better efficiency of Algorithm 4 against Algorithm 3.

Subfigure (3) presents angle histogram for subcase $v = 5$ from computations by Algorithm 3. Subfigure (4) presents angle histogram for same subcase from computations by Algorithm 4. Both results are comparable in terms of accuracy. The theoretical histogram is uniform, which conforms to these results.

As a conclusion here, the performance of the sampler is likely to increase with the number of constraints, and this is a useful quality. Incremental Algorithm 4 is more efficient than Algorithm 3, while ensuring equivalent accuracy. Algorithm 4 must therefore be favored for incremental constraints. In last example of Sect. 5.5 on Bayesian optimisation, implementation of Algorithm 4 was even necessary for an acceptable computation time.

5.4 Simulation: Case (c)

Figure 7 presents computed radial histograms and angular histograms (for coordinates x_1 and x_2 – there is only one) for subcases $v = 6$ and $v = 9$.

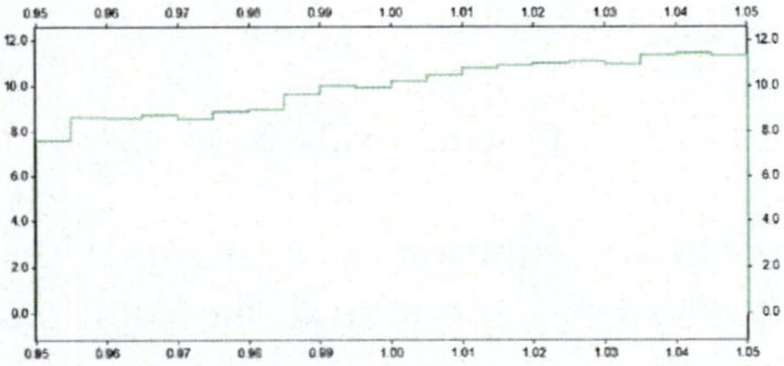

(1) Radius histogram – $v = 6$

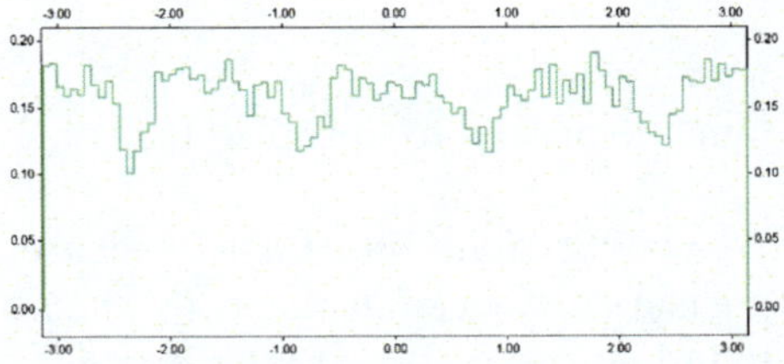

(2) Angle histogram – $v = 6$

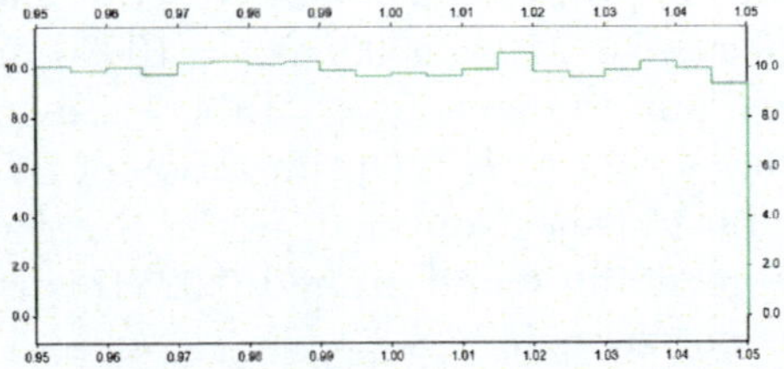

(3) Radius histogram – $v = 9$

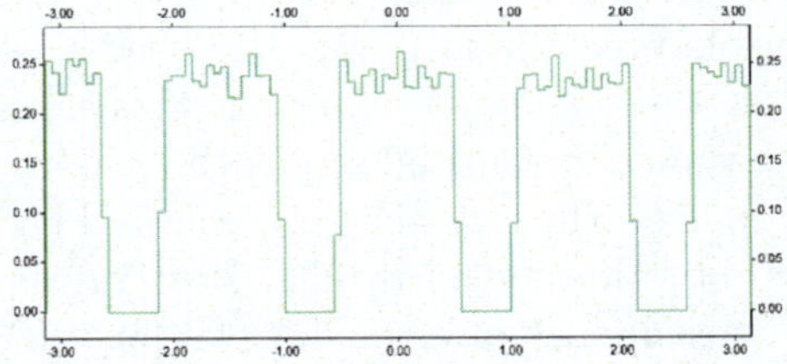

(4) Angle histogram – $v = 9$

Fig. 7. Case **(c)** – histograms – radius: 20 divisions; angle: 100 divisions

Subfigure (3) presents radial histogram for final case $v = 9$. Subfigure (4) presents angle histogram for final case $v = 9$. The quality of the histograms is comparable to what was observed previously. Due to the constraint configuration of subcase $v = 9$, the *theoretical radius histogram* is uniform, and the *theoretical angular histogram* is uniform around each subsegment, $\mathcal{A}_1$ to $\mathcal{A}_4$, with a gradual decrease on the borders. The generated histograms actually comply with these properties.

Subfigures (1) and (2) present intermediate subcase $v = 6$. Since some other coordinates than x_1 and x_2 are relaxed, contraint $x_1^2 + x_2^2 \simeq 1$ is also relaxed. Actually, theoretical angular histogram is even no longer discontinuous around subsegments, $\mathcal{A}_i$, for this reason.

As a preliminary conclusion, we consider that our sampling method is globally performant in sampling conditionally to subvarieties. A future issue will be to increase the dimension of the sampling space.

5.5 Bayesian Optimization

A Simple Geometric Problem. We intend to find the isobarycenter $\gamma = (a, b)$ of 4 *unknown points* $M_i = (x^o_{2i-1}, x^o_{2i}) \in [-5, 5]^2$ with $i \in \{1, 4\}$. The only approach that is possible for us is to test some solutions by requesting for a costly measurement; for requirement $\gamma_u = (a_u, b_u)$, this measurement evaluates:

$$e_u = g(\gamma_u, \boldsymbol{x}^o) = ||\gamma_u - h(\boldsymbol{x}^o)||_2 \text{ with } h(\boldsymbol{x}^o) = \frac{1}{4}\sum_{i=1}^{4} M_i = \frac{1}{4}\sum_{i=1}^{4}(x^o_{2i-1}, x^o_{2i}). \tag{29}$$

Our purpose is to optimize (a, b) by minimally requesting evaluation $g(\gamma, \boldsymbol{x}^o)$.

Geometric Solution. Each measure restricts the solution to a circle. After 2 measures, we usualy have to choose between two points, and the solution is found equiprobably at step 3 or 4.

Tests and Results. Points $M_1, \dots, M_4$ are $(2, -1)$, $(3, 2)$, $(-\frac{3}{2}, 4)$, $(\frac{1}{2}, 3)$. Their isobarycenter is $(1, 2)$. We used a sampler with $M = 5000$, $N = 10000$ and $[\epsilon] = [-\frac{1}{100}, \frac{1}{100}]^u$. Variable $\boldsymbol{X}$ is considered uniform on $[-5, 5]^8$. Table 1 presents a typical optimized sequence for parameters $\gamma_u = (a_u, b_u)$. Optimization of EI is done by the cross-entropy method [12] and is thus near-optimal. In this example, best value for e_u is 0.02. Best values are generally found around 0.02: this is a consequence of error interval $[\epsilon]$, which is not zero size.

Table 1. Exemple of optimization sequence

u	1	2	3	4	5	6	7	8	9	10	11	12	13
a_u	0.01	−0.41	0.84	0.97	0.99	1.00	0.98	1.02	−3.79	0.98	0.98	−0.39	1.03
b_u	0.02	−1.94	2.05	1.99	2.02	1.98	2.02	1.99	6.49	2.00	1.99	5.40	2.00
e_u	2.21	4.19	0.16	0.03	0.02	0.02	0.03	0.02	6.57	0.02	0.03	3.67	0.03

Table 2 presents the results for 100 runs. For these runs, the mean cpu time is 2627s, where *non-incremental* Algorithm 3 has been used for simulation. Three convergence conditions, $e_u \leq e_{cond}$ with $e_{cond} \in \{0.02, 0.04, 0.09\}$, are considered for these runs. Value $u_o = \min\{u/e_u \leq e_{cond}\}$ is the final step for which convergence condition is met. Percentage is given for each final step value u_o. In all cases, we notice that most optimal values are found at steps 3 or 4. The result tends to be equilibrated on 3 and 4 when stopping criterion is relaxed.

Table 2. Results for 100 runs

u_o	1	2	3	4	5	6	7	8	9	10	11	12	13
$e_{u_o} \leq 0.02$	0%	0%	26%	40%	10%	9%	2%	3%	3%	3%	1%	1%	1%
$e_{u_o} \leq 0.04$	0%	0%	36%	57%	6%	1%	0%	0%	0%	0%	0%	0%	0%
$e_{u_o} \leq 0.09$	0%	0%	40%	57%	3%	0%	0%	0%	0%	0%	0%	0%	0%

EGO Method. For the sake of comparison, we apply EGO to our problem. Gaussian process G with covariance $\text{cov}(G(\gamma), G(\gamma')) = \frac{1}{4}\exp\left(-\frac{1}{2}||\gamma - \gamma'||_2\right)$ and mean 0 is used. Table 3 summarizes the results of 100 tests. Each test implements 100 successive evaluations. Convergence conditions are $e_u \leq e_{cond}$ with different values for e_{cond} ranging from 0.5 to 0.001. Again, value u_o is the final step for which convergence condition is met. For each value e_{cond}, the table indicates the percentage of tests which succeeded to reach convergence with less than 100 evaluations. In case of convergence, the worse, mean and best final steps are given. EGO is outperformed here. However, EGO is able to achieve more refined results – *e.g.* $e_{cond} = 0.001$ – if the evaluation budget is relaxed: indeed, EGO is based on exact mathematical computation of the expected improvement.

Table 3. Results for EGO method

e_{cond}	0.5	0.2	0.1	0.05	0.02	0.01	0.005	0.002	0.001
$\% : u_o \leq 100$	91%	74%	69%	64%	55%	46%	42%	29%	14%
worse(u_o)	55	78	96	97	97	84	93	93	99
mean(u_o)	11.8	19.4	23.1	25.5	30	31.3	37.5	56.3	60.5
best(u_o)	1	1	8	8	8	15	15	19	19

Decision Space of Higher Dimension. Variable $\boldsymbol{X}$ is uniform on $[-5, 5]^8$ with realization $\boldsymbol{x}^o = (2, -1, 3, 2, -\frac{3}{2}, 4, \frac{1}{2}, 3)$. Points $M_{i,j} = (m_{i,j,k})_{1 \leq k \leq 28}$ with $1 \leq j < i \leq 8$ are built from $\boldsymbol{X}$ by setting $m_{i,j,k} = X_i X_j$ if $k = j + \frac{(i-2)(i-1)}{2}$ and $m_{i,j,k} = 0$ else. Function $g(\gamma, \boldsymbol{X})$ is defined by $g(\gamma, \boldsymbol{X}) = ||\gamma - h(\boldsymbol{X})||_2$ where $h(\boldsymbol{X}) = \frac{1}{28} \sum_{i=2}^{8} \sum_{j=1}^{i-1} M_{i,j}$. The problem is thus of much higher dimension.

We had some theoretical issues on this example. The Bayesian estimation was unable to resolve some remaining ponctual cases for zero dimension subvarieties. This difficulty has been overcome by adding exclusion constraints around already tested cases. Table 4 gives an example of run (progress is in bold). Incremental Algorithm 4 were needed for simulation, and optimization has been performed within a day. On the other hand, Algorithm 3 was far too slow in this case for optimizing in reasonable time.

Table 4. Example in a space of higher dimension

u	**1**	2	3	**4**	**5**	6	**7**	8	**9**	**10**	11	12	**13**	14	15	16	**17**	18	19	20
e_u	1.02	1.04	1.18	0.77	0.68	0.68	0.61	1	0.33	0.12	0.18	0.21	0.09	0.09	0.09	0.09	0.05	0.10	0.09	0.12

6 Conclusion

We proposed an original dichotomous method for sampling a random vector conditionally to a subvariety. This method has been parallelized, and we proposed non-incremental and incremental implementations. This generic approach, inspired from interval analysis, is accurate and efficient up to a space of dimension 11. We have shown how it could be applied efficiently to Bayesian optimization problems. The work is promizing from theoretical and applicative point of view and offers some improvement perspectives. A main issue is to enhance the efficiency of the approach with respect to higher dimensions. With this perspective in mind, some relaxation techniques applied to the subvariety may be considered.

References

1. Alefeld, G., Mayer, G.: Interval analysis: theory and applications. J. Comput. Appl. Math. **121**(1), 421–464 (2000)
2. Bonwick, J.: The slab allocator: an object-caching kernel memory allocator. In: USENIX Summer, pp. 87–98. Sun Microsystems (1994)
3. Dambreville, F.: Optimizing a sensor deployment with network constraints computable by costly requests. In: Le Thi, H.A., Pham Dinh, T., Nguyen, N.T. (eds.) Modelling, Computation and Optimization in Information Systems and Management Sciences. AISC, vol. 360, pp. 247–259. Springer, Cham (2015). https://doi.org/10.1007/978-3-319-18167-7_22
4. Dambreville, F.: Simulating a random vector conditionally to a subvariety: a generic dichotomous approach. In: SIMULTECH 2021, 11th International Conference on Simulation and Modeling Methodologies, Technologies and Applications, pp. 109–120 (2021)
5. Dezert, J., Musso, C.: An efficient method for generating points uniformly distributed in hyperellipsoids. In: Workshop on Estimation, Tracking and Fusion: A Tribute to Bar-Shalom, Monterey, California (2001)
6. Hebbal, A., Brevault, L., Balesdent, M., Taibi, E.G., Melab, N.: Efficient global optimization using deep gaussian processes. In: 2018 IEEE Congress on Evolutionary Computation (CEC), pp. 1–8 (2018)
7. Jaulin, L., Kieffer, M., Didrit, O., Walter, E.: Applied Interval Analysis with Examples in Parameter and State Estimation, Robust Control and Robotics. Springer, London (2001). https://doi.org/10.1007/978-1-4471-0249-6
8. Jones, D., Schonlau, M., Welch, W.: Efficient global optimization of expensive black-box functions. J. Glob. Optim. **13**, 455–492 (1998). https://doi.org/10.1023/A:1008306431147
9. Mockus, A., Mockus, J., Mockus, L.: Bayesian approach adapting stochastic and heuristic methods of global and discrete optimization. Inform. (lithuanian Acad. Sci.) **5**, 123–166 (1994)

10. Močkus, J.: On Bayesian methods for seeking the extremum. In: Marchuk, G.I. (ed.) Optimization Techniques 1974. LNCS, vol. 27, pp. 400–404. Springer, Heidelberg (1975). https://doi.org/10.1007/3-540-07165-2_55
11. Morio, J., Balesdent, M., Jacquemart, D., Vergé, C.: A survey of rare event simulation methods for static input-output models. Simul. Model. Pract. Theory **49**, 287–304 (2014)
12. Rubinstein, R.Y., Kroese, D.P.: The Cross Entropy Method: A Unified Approach To Combinatorial Optimization, Monte-Carlo Simulation (Information Science and Statistics). Springer, New York (2004). https://doi.org/10.1007/978-1-4757-4321-0
13. Zilinskas, A., Zhigljavsky, A.: Stochastic global optimization: a review on the occasion of 25 years of informatica. Informatica **27**, 229–256 (2016)

A Python-Based Mixed Discrete-Continuous Simulation Framework for Digital Twins

Neha Karanjkar[1(✉)] and Subodh M. Joshi[2]

[1] Indian Institute of Technology Goa, Ponda, India
nehak@iitgoa.ac.in
[2] Department of Computational and Data Sciences, Indian Institute of Science, Banglore, Banglore, India
subodhmadhav@iisc.ac.in

Abstract. The use of Digital Twins is set to transform the manufacturing sector by aiding monitoring and real-time decision making. For several applications in this sector, the system to be modeled consists of a mix of discrete-event and continuous processes interacting with each other. Building simulation-based Digital Twins of such systems necessitates an open, flexible simulation framework which can support easy modeling and fast simulation of both continuous and discrete-event components, and their interactions.

In this paper, we present an outline and key design aspects of a Python-based framework for performing mixed discrete-continuous simulations. The continuous processes in the system are assumed to be loosely coupled to other components via pre-defined events. For example, a continuous state variable crossing a threshold may trigger an external event. Similarly, external events may lead to a sudden change in the trajectory, state value or boundary conditions in a continuous process. We first present a systematic events-based interface using which such interactions can be modeled and simulated. We then discuss implementation details of the framework along with a detailed example. In our implementation, the advancement of time is controlled and performed using the event-stepped engine of SimPy (a popular discrete-event simulation library in Python). The continuous processes are modelled using existing frameworks with a Python wrapper providing the events interface. We discuss possible improvements to the time advancement scheme, a roadmap and use cases for the framework.

Keywords: Digital twins · Mixed discrete-continuous simulation · Python · SimPy

1 Introduction

A Digital Twin refers to a digital representation (computer model) of a real system that is continuously kept in sync with the system using periodic sensing of its health parameters and is used for prediction, optimization and control of the real system. The use of Digital Twins, aided by advancements in Internet of Things (IoT) technologies and Machine Learning (ML) based analytics is set to transform many sectors such as manufacturing, healthcare, urban planning, energy and transportation. While a data-driven model may suffice as a Digital Twin for some applications, a detailed simulation model

G. Wagner et al. (Eds.): SIMULTECH 2021, LNNS 601, pp. 204–223, 2023.
https://doi.org/10.1007/978-3-031-23149-0_11

is often used to create Digital Twins of complex processes in the manufacturing sector. Such systems often consist of a mix of discrete-event processes and continuous processes interacting with each other. For example, in [16] the modeling of food processing systems is described which requires a detailed simulation of continuous phenomena as well as discrete event processes with inter-dependencies.

This paper presents the key design aspects and implementation details for a Python-based Mixed Discrete-Continuous Simulation (MDCS) framework targeted for creating Digital Twins. A preliminary outline and motivation for this framework was first presented in [17]. This paper expands on the implementation aspects and presents a detailed example to demonstrate the event-based interface used for incorporating and simulating continuous processes in this framework.

In this section, we first present a broad overview of existing simulation approaches and frameworks used for building Digital Twins. We then summarize the motivation and design goals for a Python-based MDCS framework. In Sect. 2 we present the details of the framework and describe the events-based interface that can be used for integrating continuous processes into a discrete event simulation engine. In Sect. 3, we present a detailed example that illustrates the key aspects and use of the framework. While one of the continuous entities (a fluid tank) used in the example is similar to that described in [17], the example also incorporates a continuous process describing heat dissipation in a two-dimensional (2D) hot-plate that interacts with the fluid tank and other discrete event processes in the system. Unlike the fluid tank (where the state-updates are modeled by simple, linear equations), the simulation of the transient heat transfer in the hot-plate requires resolution of the governing Partial Differential Equations (PDEs) using a Finite Difference Method (FDM). We present detailed simulation results for a test case where both types of components interact with each other. An improved scheme for time-advancement and a roadmap for further development are discussed in Sects. 4 and 5 respectively.

1.1 A Review of Simulation Approaches for Digital Twins

The design of a simulation framework for Digital Twins is driven by the characteristics of the system to be modeled. The methodologies and approaches for numerical simulations depend on the type of the system under consideration, that is, whether the system is continuous, discrete or contains a mix of both types of entities. While some system models may necessitate a continuous simulation framework [3,24], a discrete-event simulation might suffice for other kinds of Digital Twins [1]. A summary of challenges and desired capabilities associated with the simulation of Digital Twins is presented in [26]. In the context of Digital Twins, continuous systems are the ones subjected to continuous evolution of the state variables in time. A few examples of such systems include transient transfer of heat, unsteady fluid flows such as air-flow over a wind turbine, chemical reactions etc. The state variables for such systems often undergo a continuous change in time (unless the system reaches a steady state). Accurate simulation of these systems necessitates solving the governing equations, which often take the form of ordinary/partial differential equations (O/PDE) or mixed differential-algebraic equations (DAE). This in turn requires using appropriate numerical schemes, for example Finite Volume, Finite Element or Finite Difference Methods as well as schemes for advancing the solution in time. The time-stepping methods can be explicit, implicit or mixed type such as the IMplicit-EXplicit (IMEX) method. In most of the time-stepping methods, the time step-size may be either fixed, or adjusted dynamically over the course of a simulation. A detailed description of continuous processes and their simulation aspects can be found in [7,22] and the references therein. In practice, frameworks such as FEniCS [23], Deal II [4], OpenFOAM [29] are used for continuous multiphysics simulations of complex systems. Numerical techniques such as reduced order models (ROM) [8,12] and Machine-Learning (ML) based metamodels [27] may be used instead of the high fidelity models to reduce the overall computing cost. In the recent times, Machine Learning is increasingly being used for scientific computing [2,6,18]. In this paper, we use the classical Finite Difference Method (FDM) for simulation of the continuous entities, however, we plan to explore the other simulation frameworks including Machine Learning in future work.

Unlike the continuous processes described earlier, discrete processes are characterized by changes in the state of the system occurring only at discrete (countable) instants of time, referred to as *events*. Discrete-event simulation is prominently divided into two approaches, viz. event-stepped and cycle-stepped approaches. We refer [15] for a detailed description of the approaches used for simulation of discrete event systems. Discrete Event System Specification (DEVS) and a subsequent generalization (GDEVS) are the two main formalisms used for specification and simulation of discrete-event systems [14,30]. Agalianos et al. present an overview of issues and challenges for discrete-event simulation in the context of Digital Twins [1]. There exist several proprietary as well as open source libraries and softwares for discrete-event simulations. A review of open source discrete simulation softwares is presented in [9].

Systems containing both discrete-event and continuous processes are termed as Mixed Discrete-Continuous (MDC) systems. Simulation of MDC systems is particularly challenging since it requires to satisfy the constraints imposed by both the continuous and the discrete entities involved. Different methods and techniques have been proposed in literature for simulation of MDC systems. A quantization based integration method was proposed by Kofman et al. for simulation of hybrid systems [21]. Nutaro et al. propose a split system approach in which a-priori knowledge about the discrete-continuous structural split in the model can be used for performing efficient simulation [25]. Klingener describes approaches that can be used to get a non-modular simulation framework [19,20]. An approach called Discrete Rate Simulation has been proposed in [10] for simulating linear continuous models (such as constant-rate fluid flows) within a discrete-event framework. A usecase for this approach has be demonstrated by Bechard et al. in [5]. Eldabi et al. present a detailed overview of various strategies used for MDC simulation in [11].

1.2 Motivation and Design Goals

While the ability to perform mixed discrete-continuous simulations is a key requirement for the development of the framework presented in this paper, the other design goals are as follows:

1. The ability to model heterogeneous systems containing different kinds of continuous processes, each possibly requiring a different numerical method for its solution and/or different characteristic time-step sizes.

2. Support for capturing the effect of periodic sensor updates from the real system on the model's state.
3. The ability to perform real-time simulation.
4. The framework should be open-source and flexible. It should be easy to integrate existing libraries for enabling analytics and visualization (e.g. optimization, machine learning, data handling, scientific computing and plotting libraries) into the framework.
5. The language used by the framework should support modular descriptions and the use of object oriented features for modeling complex systems with many interconnected components.

While there exist a number of frameworks that are targeted separately for either continuous simulation or discrete-event simulations, the requirement of simulating both discrete and continuous processes together, possibly interacting with each other, introduces some challenges. A majority of the existing frameworks for MDC simulations are either commercial or domain-specific, with exceptions such as OpenModelica [13]. However, there still exists the need for a mixed simulation framework that is written in a general-purpose, object-oriented language which allows integration with existing continuous simulation frameworks. Python is an attractive choice for implementing such a framework because of its wide user base, ease of use and the availability of a number of libraries for analytics and visualization.

2 Framework for Mixed Discrete-Continuous Simulation

In this section, we present the basic definitions and assumptions, describe the main aspects of the framework (including the events interface) and discuss the implementation details. The system to be modeled can be thought of as a mix of continuous and discrete entities that interact with each other.

- An **entity** in this context is a collection of state variables, methods and processes representing a particular object to be modeled in the system.
- A **discrete entity** refers to a process whose state can change only at discrete time instants (events).
- A **continuous entity** is an entity whose state may be considered to change continuously with time and may require continuous simulation/monitoring.

When simulating discrete and continuous processes interacting with each other, the fundamental questions related to the advancement of time have been addressed by formalisms proposed for hybrid simulations, for e.g. [25]. However, from an implementation perspective, the simulation approaches can be broadly classified into two categories as summarized below:

(A) In the first approach, the advancement of time is controlled by a single time-stepped continuous simulation loop. The discrete events to be modeled are embedded into the continuous simulation code as conditional updates to the state variables or boundary conditions during simulation. These updates may occur at pre-defined time-steps or whenever a certain condition on the state variables is met (for example, when the value of the state variable crosses a particular threshold). The step-size for advancing time is fixed and determined by the stability considerations of the numerical scheme used for continuous simulation. If the required step-size differs across multiple continuous entities in the system, the smallest of the step-sizes needs to be used for updates in *all* of the continuous entities. Such an approach is well suited for systems mainly consisting of tightly coupled continuous entities.

(B) In the second approach, time advancement is handled by an event-stepped discrete-event simulation algorithm. Simulation of continuous entities is performed by invoking their state-update functions periodically or at selected time-instants from within the event-stepped loop. Interactions of a continuous entity with other entities in the system are modeled via **events**. The continuous entity must generate and advertise events when certain conditions on its state variables are met (for example, when the state value crosses a certain threshold), so that external activities can be triggered on the occurrence of this event. Similarly, external events may cause a sudden change in the state values or trajectory of the continuous entity. Therefore, whenever such external events occur, there must be a mechanism to update the state/boundary conditions of the continuous entity and invoke its state update function. This approach is well-suited when the continuous entities in the system are few and loosely coupled.

Approach **(B)** is particularly suited for simulation of Digital Twins in manufacturing and process engineering domains since the systems to be modeled are heterogeneous, typically consisting of a larger number of discrete entities and a few continuous entities that are loosely coupled and interact in well-defined ways. From a modeling perspective, describing interactions between components through events can lead to modular descriptions that are easier to read, maintain and debug. We present a framework based on this approach and describe how the interactions can be modeled and time advancement can be performed.

2.1 The Events Interface

In our framework, time advancement is performed using the event-stepped algorithm of a general purpose discrete-event simulation framework (such as SimPy). To incorporate continuous entities into the simulation, a wrapper for each continuous entity is made, which provides an interface as summarized in Fig. 1. A continuous entity is characterized by its state variables and a state-update function. The modeler also needs to provide a definition of all events that may be generated by the entity and affect the external world and vice-versa. These events can be classified into four types as follows:

1. **Perturbation Event:** An externally-created event that may affect the state/trajectory of the continuous entity.
2. **Probe Event:** An externally-created event which involves querying the state of the continuous entity and thus necessitates updating its state up-to a given time.
3. **Generated Event:** An event triggered by the continuous entity itself as a consequence of its state update which may affect other entities in the system.

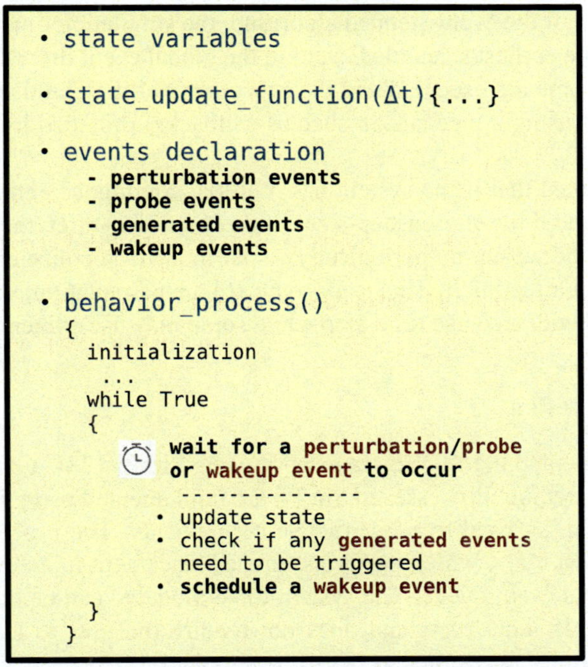

Fig. 1. Wrapper for a continuous entity with the events serving as its interface [17].

4. **Wakeup Event:** An event scheduled by the continuous entity itself for performing state updates after a fixed time step or for creating generated events whose time of occurrence can be predicted in advance.

The behaviour of the continuous entity is modeled as a discrete-event process. This process is activated whenever a perturbation, probe or a wakeup event for the entity occurs. When activated:

1. The state updates for the entity up-to the current time are computed.
2. If any condition for generating output events is met (for example, the state variable crossed a threshold), the generated events are triggered.
3. The continuous entity schedules a wakeup event for itself after a particular time interval. This time interval is determined based on one of two approaches as follows:
   - **(a) Predictive time-stepping:** If the current trajectory of the state values in the entity is known, and if all of the output events can be predicted ahead of time based on this trajectory, the wakeup can simply be scheduled at the time the earliest output event is predicted to occur. (This is illustrated via the example of a fluid tank described in Sect. 3.2.
   - **(b) Fixed time-stepping:** If the exact time instants of events that will be generated by this entity cannot be predicted ahead of time, the state needs to be updated after regular time intervals that are small enough, by scheduling a wakeup event periodically. The step-size may need to be chosen based on numerical stability requirements. (This is illustrated via the example of a heater entity described in Sect. 3.1.

At each iteration of the event-stepped algorithm, the simulation time is advanced to the time-stamp of the earliest scheduled event in the global event list. All events scheduled to occur at this time are executed and the processes waiting on this event are automatically triggered (using a mechanism such as callbacks) provided by the discrete-event framework.

It is to be noted that if the system has multiple continuous entities requiring fixed time-stepping, and if the chosen step-sizes of these entities differ, then in this approach, each entity will be woken up periodically as per its own step-size only, unless there is an external event affecting it. This leads to an efficient simulation, where some entities may be updated with a coarse time-step while some may use a finer time-step.

2.2 Implementation

We have implemented the MDCS framework using SimPy [28], a discrete-event simulation library in Python. Processes in SimPy are implemented using Python's generator functions and can be used to model active components. The processes are managed by an *environment* class, which performs time advancement in an event-stepped manner using a global event queue. The system to be modeled can be described in Python using a few SimPy constructs and does not require the user to learn a new modeling language. SimPy also supports real-time simulation. However, SimPy is designed for discrete-event simulation and currently offers no features for modeling continuous systems [28]. The events-based interface presented in this paper can be implemented as an abstract wrapper class to allow integration of continuous entities. The modeler needs to create the wrapper class for each unique type of continuous entity describing the aspects summarized in Fig. 1. When the simulation of the continuous entity needs to be performed using an external solver (such as Deal II or OpenFOAM), the state update function in the abstract class serves as a wrapper for invoking the state updates via the external continuous solver. Similarly, the continuous solver code needs to be instrumented to detect conditions that must trigger events, and set a flag variable in the wrapper. State or boundary value updates to the continuous entity can be implemented via global shared variables that may be updated by external processes, but are read at each iteration in the continuous solver code. The implementation is straightforward if the simulation and state updates for the continuous entity are also described as Python code. In the example presented in Sect. 3 the state updates of continuous entities are performed directly by including their description as Python modules.

3 A Modeling Example

We present a modeling example to illustrate the key aspects of the MDCS framework. The system to be modeled consists of discrete-event processes as well as two types of entities that are modeled in the continuous domain. The first type of entity (a fluid tank) has simple, linear state-update equations. Simulation of this entity can be performed using the predicted time-stepping approach. The second entity is a heater with a square-shaped plate which can be heated from two opposite sides. The temperature within the plate varies continuously with time and also across the length and breadth of the plate.

The diffusion of heat in the plate is simulated using fixed time-stepping approach, with the time-step size determined by numerical stability requirements. The complete system to be simulated consists of both types of components interacting with each other.

We first describe each of these components and their simulation approaches in detail. We then describe the complete system consisting of the fluid tank and the heater instances interacting with each other as well as with other discrete-event processes. We show how the interactions can be described in the framework, and present detailed simulation results.

3.1 Heater

Description. The physical system consists of a square-shaped hot-plate with each side 1m long and made of a composite material. Two opposite edges of this plate can be heated with the help of heating coils. We make the simplifying assumption that the heating coils instantly reach the steady-state high temperature value without any lag when switched on. The remaining two edges of the hot-plate are maintained at a constant lower temperature at all times. After the heating coils are turned on, heat dissipates in the hot-plate and eventually the temperature distribution attains a steady state profile. How quickly the heat dissipates depends on the thermal conductivity, the specific heat capacity and the density of the material of the hot-plate. A physical parameter called as thermal diffusivity (α, units m^2/s) is often used to describe the resistance offered by any material for heat dissipation. It takes into account the collective effects due to all the material properties mentioned above. We assume the hot-plate to be made of a highly conductive composite material with thermal diffusivity equal to 0.0005 m^2/s. When the heater is turned off, the heat starts flowing out via all the four edges. We assume the convective and the radiative losses to be negligible and the entire heating and cooling of the hot-plate takes place only through the boundaries. We measure the temperature of the hot-plate with the help of a temperature probe. The framework allows specifying the location of the probe at any location. We assume the probe to locate at the center of the plate in this example. For this test case, we specify the heater temperature (when turned `ON`) as 100 °C (`HIGH_temp`) and the room temperature, which is also the temperature maintained constantly at the other two edges, as 25 °C (`LOW_temp`). Further, when the heating is switched `OFF`, all the four boundaries are assumed to be at the uniform temperature equal to `LOW_temp`. Figure 2 shows a schematic of the hot-plate along with the state-variables and the state-update equations governing the conductive heat transfer.

Numerical Simulation. Let the hot-plate be modeled by a square shaped computational domain $\Omega := [0,1] \times [0,1]$. Let the temperature in the plate be described by $u(x, y)$, where (x, y) is any physical location. Further, time is denoted by t. The thermal diffusivity α is a material-specific property, which depends on the thermal conductivity of the material, its specific heat capacity and its density. The heat conduction through a homogeneous material is described mathematically by the following Partial Differential Equation (PDE).

$$\frac{\partial u}{\partial t} = \alpha \nabla^2 u, \ \forall (x, y) \in \Omega \tag{1}$$

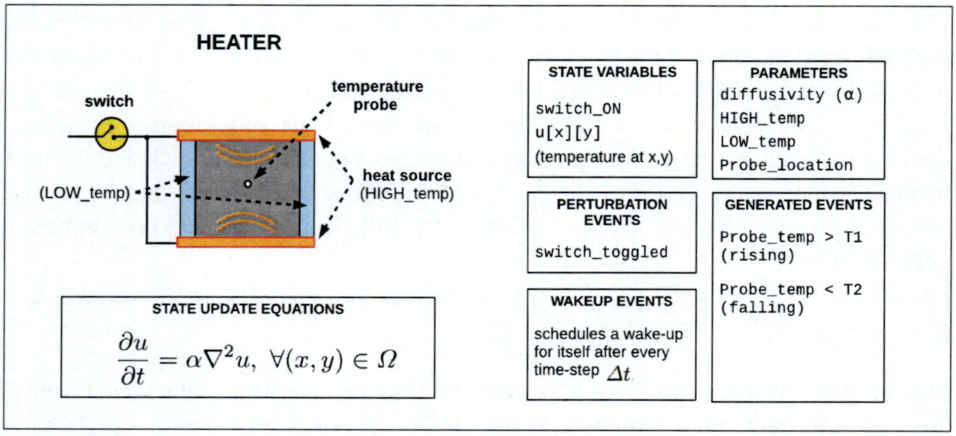

Fig. 2. Description of the Heater model.

The boundary conditions are as follows. $u(x,0) = u(x,1) =$ Heater temperature: i.e. HIGH_temp (heater ON) or LOW_temp (heater OFF) and $u(0,y) = u(1,y) =$ LOW_temp $\forall$ t.

To solve this equation numerically, we use the Finite Difference Method (FDM). First, the domain Ω is discretized in a structured, uniform grid with the grid-spacing h. Let the number of grid-points in each direction x and y be N, i.e. N^2 in total. Therefore, the location of any grid-point (i,j) is given as $(x_i, y_j) = (ih, jh),\ i,j = 0,1,2,...,N-1$. Similarly, let the time-step for advancing the numerical simulation be Δt. The temperature at any grid-point (i,j) at time level k is denoted as $u^k_{i,j}$. The difference equation corresponding to the governing equation (1) can be written as (derived from Taylor's series expansion, i.e. standard FDM formulation):

$$\frac{u^{k+1}_{i,j} - u^k_{i,j}}{\Delta t} = \alpha \left(\frac{u^k_{i+1,j} - 2u^k_{i,j} + u^k_{i-1,j}}{h^2} + \frac{u^k_{i,j+1} - 2u^k_{i,j} + u^k_{i,j+1}}{h^2} \right) \tag{2}$$

with the Dirichlet boundary conditions given as

$$u^k_{0,j} = u^k_{N-1,j} = \texttt{LOW\_temp}$$

and

$$u^k_{i,0} = u^k_{i,N-1} = \texttt{HIGH\_temp} \text{ (heater ON) or } \texttt{LOW\_temp} \text{ (heater OFF)}$$

Equation (2) can be further simplified as

$$u^{k+1}_{i,j} = \gamma \left(u^k_{i+1,j} + u^k_{i-1,j} + u^k_{i,j+1} + u^k_{i,j+1} - 4u^k_{i,j} \right) + u^k_{i,j} \tag{3}$$

where, $\gamma = \alpha \Delta t / h^2$. In the beginning of the simulation, the initial conditions $u^0_{i,j} =$ LOW_temp, $\forall i,j$ hold true for the entire domain. Equation (3) is solved iteratively to advance the simulation in time. This time-stepping scheme is also known as the Forward-Euler (FE) explicit timestepping scheme. It is to be noted that, the FE timestepping scheme is only conditionally stable, i.e. the time-step value needs to be 'small' enough to yield stable computations. According to the Von-Neumann stability analysis, the time-step value comes out to be $\Delta t \leq h^2/(4\alpha)$. i.e. this is the largest value of Δt that can be safely used for advancing the solution in time. This in turn also dictates the frequency of wake-up events that needs to be scheduled. An efficient, vectorized implementation of Eq. (3) is performed using Numpy, a Python library for scientific computing. Python also makes it convenient to incorporate the numerical implementation of the heater in the discrete event framework.

Interface with the Discrete-Event Simulation Framework. The heater entity is implemented as a Python class. The state variables, parameters and the state-update function (i.e. the numerical solution of the governing PDE given by Eq. 1) become the members of this class. The interaction of the heater with the environment happens via the following types of events:

1. **Perturbation Events:** These are the events which can cause the heating coils to turn on or off. This in turn results in heat transfer into or out of the system affecting the temperature distribution.
2. **Probe Events:** We can specify the location of the probe on the hot-plate. External components can probe the temperature at this location at the current simulation time. This necessitates the updating the heater state up-to the current time. If time at which the probing is performed is sooner than the next wake-up governed by the time-step Δt corresponding to the Forward Euler method, an update at a smaller time-step is performed. Since this doesn't affect the stability of the numerical method, the update can be safely performed.
3. **Generated Events:** User can specify a threshold value of temperature (or multiple values) at the probe location. Whenever the *rising* temperature at the probe location crosses the threshold values, an event (probe-temperature crossed (rising)) is generated. Similarly whenever the *falling* temperature at the probe location crosses the threshold value, another event (probe-temperature crossed (falling)) is generated. External processes waiting for these events to occur are then automatically notified (using the `yield<event>` construct of SimPy).
4. **Wake-up Events:** Since numerical solution of the governing PDE requires an iterative time-stepping scheme, a wake-up event is generated after every time-step which schedules the state-update for the next time-step. The time-step size is governed by the stability requirements of the numerical method. If any other event (probe/perturbation/generated) occurs before the scheduled wake-up takes place, a state-update is scheduled by those events as described earlier. Once those events are executed, the system resumes the wake-up cycle for advancing in time.

Numerical Results. We validate the heater model as follows. A heating and cooling schedule is designed to validate the heater model. The schedule consists of two identical cycles of heating and cooling in the time span of first 50 min, and a random heating and cooling schedule from $T = 50$ min to $T = 120$ min. Both cycles till $T = 50$ min are of identical time duration and consist of a heating process followed by a cooling process. Figure 3 shows a time-series plot of the temperature at the center of the hot-plate (measured by the probe). The framework allows specifying threshold temperature values to

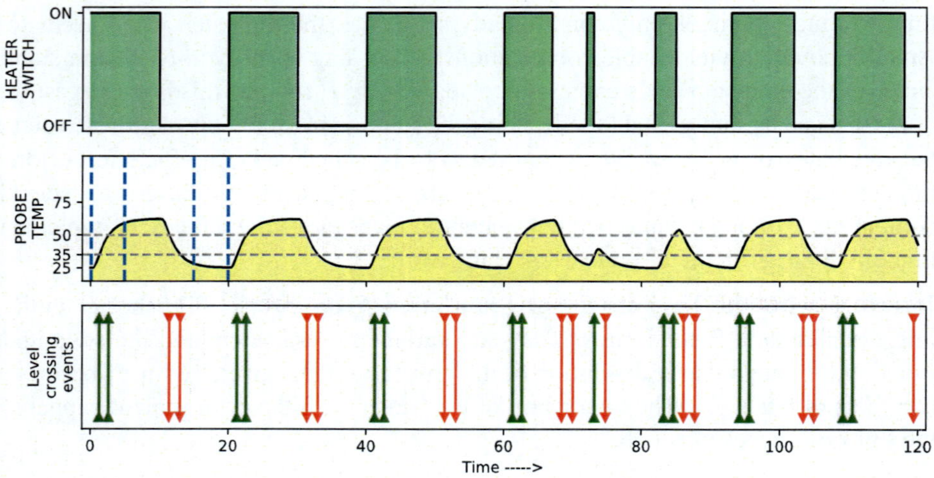

Fig. 3. Time evolution of the probe temperature showing events generated during simulation. The level crossing events (rising and falling) are shown with upward and downward arrows respectively. The threshold temperatures were set at 35 °C and 50 °C.

schedule Generated Events. In this example, we have set two temperature values, i.e. u = 35 °C and u = 50 °C as shown in the figure. Whenever the *rising* temperature at the probe location crosses the threshold values, a Generated Event (probe-temperature rising, denoted by green upward pointing arrow) is created. Similarly whenever the *falling* temperature crosses the threshold values, another generated event (probe-temperature falling, denoted by red downward pointing arrow) is created. These generated events can affect either the discrete event schedule or other continuous systems in the framework. The entire simulation is run through a discrete event scheduler, which schedules a wake-up event for each time step to advance the simulation in time. It can be seen that, the temperature at the center of the hot-plate starts increasing when the heater is turned `ON` and asymptotically reaches a steady-state value of 62.8 °C. At time $T = 12.5$ min, the heater is turned `OFF` and the temperature starts dropping exponentially until it reaches the `LOW_temp` value of 25 °C. As the heat loss takes place only at the edges, a higher residual temperature remains at the interior parts of the hot-plate for a short time even after the heater is turned `OFF`. Eventually the heat loss results in a uniform temperature of 25 °C over the entire area of the hot-plate. Figure 4 shows snapshots of the hot-plat at time Δt, 5, 15 and 20 minutes, where, Δt is the time after the first time-step is performed. Corresponding time-values are highlighted in 3 with dashed blue lines. It can be clearly seen that, at Δt the two opposite edges are at `HIGH_temp` while the rest of the hot-plate is at `LOW_temp`. The value of Δt is dictated by the stability conditions as stated earlier. At $T = 5$ min, heat has dissipated inside the domain raising the temperature differentially at different parts. After the heating is turned `OFF`, a residual high temperature is observed for a short while in the interior parts of the plate, for example as shown in the figure at $T = 15$ min. This residual temperature returns to `LOW_temp` as the heat loss takes place through all four boundaries. Sometime before $T = 20$ min, the temperature over the entire domain asymptotically returns to 25 °C.

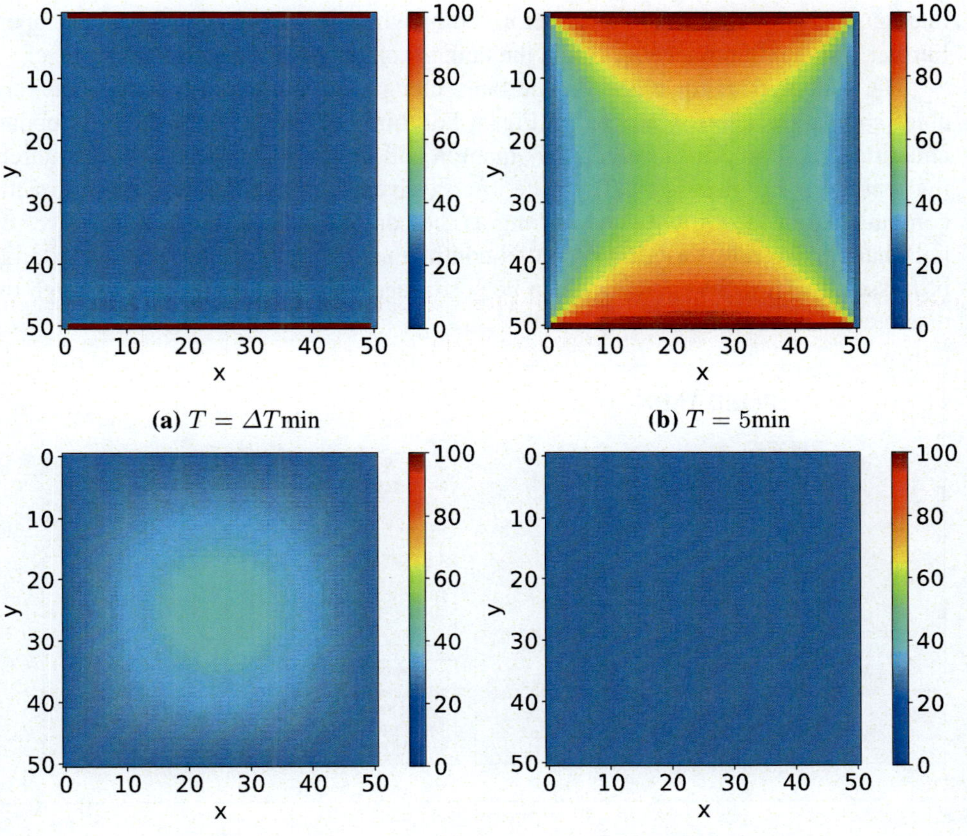

(a) $T = \Delta T$min (b) $T = 5$min

(c) $T = 15$min (d) $T = 20$min

Fig. 4. Temperature profile snapshots at time $T = \Delta t, 5, 15, 20$ mins.

The cycle repeats after $T = 25$ min till $T = 50$ min. After $T = 50$ min, the heating and cooling schedule is kept random. This is to demonstrate that the heating and cooling can be randomly applied in the ongoing simulation and an a-priori knowledge of the same is not required for running the simulation. This has been made possible due to our approach to time-stepping the simulation via a discrete event scheduler, including for the embedded continuous processes. In the subsequent sections, we demonstrate an example where the heater works along with another continuous simulation entity (water tank) in the discrete event framework, such that both the discrete event framework and the continuous entities can potentially have an effect on each other.

3.2 Fluid Tank

Description. Consider a fluid tank whose level is to be simulated and monitored continuously with respect to time. The tank has inlet and outlet valves that can be opened and closed based on external triggers. Opening/closing of these valves changes the state trajectory of the tank. The flow rate through the inlet/outlet are assumed to be fixed parameters in the model. They are specified in units of length per-time and defined as maximum change in the fluid level per unit time when the corresponding valve is open. Further, the maximum level at which the tank is considered full, is also a parameter.

The state-update equations for the tank, along with a summary of the state variables and parameters is presented in Fig. 5. For this continuous entity, the state-update equation is a simple linear algebraic equation and therefore, an iterative time marching method is not necessary. The state after a given time interval Δt can be directly computed from the current state as long as the condition of the inlet/outlet valves do not change. We now describe how this continuous entity interacts with the external processes, and how its simulation can be performed by time-advancement through the discrete-event engine.

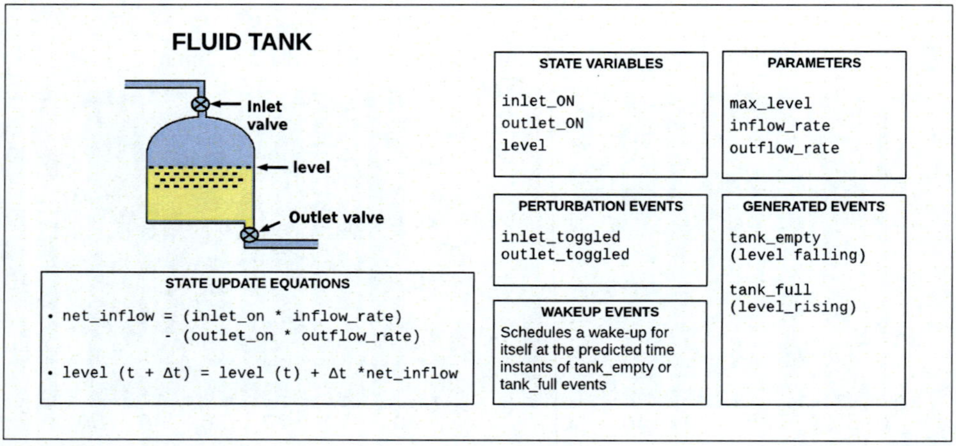

Fig. 5. Model of a fluid tank.

Discrete-Event Interface and Simulation: The tank entity can be implemented as a Python class with an interface similar to that in Fig. 1. The state variables, parameters and the state-update function become members of this class. The tank interacts with the environment via the following events serving as an interface:

1. **Perturbation Events:** External events can cause the tank's inlet or outlet valves to toggle their state, which can change the trajectory of the level.
2. **Probe Events:** External components can probe the level of the tank at the current simulation time. This necessitates updating the tank state up-to the current time.
3. **Generated Events:** Whenever the tank level falls, and the tank becomes empty, a `tank_empty` event is generated and triggered by the tank entity itself. External processes waiting for this event to occur are then automatically notified (using the `yield` <`event`> construct of SimPy). Similarly, when the tank level is rising and reaches the maximum value, a `tank_full` event is generated.
4. **Wake-up Events:** Owing to the linear state-update equations, the state updates need not be performed periodically at fixed time-steps. Rather they can be performed directly at time-instants of interest. Whenever a state-update is performed, the future time instant at which the tank is expected to become empty (if the level is falling) or full (if the level is rising) is computed, and the entity schedules a wake-up event

for itself at this precise time instant. When the wake-up event (or any perturbation/probe/generated events) occur in the tank, the state-update is performed, empty/full events are triggered if the tank has become empty/full at this instant, and the next wake-up event is scheduled, based on the current trajectory.
If it so happens that tank's state trajectory changes sometime before the next wake-up event (for instance, due to the toggling of a valve), the state is updated and a new wake-up event is scheduled based on the updated trajectory. The old wake-up event however is not cancelled. It simply causes the tank state to be updated up-to the time instant of the old wake-up event and does not have any side effects on the state.

Validation and Simulation Results. Figure 6 shows the simulation results obtained for a simple validation exercise involving a single tank instance. Here, an external SimPy process toggles the tank's inlet and outlet valves after random time intervals. The generated `tank_empty` and `tank_full` events are also indicated in the plot.

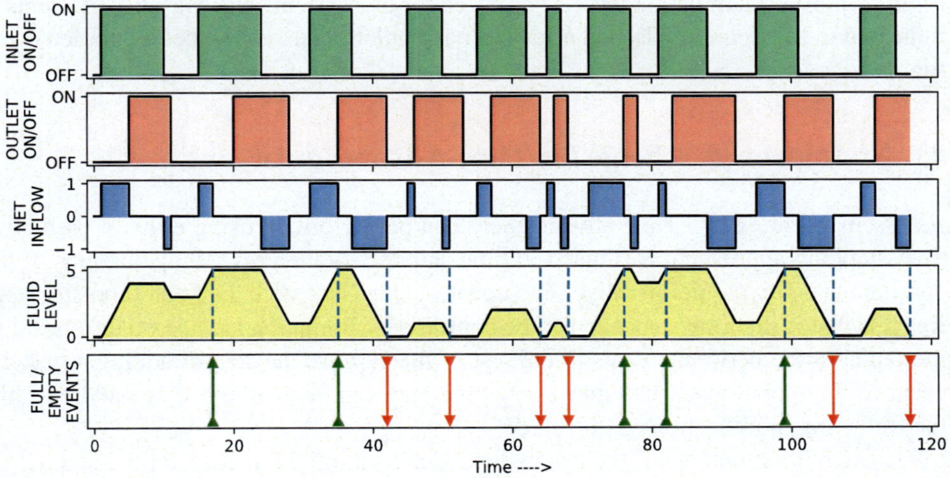

Fig. 6. A plot of the time evolution of the tank's state showing the empty/full events generated during simulation. The inlet/outlet values are toggled after random time intervals.

3.3 The System

We now describe the complete system where the fluid tank and heater instances interact with each other and with the environment through the *events interface*. To illustrate a two-way dependency between the fluid tank and heater instances, we model the following interactions in the system:

1. At the start of simulation, the heater is assumed to be OFF and the temperature of the entire heating plate is set to the `LOW_temp` value of 25 °C. The fluid tank is assumed to be full. The tank's inlet and outlet valves are both closed.
2. The heater is turned ON. As soon as the probe temperature crosses a certain threshold (in this case 50 °C), the system can start processing external jobs one-by-one. The arrival of jobs is modeled by a stochastic discrete-event process.

3. Each arriving job has a duration which is random and uniformly distributed between 0.5 to 1 min. Also the time interval between arrival of successive jobs is also uniformly distributed between 0.5 to 1 min. The tank outlet valve needs to be kept open for the duration of each job. Thus the tank gradually empties as successive jobs are processed.
4. As soon as the tank becomes empty, the heater is turned OFF, and the tank refill process is initiated. The tank inlet valve is opened and the outlet valve is closed. The tank gradually becomes full again.
5. As soon as the tank becomes full, the entire loop is repeated, starting from step 2.

To implement these interactions, an additional Python class (`Controller`) is created. The behavior of the controller can be described in a straightforward manner as a SimPy process. Listing 1.1 is an excerpt from the controller's behavioral loop described above. The code listing serves to highlight the ease with which complex interactions between continuous and discrete-event entities can be described by the modeler. Figure 7 presents simulation results generated for this system. The causality of events is indicated in the figure by dashed arrows. The simulation also produces a detailed event log.

4 An Improved Scheme for Time Advancement

As discussed in Sect. 2, time advancement can be performed using either a predictive time-stepping approach, or using fixed time-steps when the governing equations of the continuous entity require iterative state updates. One issue with the fixed time-step approach is that if the time step size for the continuous simulation is very small relative to the typical time between events in the rest of the system, the cost of adding a wakeup event to the global event list after every time-step can be prohibitive. To address this, the following modification can be used:

At each iteration (K) of the event-stepped algorithm, the *tentative* time-step (Δ_K) is taken to be the difference between the scheduled time of the next event in the global event list ($t_{\text{next\_event}}$) and the simulation time for the current iteration(t_K).

$$\Delta_K = t_{\text{next\_event}} - t_{\text{K}}$$

Each continuous entity is then asked to *peek-ahead* in time by a total period of Δ_K by executing its state-update equations. This can be done either using a single step of size Δ_K or by dividing this period into finer time-steps as dictated by the time marching scheme. The computation of state updates for multiple continuous entities can potentially be executed in parallel. If no output events of interest are predicted to be generated by any of the continuous entities in this period, the time can be advanced by Δ_K and the computed state-updates in each of the continuous entities can be applied before proceeding to the next iteration. However, if it is found that for a continuous entity i, an event of interest is generated at time $t_i < t_{\text{next\_event}}$, then it may be possible that this event could affect the state or trajectories of the other entities in the system. Thus the actual time-step taken must be the one that advances time to the earliest predicted event

```
# Python code excerpt describing the control loop
# note: env is an instance of SimPy's environment object

class Controller():
  # initialization, and other member functions
  # .....

  def controller_behavior(self,env):

    # initially the tank is full and its valves are closed.
    tank.level=tank.max_level
    tank.inlet_ON=False
    tank.outlet_ON=False
    # initially the heater is off. check this.
    assert(not heater.is_heater_ON())

    while(True):

      # turn ON the heater
      heater.turn_ON_heater_switch()
      # wait until the probe temperature goes
      # above a set threshold.
      if(heater.probe_temp() < 50):
        yield heater.threshold_crossed_rising_event

      # now, start processing jobs by periodically
      # toggling the tank outlet valves. Do this until
      # the tank level falls below a threshold.

      while(not tank.is_empty()):
        # turn ON the tank valve
        tank.toggle_outlet_valve()
        job_duration = random.uniform(0.5, 1)
        job_done   = env.timeout(job_duration)
        tank_empty = tank.empty_event

        # wait until the job is done or the tank becomes empty
        what_occured = yield (job_done | tank_empty)

        if job_done in what_occured:
          # wait for some time and then repeat with next job
          tank.turn_OFF_outlet_valve()
          idle_duration = random.uniform(0.5, 1)
          yield env.timeout(idle_duration)
        else:
          # tank became empty.
          break
      # tank became empty. Now turn OFF the heater,
      # and start the tank refilling process.
      # then, wait until the tank becomes full again.
      heater.turn_OFF_heater_switch()
      tank.turn_OFF_outlet_valve()
      tank.turn_ON_inlet_valve()
      yield tank.full_event

      # now the tank has become full.
      # turn off the tank valves.
      tank.turn_OFF_inlet_valve()
      tank.turn_OFF_outlet_valve()
      # done. repeat the loop.
```

Listing 1.1. Python excerpt for implementing the system control loop described in Sect. 3.3.

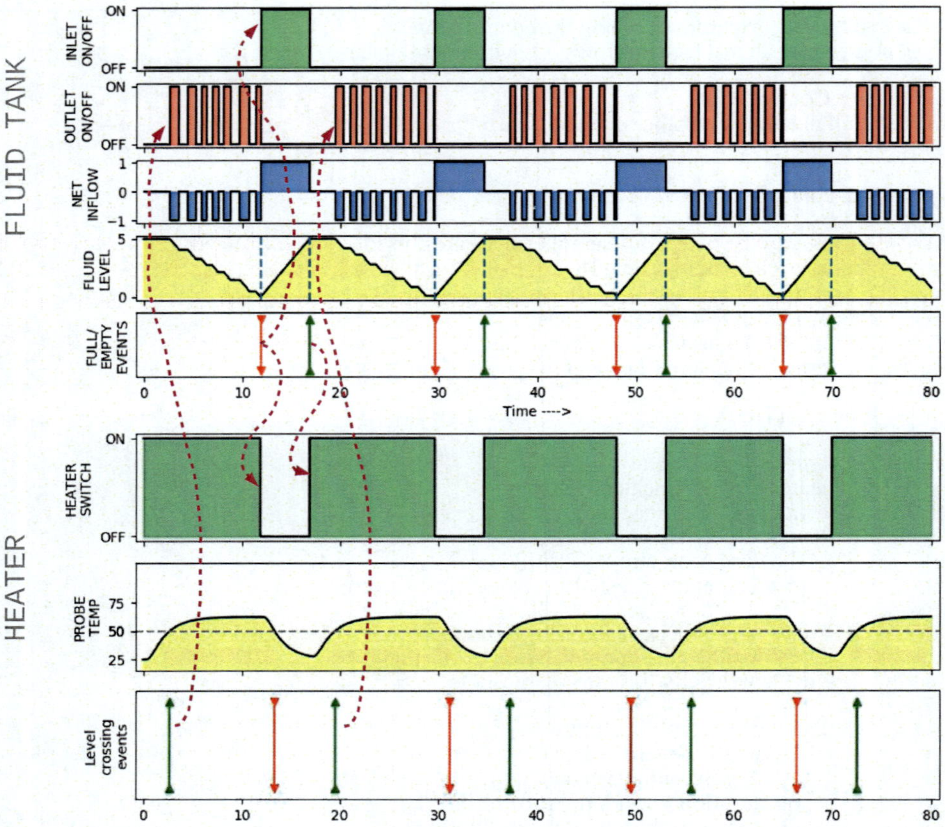

Fig. 7. Simulation results for a system containing interacting tank and heater instances. The causality of events is indicated by dashed arrows.

across all of the continuous entities. That is, the simulation time should be advanced to $t_{K+1} = \min_i(t_i)$ in the next iteration.

The earliest event predicted to occur at time t_{K+1} can then be inserted into the global event-list, so that its effect on other entities can be propagated as usual in a discrete-event framework, and the state updates in all of the continuous entities computed up to time t_{K+1} can be applied before advancing simulation time to t_{K+1}. A further optimization is to adaptively adjust the tentative time step Δ_K for improved performance. Implementing this requires the continuous simulation framework to support a peek-ahead or roll-back feature.

A second approach is to use meta-modeling (in the form of regression-based models, neural networks with supervised learning, or physics-informed neural networks) for predicting the time instants of generated events in advance. In fact, detailed models of continuous processes can often be replaced by reduced order surrogate models when accuracy needs to be traded for evaluation speed. In such models, the time instants of generated events can be predicted ahead of time, and the predictive time-stepping approach can be used for efficient simulation. Exploring the use of these approaches for building digital twins is a promising direction.

5 Future Work and Conclusions

In this paper we presented a Python based Mixed Discrete-Continuous Simulation (MDCS) framework specifically targeted for Digital Twins applications. The framework is based on Python's SimPy library and uses its event-stepped algorithm for coordinating the time advancement. We presented a detailed example of interacting continuous entities simulated using this framework. The design aspects of the framework and the simulation approach make it well-suited for digital twin applications for the following reasons:

- The event-stepped approach can result in a more efficient simulation for scenarios where only a few kinds of events affect the trajectory of continuous entities in the system.
- In this approach, it is possible for different continuous entities in the system to use different continuous solvers and internal time step-size values.
- The loose coupling between the continuous entities presents opportunities for executing their behavior in parallel within a single time-step for real-time simulation.
- For modeling entities where a high level of accuracy may not be necessary, coarse surrogate models can be used to predict the trajectory and time of generated events and schedule a wakeup ahead of time.
- Sensor value updates from the real system can be easily incorporated into the simulation as perturbation events affecting the state.

Future improvements to the framework are planned along the following directions:

1. **Integration with Existing Continuous Simulation Frameworks**
   For fast simulation of continuous processes, integration with established continuous simulation frameworks is necessary. This requires building wrappers for invoking state update functions of continuous solvers such as OpenFOAM. We also plan to integrate Dolfin [23], a Python based finite-element library for multiphysics modeling and simulation.
2. **Incorporating Analytics**
   For Digital Twin applications, analytics modules need to be incorporated into the framework for parameter extraction from sensor data, prediction, optimization and for building surrogate models in run-time.
3. **Acceleration for Real-time Simulations**
   The requirement for real-time simulation creates a need for simulation acceleration that is possible using hardware platforms such as GPGPUs, FPGAs or parallel execution on multi-core systems. It is possible to explore architectures that can take advantage of these technologies for simulations.
4. **Support for Sensing and Control**
   Sensing and control are integral aspects of a Digital Twin. Features that support these aspects need to be explored and integrated into the framework.

References

1. Agalianos, K., Ponis, S.T., Aretoulaki, E., Plakas, G., Efthymiou, O.: Discrete event simulation and digital twins: review and challenges for logistics. Proc. Manuf. **51**(2019), 1636–1641 (2020). https://doi.org/10.1016/j.promfg.2020.10.228

2. Aimone, J., Parekh, O., Severa, W.: Neural computing for scientific computing applications. In: ACM International Conference Proceeding Series, vol. 2017-July (2017). https://doi.org/10.1145/3183584.3183618
3. Aversano, G., Ferrarotti, M., Parente, A.: Digital twin of a combustion furnace operating in flameless conditions: reduced-order model development from CFD simulations. Proc. Combust. Inst. **38**, 1–9 (2020). https://doi.org/10.1016/j.proci.2020.06.045
4. Bangerth, W., et al.: The `deal.II` library, version 8.4. J. Numer. Math. **24**, 135–141 (2016). https://doi.org/10.1515/jnma-2016-1045
5. Bechard, V., Cote, N.: Simulation of mixed discrete and continuous systems: an iron ore terminal example. In: 2013 Winter Simulations Conference (WSC), pp. 1167–1178 (2013). https://doi.org/10.1109/WSC.2013.6721505
6. Brunton, S.L., Noack, B.R., Koumoutsakos, P.: Machine learning for fluid mechanics. Annu. Rev. Fluid Mech. **52**, 477–508 (2020). https://doi.org/10.1146/annurev-fluid-010719-060214
7. Cellier, F.E., Kofman, E.: Continuous System Simulation. Springer, USA (2006). https://doi.org/10.1007/0-387-30260-3
8. Chinesta, F., Ladeveze, P., Cueto, E.: A short review on model order reduction based on proper generalized decomposition. Arch. Comput. Methods Eng. **18**, 395–404 (2011). https://doi.org/10.1007/s11831-011-9064-7
9. Dagkakis, G., Heavey, C.: A review of open source discrete event simulation software for operations research. J. Simul. **10**(3), 193–206 (2016). https://doi.org/10.1057/jos.2015.9
10. Damiron, C., Nastasi, A.: Discrete rate simulation using linear programming. In: Winter Simulation Conference Proceedings, pp. 740–749 (2008). https://doi.org/10.1109/WSC.2008.4736136
11. Eldabi, T., Tako, A.A., Bell, D., Tolk, A.: Tutorial on means of hybrid simulation. Proceedings of the 2019 Winter Simulation Conference, pp. 273–284 (2019)
12. Feng, L.: Review of model order reduction methods for numerical simulation of nonlinear circuits. Appl. Math. Comput. **167**, 576–591 (2005). https://doi.org/10.1016/j.amc.2003.10.066
13. Fritzson, P., et al.: The OpenModelica integrated environment for modeling, simulation, and model-based development. Model. Identif. Control **41**(4), 241–295 (2020). https://doi.org/10.4173/mic.2020.4.1
14. Giambiasi, N., Escude, B., Ghosh, S.: GDEVS: a generalized discrete event specification for accurate modeling of dynamic systems. In: Proceedings - 5th International Symposium on Autonomous Decentralized Systems, ISADS 2001, pp. 464–469 (2001). https://doi.org/10.1109/ISADS.2001.917452
15. Hill, R.: Discrete-event simulation: a first course. J. Simul. **1**(2), 147–148 (2007). https://doi.org/10.1057/palgrave.jos.4250012
16. Huda, A.M., Chung, C.A.: Simulation modeling and analysis issues for high-speed combined continuous and discrete food industry manufacturing processes. Comput. Industr. Eng. **43**(3), 473–483 (2002). https://doi.org/10.1016/S0360-8352(02)00120-1, https://www.sciencedirect.com/science/article/pii/S0360835202001201
17. Karanjkar, N., Joshi, S.M.: Mixed discrete-continuous simulation for digital twins. In: Proceedings of the 11th International Conference on Simulation and Modeling Methodologies, Technologies and Applications, SIMULTECH 2021, Online Streaming, 7–9 July 2021, pp. 422–429. SCITEPRESS (2021). https://doi.org/10.5220/0010580804220429
18. Karniadakis, G.E., Kevrekidis, I.G., Lu, L., Perdikaris, P., Wang, S., Yang, L.: Physics-informed machine learning. Nat. Rev. Phys. **3**(6), 422–440 (2021). https://doi.org/10.1038/s42254-021-00314-5
19. Klingener, J.F.: Combined discrete-continuous simulation models in ProModel for windows. In: Winter Simulation Conference Proceedings, pp. 445–450 (1995)

20. Klingener, J.F.: Programming combined discrete-continuous simulation models for performance. In: Winter Simulation Conference Proceedings, pp. 833–839 (1996). https://doi.org/10.1145/256562.256824
21. Kofman, E.: Discrete event simulation of hybrid systems. SIAM J. Sci. Comput. **25**(5), 1771–1797 (2004). https://doi.org/10.1137/S1064827502418379
22. LeVeque, R.J.: Numerical Methods for Conservation Laws. Birkhauser, Basel (1990)
23. Logg, A., Wells, G.N.: DOLFIN: automated finite element computing. ACM Trans. Math. Softw. **37**(2), 1–28 (2010). https://doi.org/10.1145/1731022.1731030
24. Molinaro, R., Singh, J.S., Catsoulis, S., Narayanan, C., Lakehal, D.: Embedding data analytics and CFD into the digital twin concept. Comput. Fluids **214**, 104759 (2021). https://doi.org/10.1016/j.compfluid.2020.104759
25. Nutaro, J., Kuruganti, P.T., Protopopescu, V., Shankar, M.: The split system approach to managing time in simulations of hybrid systems having continuous and discrete event components. Simulation **88**(3), 281–298 (2012). https://doi.org/10.1177/0037549711401000
26. Shao, G., Jain, S., Laroque, C., Lee, L.H., Lendermann, P., Rose, O.: Digital twin for smart manufacturing: the simulation aspect. In: Proceedings - Winter Simulation Conference 2019-Decem (Bolton 2016), pp. 2085–2098 (2019). https://doi.org/10.1109/WSC40007.2019.9004659
27. Simpson, T.W., Peplinski, J.D., Koch, P.N., Allen, J.K.: Metamodels for computer-based engineering design: survey and recommendations. Eng. Comput. **17**, 129–150 (2001)
28. SimPy-Team: Simpy: Discrete-event simulation for python (2020). https://simpy.readthedocs.io/en/latest/
29. Weller, H.G., Tabor, G., Jasak, H., Fureby, C.: A tensorial approach to computational continuum mechanics using object-oriented techniques. Comput. Phys. **12**(6), 620–631 (1998). https://doi.org/10.1063/1.168744, https://aip.scitation.org/doi/abs/10.1063/1.168744
30. Zeigler, B.P.: Devs representation of dynamical systems: event-based intelligent control. Proc. IEEE **77**(1), 72–80 (1989). https://doi.org/10.1109/5.21071

Author Index

G. Wagner et al. (Eds.): SIMULTECH 2021, LNNS 601, p. 225, 2023.
https://doi.org/10.1007/978-3-031-23149-0